*Naval Strategy
and National Security*

International Security Readers

Naval Strategy and National Security

AN *International Security* READER

EDITED BY
Steven E. Miller
and Stephen Van Evera

PRINCETON UNIVERSITY PRESS
PRINCETON, NEW JERSEY

Published by Princeton University Press, 41 William Street, Princeton, New Jersey 08540

In the United Kingdom· Princeton University Press, Guildford, Surrey

First Princeton Paperback printing, 1988
First hardcover printing, 1988

LCC 87-36108   ISBN 0-691-07775-4   ISBN 0-691-02272-0 (pbk)

Clothbound editions of Princeton University Press books are printed on acid-free paper, and binding materials are chosen for strength and durability. Paperbacks, while satisfactory for personal collections, are not usually suitable for library rebinding

Printed in the United States of America by Princeton University Press, Princeton, New Jersey

Contents

The Contributors

STEVEN E. MILLER is an editor of *International Security*. He is a member of the MIT Defense and Arms Control Studies Program and teaches in the Department of Political Science at MIT.

STEPHEN VAN EVERA is an Adjunct Fellow at the Center for Science and International Affairs, Harvard University.

R. JAMES WOOLSEY is a partner in the law firm of Shea and Gardner in Washington, D.C. He served as Undersecretary of the Navy in the Carter administration.

LINTON F. BROOKS is a captain in the U.S. Navy now serving as Director of Defense Programs on the staff of the National Security Council.

JOHN J. MEARSHEIMER, Professor of Political Science at the University of Chicago, is completing a book on B. H. Liddell Hart.

JOSHUA M. EPSTEIN is a Senior Fellow at the Brookings Institution.

MICHAEL MCCGWIRE is a Senior Fellow at the Brookings Institution.

KARL LAUTENSCHLÄGER is a staff defense analyst at the Los Alamos National Laboratory. He has been an advanced research scholar at the Naval War College and a visiting faculty member at the Fletcher School of Law and Diplomacy, and was a naval officer for five years with two combat deployments to the Tonkin Gulf.

RICHARD L. GARWIN is IBM Fellow at the Thomas J. Watson Research Center, Yorktown Heights, New York, Adjunct Professor of Physics at Columbia University, Adjunct Research Fellow at Harvard University, and Andrew D. White Professor at Cornell University.

HAROLD A. FEIVESON is a senior research analyst at the Center for Energy and Environmental Studies at Princeton University. JOHN DUFFIELD is a Ph.D. candidate at the Woodrow Wilson School, Princeton University.

DESMOND BALL is Head of the Strategic and Defence Studies Centre, The Australian National University.

BARRY R. POSEN is Associate Professor of Political Science at MIT and a member of the MIT Defense and Arms Control Studies Program.

SEAN M. LYNN-JONES is Managing Editor of *International Security* and a Research Fellow at the Center for Science and International Affairs, Harvard University.

Preface | *Stephen Van Evera*

During the 1980s an intense debate has arisen over American naval policy, focused on the Reagan administration's "Maritime Strategy." According to official statements, this strategy posits an aggressive global role for the U.S. Navy in the event of a conventional war with the Soviet Union, including early attacks on Soviet ballistic missile submarines (SSBNs) in arctic waters, offensive strikes against Soviet bases on the Soviet mainland, and attacks against Third World countries allied to the Soviet Union.[1] Maritime Strategy proponents have claimed that such a strategy would help deter war, would encourage the Soviets to make peace should war break out, and would leave the United States in a strong position at the war's end. Opponents have argued that the strategy would require too many forces, would do little to deter war, and would increase the risk of nuclear escalation should conventional war break out.

This dispute over strategy has been echoed in a related controversy over the proper size of the navy fleet and the navy budget. Big-navy proponents argued in the early 1980s that the Maritime Strategy required a bigger fleet; specifically, Reagan administration officials favored an increase from the 480-ship navy of the late 1970s to a 600-ship navy, and from 13 aircraft carrier battle groups to 15 battle groups. Critics answered that the Soviet threat to Europe was the principal threat to American security, that the navy was peripheral to the defense of Europe, and that the navy therefore was already large enough, or perhaps too large. These critics suggested that spending on ground and air forces for the defense of central Europe, or on strategic forces, should take priority over naval increases. Their arguments failed in the short run, as the Congress eventually approved the 600-ship program, but the concept of a 600-ship program remains controversial, and some analysts now favor scaling the navy back.

Related disputes have also developed over the survivability and capability of American aircraft carriers and submarines. Critics of carriers have charged that large surface ships are increasingly vulnerable to new antiship technologies such as cruise missiles, and that the United States should shift to smaller ships or to land-based forces. Some observers have also warned that the American SSBN fleet, previously the most secure element of the American strategic force, may be growing vulnerable to Soviet antisubmarine warfare.

---

1. The Maritime Strategy was set forth most fully in an article by Admiral James D. Watkins, "The Maritime Strategy," Supplement to U.S. Naval Institute *Proceedings*, January 1986, pp. 2–17. For more literature on the Maritime Strategy see works cited in the articles by Linton Brooks and John Mearsheimer reprinted below.

The articles in Part I of this book explore the principal issues in the naval strategy debate, while articles in Parts II and III explore related technical, operational, and arms control issues. James Woolsey's essay "Planning a Navy" sets the stage by discussing how naval requirements should be assessed. Woolsey argues that because we cannot easily foresee the nature of the problems that we will someday ask our navy to solve, and because naval forces have such a long service lifetime, we should build a flexible navy that can perform a wide range of missions, including some that are not now required. He also argues that aircraft carriers are less vulnerable than their critics claim, and can be adequately secured from attack.

Linton Brooks, John Mearsheimer, and Joshua Epstein develop the case for and against the Maritime Strategy. Captain Brooks' article "Naval Power and National Security" is the most comprehensive argument for the Maritime Strategy published to date. He asserts that the Maritime Strategy plays a vital role by shoring up the defense of NATO's northern and southern flanks, by giving the United States the capacity to continue fighting even if Europe is lost, and by raising the cost to the Soviets of launching a conventional war, since it would put Soviet SSBNs at a risk in such a war. He notes that the aggressive aspects of the strategy are not without risk, but argues that this risk is smaller than the dangers it averts.

In contrast, Mearsheimer, in "A Strategic Misstep: The Maritime Strategy and Deterrence in Europe," argues that the naval operations envisioned by the Maritime Strategy require unduly large naval forces, that these operations are not essential to the defense of Western Europe, and that they would increase the risk of nuclear war. Epstein criticizes in detail an important component idea of the Maritime Strategy in "Horizontal Escalation: Sour Notes of a Recurrent Theme."

Michael MccGwire's "Naval Power and Soviet Global Strategy" describes and assesses the naval thinking of America's principal adversary. He notes that Soviet military doctrine envisions a decisive fight to the finish with the West which culminates in the extirpation of the capitalist system. The Soviet Navy has five principal missions in this decisive battle: strikes against shore targets in the West; destruction of Western navies; interdiction of Western sea lines of communication; protection of Soviet sea lines of communication; and support for Soviet ground force operations. Within these missions, the Soviets have recently emphasized defense of their own nuclear deterrent at the expense of attacking Western nuclear submarines or sea lanes.

The articles in Part II discuss aspects of naval technology that bear on naval policy questions. Karl Lautenschläger, in "Technology and the Evolution of Naval Warfare," surveys the history of naval technology over the past several centuries. He concludes that fears of technological surprise are largely misplaced; militarily significant technological surprise has been relatively rare, and has arisen from the integration of existing technologies, not the invention of new technologies. Richard Garwin's article "Will Strategic Submarines Be Vulnerable?" concludes that American SSBNs are not vulnerable to Soviet attack, and will probably not become vulnerable in the near future. Lautenschläger's survey "The Submarine in Naval Warfare" traces the dramatic evolution in the missions performed by submarines since 1900, and notes past mistakes in adapting submarines for their new roles. Lautenschläger also punctures myths about the current capabilities of submarines. Harold Feiveson and John Duffield's contribution, "Stopping the Sea-Based Counterforce Threat," outlines the growing threat to strategic bombers and land-based intercontinental ballistic missiles (ICBMs) posed by the increasing accuracy and MIRVing of Soviet and American submarine-launched ballistic missiles (SLBMs), and suggests solutions that the authors argue could still avert this threat.

The contributions in Part III address special topics in controlling the risks of naval operations in peacetime and wartime. Desmond Ball's "Nuclear War at Sea" surveys the ways that a nuclear war at sea might arise, assesses the scope of these dangers, and suggests possible solutions. Barry Posen's "Inadvertent Nuclear War?" explores the risk of escalation that would arise if, during a conventional war, American submarines attacked Soviet attack submarines (SSNs) in a manner that also destroyed Soviet SSBNs. (After Posen's article appeared the Navy acknowledged that in a conventional war it will deliberately search out Soviet SSBNs. If such operations create risk, therefore, this risk will arise deliberately, not inadvertently.) Finally, Sean Lynn-Jones' article "A Quiet Success for Arms Control" assesses the achievements of the 1972 Agreement on the Prevention of Incidents at Sea between the United States and the Soviet Union. Lynn-Jones argues that the Agreement has significantly reduced the number of dangerous incidents and cases of harassment at sea.

Part I:
Naval Strategy

Planning a Navy: The Risks of Conventional Wisdom

R. James Woolsey

The Spring 1978 issue *of* International Security *presented two essays addressing the contentious political and technical problems central to shaping the U.S. Navy in the 21st century. Senator Gary Hart and ship designer Reuven Leopold examined respectively Congressional sentiment and state of the art weapons acquisitions as they each confronted the certainty of a numerically constrained fleet.*

This debate over force structure—and the national defense policies reflected by the structure—inherently embraces a number of assumptions, as well as a variety of analytical approaches. We have asked Under Secretary of the Navy R. James Woolsey to assess the competing arguments that have surrounded discussions of alternative naval futures for the past months. What results is both a personal evaluation of now standard criticism, and a comprehensive recommendation for weathering what the Navy sees as difficult days ahead.

Forty-three years ago John Maynard Keynes closed his *General Theory* with the following words:

... the ideas of economists and political philosophers, both when they are right and when they are wrong, are more powerful than is commonly understood. Indeed the world is ruled by little else. Practical men, who believe themselves to be quite exempt from any intellectual influences, are usually the slaves of some defunct economist. Madmen in authority, who hear voices in the air, are distilling their frenzy from some academic scribbler of a few years back. I am sure that the power of vested interests is vastly exaggerated compared with the gradual encroachment of ideas. Not, indeed, immediately, but after a certain interval; for in the field of economic and political philosophy there are not many who are influenced by new theories after they are twenty-five or thirty years of age, so that the ideas which civil servants and politicians and even agitators apply to current events are not likely to be the newest. But, soon or late, it is ideas, not vested interests, which are dangerous for good or evil.

Keynes's principle—let us call it the principle of intellectual stagnation in the public service—has some important implications for naval forces, or at least for one conventional wisdom about naval forces. I say "one conventional wisdom" advisedly, because if there is one thing that has become clear in recent years about

---

R. James Woolsey is Under Secretary of the Navy. From 1974 until March 1977 he was associated with the law firm of Shea and Gardner in Washington, DC. He was General Counsel to the U.S. Senate Committee on Armed Services, 1970–73.

naval forces it is that the number of offices, institutions, and influential individuals in the government with different but firmly held views about the proper future of the Navy is beginning to approach the number of ships in the fleet.

The Conventional Wisdom

The particular conventional wisdom about the Navy that I want to examine, with Keynes's principle in mind, is a rather widely-held one. It is bottomed on two views. The first is an assumption about the future of technology. It is that, due to projected improvements in anti-ship missiles and the difficulty of defending against them, surface ships over the next twenty or thirty years will become increasingly unable to survive at sea. The second basis of this particular conventional wisdom is harder to describe. It is a set of views that might be called quantitative policy analysis. It is the notion that military forces should be designed almost exclusively to yield favorable results from computer calculations of the outcomes of very specific military engagements, using complex models with many assumptions. The particular conventional wisdom produced by these views runs something like this:

> Naval forces, particularly surface ships, are becoming increasingly obsolete. Ever since the Israeli destroyer, *Eilat*, was sunk in 1967 by a Soviet-made cruise missile, this trend has been clear and it becomes clearer each year. Cruise missiles make surface ships increasingly vulnerable to attack by all sorts of platforms—submarines, other surface ships, and aircraft.
>
> The U.S. Navy has compounded this problem because it has become used to placing all its offensive power entirely in a single platform, the large-deck aircraft carrier. These ships are becoming so expensive that it is going to be difficult for the Navy to maintain very many of them, and they create the difficulty of having all of one's eggs in very few baskets.
>
> Vulnerability and limited numbers mean that carriers and other surface ships could not prudently be risked in a major war in the future. This means that surface ships should primarily be used for specific purposes: showing the flag in peacetime and projecting power ashore in contingencies such as Korea or Vietnam, where they can operate from an ocean sanctuary against third world countries that lack the sophisticated naval forces of the Soviet Navy. But, for these peacetime and minor contingency purposes, the expense of operating carriers is unsupportably high. The United States could probably afford some reduction in large carriers and substitute, e.g., amphibious ships. Since many of them are large, look impressive, and roughly resemble carriers, they could be used for port visits and peacetime deployments.

The Navy's main, and only vital, mission in a U.S.-Soviet war is to protect the sea lines of communication from the East Coast of the United States to Western Europe. A war in Europe would likely be over quickly since it would either turn nuclear or one side would suffer significant defeat within the first thirty-odd days. Sealift might thus not be a particularly significant factor, but to hedge against a longer war, some Navy for sea lane protection is necessary. This could be done primarily by land-based anti-submarine warfare aircraft such as the P-3, small surface ships such as frigates, and a few nuclear attack submarines. In a big war, any other mission for naval forces—such as conducting operations within range of Soviet land-based aircraft—is too dangerous, or at least too expensive, to plan.

This particular conventional wisdom is not wholly in error, but it has serious flaws for the reasons Keynes described. First is its key technical judgment: that surface ships will become ever more vulnerable to anti-ship cruise missiles over the next two to three decades.

It is instructive to note that the people who lost the Eilat in 1967 have learned something. I have recently been on an Israeli patrol boat and reviewed their 100 percent successful tactics for avoiding hits by the large number of anti-ship missiles that were fired against them in the 1973 war. Suffice it to say this is an area in which their, and the United States' advantages and potential advantages, are far from irrelevant—e.g., skill with electronic countermeasures, innovative tactics, and intelligent and well-trained crews.

Moreover, defending ships that can move and, in a conventional war, that can tolerate a small number of hits, is inherently an easier problem than perfectly defending a fixed land-based site in a strategic nuclear exchange—where the penalty for leakage is greater. Even if a surface ship, particularly a large one, is hit by a conventionally-armed anti-ship missile, the probability of the ship being put out of action, much less sunk, is certainly not unity. It is true that, for a number of years in the 1950s and 1960s, the United States constructed ships in part under the assumption that any large-scale war would be nuclear and that hardening or passive protection was not, ordinarily, worth the money. However, a new Sea-Based Air Platforms Assessment, done by the Navy at the request of Congress, indicates that substantial improvement in the survivability of carriers, large or small, is possible by constructing new ships with more armor and passive protection than has been used in recent years. Further, vertical and short take-off and landing (V/STOL) aircraft could significantly improve the capability of any carrier, of whatever size, to conduct its mission after suffering damage, because such aircraft can continue to operate even if the ship has been slowed (for that matter, if it is dead in the water) and even if the deck area, where catapaults or arresting gear would normally be located, has been damaged.

Most importantly, however, systems now coming into U.S. naval forces have been in development over the last decade that will make significant contributions toward reducing the number of missiles that might be able to penetrate defenses—electronic warfare suits, the PHALANX Close-In-Weapon-System, and particularly the AEGIS area air defense system.

Furthermore, there are either in the fleet, entering it soon, or in development, systems capable of destroying the platforms launching the incoming missiles to guard against gradual attrition. The F-14 fighter and E-2C early warning aircraft provide a good capability against aircraft launching anti-ship missiles, although that difficult job is one which needs substantial work in the future. Improvements in anti-submarine warfare forces—towed arrays, and others—make the job of engaging the missile-launching submarine a more reasonable enterprise than was the case a few years ago. And hostile surface ships can now be engaged by carrier-based aircraft and soon by the HARPOON missile just coming into the fleet.

Torpedoes continue to be a problem for any surface ship that permits a hostile submarine to move sufficiently close, but in the open ocean this can be made difficult. In any case the death of the surface navy from the cruise missile—as Mark Twain said of his own premature obituary—has been greatly exaggerated.

What about the problem of all U.S. offensive eggs being locked into a few large carrier baskets? There are two ways open for packing some offensive punch into a much larger number of platforms. They are not mutually exclusive. The first, of comparatively low cost, is to equip as many as possible of combatant platforms with cruise missiles. The program is already under way to equip all surface combatants as well as a number of aircraft and submarines with the HARPOON. Over the longer haul, cruisers, destroyers, and attack submarines could be equipped with the considerably longer range TOMAHAWK cruise missile both for the anti-surface-ship mission and for land-attack missions. In some of the latter it is possible to have a significant military impact by delivering only a relatively small number of very accurate conventional warhead weapons.

The second way to spread out offensive punch is to disperse the aviation eggs in more baskets by developing V/STOL aircraft for deployment on small carriers and other air capable surface combatants. This may prove expensive, but it is a road that should be explored vigorously in research and development for a few years. Even if no more large-deck carriers are built, there will be at least twelve of them in our naval forces nearly into the 21st century. The only investment required to ensure this is to conduct a set of very thorough overhauls, called the Service Life Extension Program, beginning in 1981. Because of this, any transition to V/STOL would not, and need not, be a sudden proposition. It is more akin to an evolving reliance on solar power in place of fossil fuels than it is akin to, e.g., replacing one rifle with another.

"Well, all right," the answer might be, "suppose surface ships are not becoming as vulnerable as the conventional wisdom might indicate, and suppose there are ways to spread our firepower on to more platforms than twelve large-deck carriers. You still haven't told me what you want to use the Navy for. Are you interested in power projection? Are you interested in sea control? You can't afford to do everything. Shouldn't the United States be concentrating, for example, on protecting the sea lines of communication in the event of a NATO war, using the most cost-effective systems possible? Tell me the scenario you want to operate in and I will help you design and size your force. I will help you discover 'how much is enough.' The Navy has to get its act together and decide what war it wants to fight."

The Limitations of Quantitative Policy Analysis

These sorts of questions, claims, and advice are the bread and butter of what I described earlier as "quantitative policy analysis": a method of decisionmaking that relies heavily, in the military field, on designing forces to cope with very specific scenarios, utilizing complex computer models dependent on numerous detailed assumptions.

I am not using "systems analysis," because that term—the name of both the original office established in the Defense Department by Secretary McNamara and the decisionmaking methods it spawned throughout government—has become identified in many people's minds with a number of specific issues. Further, I am not referring to that office itself, nor to the role of particular individuals who worked or work there. Having worked there myself, I have a high regard for many others who have, for those who have led it, and for those who lead it today. What I am describing is an admittedly rather single-minded attachment to a specific tool of decisionmaking: a single-mindedness I would attribute to no one all of the time, but to lots of us some of the time, and to some of us most of the time.

This type of analysis has made several very positive contributions.

First, it has concentrated the public debate on outputs rather than inputs. Clearly it is better that we discuss U.S. capabilities and the costs of them rather than, line item for line item, whether 5 percent more or fewer generators should be bought this year than last. Second, analysis, when properly done, can turn attention toward overall systems—toward the capability, say, of destroyers and helicopters operating together as an anti-submarine warfare system; this is clearly better than merely looking at the pieces of the puzzle. Third, analysis and the budget procedures that accompany it can help develop better balanced capabilities within budget constraints by, for example, calling attention to the fact that adequate airlift, sealift, or logistics may not be provided for the number of divisions being bought.

These alone are enough to say that analysis deserves one or two cheers—not three by any means, but not zero either. A background in analysis is especially useful additional learning for those who are schooled in the arts of leading men and women and commanding. This seasoning has produced some of the ablest members of the military.

But one of the oldest and most honorable strains in the analytic discipline is the principle that one must always tell the client or decisionmaker who asks the questions that set up the analysis, whether or not his questions are the right ones.

In that spirit, we should ask whether "how much is enough," the quintessential question for quantitative policy analysts, is the right, or at least the most important, question for defense planning.

The leading reason why "how much is enough" may not be the major question that needs to be asked in naval force structure debates is that, in answering it, analysts often assume away the important issues in the process of designing forces for the specific scenarios that they know how to work within. Designing a Navy around specific scenarios requires one to look too far into the future to be realistic. Ships are platforms—more like capital investments than like specific weapons. Over 70 percent of the ships that will be in the fleet in the year 1990 already exist or have been authorized. A carrier authorized in the next year or two can spend over half its service life in the 21st century. For an example, in the year 2010, such a carrier (which would enter the fleet in the mid-1980s) would just be entering middle age—that is, it would be coming out of its major service life extension overhaul and would be looking forward to another fifteen years in the fleet. Now, we have no better idea today what specific wars or crises will occur between now and 2010—thirty-two years from now—than we had in 1946—thirty-two years ago —about the crises of today. Who foresaw in 1946 that in 1978 our thinking about when and how we might need naval forces could be significantly influenced by, e.g., a commitment to Israel, the need to protect sea lines of communication to Persian Gulf oil, a split between a Communist China and the Soviet Union, or U.S.-Soviet parity in strategic nuclear weapons? In 1946 the State of Israel did not exist, Persian Gulf oil was just being discovered, the People's Republic of China was just being born, and neither the United States nor the Soviet Union had heard of an ICBM or an SLBM. What makes anyone even remotely confident that the national security problems of the early 21st century are any clearer today than the forces that drive naval needs in 1978 were clear in 1946?

It is neither lack of effort nor a temporary and remediable lack of will power that leads to the Navy not being able to tell now just what sort of war it wants to fight in thirty years. Designing a Navy is not like planning a Cook's Tour—next week I'll be in Belgium, it will be March, so I'll need a raincoat. It is far more like forming the Lewis and Clark Expedition: preparations for a wide range of problems are necessary, so flexibility is vital.

How this can best be done is a massive subject. But three points are relevant. First, since ships are capital investments and last such a long time in peacetime, and since the pace of technological change for weapons and sensors—such as radars and sonars—is so rapid ships must be able to accommodate change and modernization readily. Carriers are inherently capable of doing this. As the suits of aircraft are changed, the ship can thereby accommodate major alterations in missions. The carriers have done this many times since World War II. Another way to promote flexibility is to build even small surface combatants to take some aircraft. U.S. destroyers and frigates took on important new capabilities when anti-submarine warfare helicopters were put aboard them a few years ago. They will evolve further as more advanced helicopters replace these in the 1980s. This interest in the flexibility provided by multiplying the number and type of platforms that can use aviation at sea is another key reason to support the development of V/STOL aircraft. A third way to make it easier for ships to accommodate change is to build surface ships in such a way that their weapons suits—guns, radars, missiles, etc.—can be changed far more readily and cheaply than is now possible as new weapons and sensors become available, almost as if one were changing modules. The United States is currently working hard on this promising concept, but there is much to do.

The Spectrum from Peace to War

In addition to the need to design a Navy for flexibility over a long period of time, there is another major reason to consider rejecting the precise quantitative and scenario-dependent method of designing a Navy. It is that naval forces are particularly vital for dealing with the *dynamic* problem of the transition, a transition we want to prevent, from peace to war. Any specific scenario, whether it is protecting the sea lines of communication to Europe in the event of a NATO war, projecting power in a certain type of crisis in the third world in the absence of Soviet intervention, or any other, has to be a snapshot in time. One does not want to ask only the question convenient for quantitative policy analysis to answer—"What might things look like at a specific time?"—but the much harder question: "What are the problems of going to or being forced from one position to another, and how can I control that process to avoid risk?" The President and his advisers need to know what sort of forces can best help them manage the perennial danger of crises escalating into war, and to manage them in such a way that the other side will know that at each step of the road the United States is in control. Unfortunately for the analyst, it is in the complexity of evolving and dynamic situations, of peace threatening to evolve into war, that naval forces are most relevant. That makes them difficult to analyze in one- or two-mission or one- or two-scenario formats. Naval forces are highly relevant to this delicate transition because they can do three things: (1) they can help maintain stability in peacetime through forward deployments and perceptions of their potential power if used; (2) they can help contain or manage crises as they evolve; and (3) they can help deter general war by being clearly better able to fight it than their foreseeable adversaries. These tasks are, in a very real sense, a seamless web. They have recently been reviewed in a major study for the Navy, Sea Plan 2000. Let me summarize its message at the risk of radical oversimplification.

Naval forces can help maintain stability in peacetime by forward deployment. The United States maintains today two carrier task groups in the Mediterranean and two in the Western Pacific; several Marine amphibious units are deployed in both areas as well. Since 1945, the United States has used such sea power as a means of affecting the behavior of decisionmakers in other nations in peacetime. These forward deployments are intended to demonstrate U.S. interest and resolve, to reassure allies, to deter enemies, and to ensure quick response. So one important question in designing a Navy is, how do force structure decisions have an effect on these forward deployments? For the issue is not whether a permanent reduction in U.S. naval forces would constrain forward deployment and therefore foreign policy, but rather to what degree it would do so.

Another part of the picture of assessing the overall contribution of naval forces to peacetime stability is the perception of the Soviet-U.S. naval balance. It is not inevitable that the United States concede to the Soviets parity in all military capabilities. They do not enjoy it now. The forward strategy linking the United States to other continents requires use of the seas, and makes any perception that the Soviets could deny the United States control of the seas particularly damaging. Such perception is not warranted by the projected trends in technology, if the United States has the will, the skill, and the money to proceed to deploy what has been developed.

Another major task that national decisionmakers must face in the spectrum between peace and war is the containment or management of local crises. In some crises a President may wish to commit U.S. troops immediately to preempt certain potential moves by an adversary, to evacuate Americans in jeopardy, or to ferry supplies rapidly to a friend or ally. Naval forces are not the only possible means. The quick response of airlift provides the President with a valuable tool, for example. But airlift has limitations and in a number of cases, naval forces may be preferred for good reason. For example, naval forces can be deployed to a crisis area without being committed to battle and without committing allies. Such demonstrations manifest both U.S. concern and capabilities. In over 200 crises, large and small, since 1945, in which the United States was involved, U.S. Navy and Marine forces were deliberately employed in 177 cases, while U.S. land-based air or ground forces alone were demonstrated in fewer than 90 cases.

Naval forces may be the most acceptable form of military presence in crisis situations. They can convey, if the policymaker chooses, calculated ambiguity and calibrated response. Their presence does not irrevocably commit the United States to a given course of action. They do, however, seriously complicate the calculations of opposing parties. U.S. fighting forces can be assembled for action without using bases in other nations. Indeed naval forces help make the United States comparatively indifferent to the vicissitudes of other nations' policies about base rights, whether for itself or for hostile countries, and Naval forces thus help make it more able to tolerate shifts in political winds without feeling that vital interests are injured. If a crisis is resolved satisfactorily, naval forces can be withdrawn with limited fanfare. In sum, naval forces provide a policy maker with important flexibility, and a tool for orchestrating events.

To be able to successfully support U.S. policy in a crisis, its naval forces require several things.

They must have the striking power to affect events ashore.

They must have local superiority over potential adversaries. The benefit of naval superiority is that it signals to the Soviets and others that their adventurism occurs against the backdrop of local U.S. forces that are capable of fighting and winning. In a crisis the United States wants it to be the other commander on the scene who is forced to tell his superiors that he must back down or risk escalation.

There must be sufficient forces to permit coverage of different crisis areas, so that responding to a crisis in one area does not involve a risk of being unable to deal quickly with a new outbreak somewhere else. This does not imply that the United States must be everywhere all the time. It does mean that reductions in U.S. force levels will increasingly constrain its credibility.

Most of all, there is the possibility that a crisis could escalate to actual fighting, with some losses to U.S. forces. These forces are becoming more, rather than less, capable of responding to a sudden attack. Nevertheless, prudent planning requires that the possibility of some initial losses be recognized. Calculations of total forces must take this possibility into account.

Another major task for national decisionmakers is the deterrence of a major war. The deterrence of conflict will depend upon a credible warfighting capability. Maintaining such a capability is complex and difficult for a whole series of reasons, but two are salient. First, U.S. Allies are overseas—many of the most important ones close to the borders of the Soviet Union. If there is effective sea denial, by *both* sides, the United States loses. It must be able to use the seas to maintain its alliances and security. Second, it is no longer the case that the threat to use of the seas occurs only in the near-coastal waters of the Soviet Union. The increasingly-capable blue-water Soviet Navy, and particularly the long-range Backfire bomber going into the Soviet Naval Aviation forces, makes all the world's waters of interest a potential theater of conflict. The Backfire, for example, can range from Soviet bases to the environs of the Azores in the Atlantic and Pearl Harbor in the Pacific.

One of the most immediate concerns to the United States in deterrence and warfighting must be the defense of vital sea lanes. No matter what the scenario— minimum warning or long warning, short war or extended war—large amounts of material must be moved by sea. This is usually viewed strictly in the context of the North Atlantic. But Hawaii and Western Pacific allies must be supported by sea, and a continuing flow of vital overseas resources—particularly petroleum products—must be delivered to the United States and its allies to sustain economy and industry.

Naval forces contribute to deterrence and to the ability to fight a global war by a clear demonstration of an ability to support allies or strategic friends on the flanks of the Soviet Union. Sea lane defense, by itself, does not protect flanks. NATO is a collective alliance, relying upon the commitment of all its members to the common defense. If any of these members doubted America's commitment or capability to support them, it could generate serious pressures on alliance cohesion.

Then, in a general U.S.-Soviet war, U.S. Naval forces must be capable of flexible options worldwide. A major conflict will almost certainly be conducted on a global scale. If that should occur, we must be able to destroy Soviet naval forces in many areas of the world. By our having that capability, Soviet planning is complicated, for they are forced into a defensive posture. Whether or not a national leader chose to exercise the option, the capability to conduct offensive operations against an enemy fleet is crucial in order for these forces to be useful to the nation. Alfred Thayer Mahan hammers this point home throughout "The Influence of Sea Power on History." An offensive capability, Mahan points out, was the central difference over the years between the British and French fleets, and the key to British success.

Conclusion

These needs for managing dynamic situations—for maintaining stability in peacetime, for providing the tools to manage crises, for deterring war by being able to fight it—require a range of types of naval forces. Managing stability, crises, and deterrence means being able to conduct military operations, where needed, on, under, above, and along the shores of 70 percent of the earth's surface. This is simply too complex a task to be accomplished by one or two specific types of platforms or systems. Before asking "How much is enough?" the balance of forces to cope with dynamic, distant, and complex needs must be determined. This latter problem is far harder, and far less amenable to quantitative policy analysis.

There is a final reason why it is unwise to design a Navy to prevail only in certain specific scenarios, relying primarily on the techniques of quantitative policy analysis. It is that such analysis, in general terms, may lead to a fundamental misunderstanding of the nature of war. It is often said that quantitative analysis focuses attention where it should be focused—at marginal changes. Now, at one level, concentrating on the margin—that is, on the costs and benefits of the next decision, not some overall historical average—is just common sense. Ignoring sunk costs is the first principle of most successful businessmen and all successful poker players. But a fixation on marginal change can be stultifying if it so narrows the analyst's imagination that he never moves his gaze from the bow-wave to the horizon. A focus on changes at the margin and on quantitative questions can too often produce an attitude that innovation is suspect and that the only changes of any interest are to buy several fewer of these or slightly increase the number of those—a sort of military instinct for the capillaries.

Military breakthroughs do not come that way. They come by approaching things from a new perspective, by devising a different way, for example, to exploit the effect of mass or shock, a way to use surprise or concealment to accomplish what was previously accomplished by ponderous force, or a way to disperse and then concentrate for battle that confounds the enemy's planning. Quantitative scenario-specific analysis often misses this fundamental truth about military matters. It does not take each element, for example, of naval warfare—anti-air warfare, anti-surface warfare, anti-submarine warfare, and so on—and ask how in each type of combat we might most readily make a potential enemy's past investments in weapons worthless. Such analysis doesn't ask, "How can I exploit my advantages?" "How can I destroy the will of an enemy commander?" As the Chinese strategist Sun Tzu wrote long ago—the least desirable way to achieve victory is to destroy an enemy's cities; the next least desirable is to kill his soldiers; better is to destroy his alliances; but best of all is to destroy his plans and never have to fight at all. Only intellectual audacity permits this most humane type of victory, and intellectual audacity is not normally found at the margin.

This capacity for asking the right question, for intellectual audacity, is not a talent that is foreign to totalitarian or aggressive societies. One needs only to recall Heinz Guderian's development of tank warfare and the blitzkreig in Germany in the late 1930s. The United States must not be the only superpower on this planet which is asking only, "How much is enough?" We must ask many other, and more audacious questions too, or risk ugly surprises. The best illustration is probably one such surprise that occurred some time ago.

In the late 18th century, during the re-evaluation that defeat always forces on a country's military establishment, French armorers discovered methods of casting cannon that improved accuracy and made artillery light enough to be pulled by horses rather than oxen. The implications of this development were not immediately clear. There was some experimenting, but most commanders used horse-drawn artillery as they had oxen-drawn—to make ponderous sorties from fixed forts and magazines and to fight in the rigid 18th century manner. Viewed in this context, horse-drawn artillery was a marginal improvement of sorts—it probably was not cost-effective. The development was only fully exploited during and after the French Revolution, in particular by a young Corsican artilleryman. Doubtless his success depended heavily on his own tactical genius and on the social effects of the French Revolution, which permitted the *levée en masse*—making possible the 19th century version of a human sea attack. But accurate horse-drawn artillery opened radical and unforeseen new possibilities of organization, mass, maneuver, and surprise that enabled his armies to shatter 18th

Century concepts of warfare and the armies that practiced them. New units called "divisions" were formed under aggressive young commanders, and each was given its own artillery. Forts and magazines were bypassed, and the Alps crossed, in rapid marches; firepower was massed quickly to destroy opposing armies before they could concentrate on the battlefield. In other men's hands, a lighter cannon barrel had been just a lighter cannon barrel. In his, it was a major element in the conquest of Europe.

So the risks of becoming rigid in thinking about military forces—of designing forces to fight the way they have always fought, of concentrating only on how much is enough at the margin—are great. Keynes must be proven wrong. The United States naval forces must not perpetually be the slave of vulnerability calculations of 1967 or analytical tools of 1963. Naval planners must learn continuously. The risks of doing otherwise are clear—the learning will take place, but much more painfully. For example, it took the rest of the European continent twenty years to learn enough from that artilleryman, turned Emperor, to defeat him. And, even then, as his ultimate conqueror said, it was a near-run thing.

Naval Power and National Security

The Case for the Maritime Strategy

For the past five years, U.S. Navy officers and their civilian colleagues have been taking to heart the centuries-old dictum of the first great theorist of conflict, Sun Tzu: they have been studying war. While military reformers have focused on the need for improved military strategy in a land campaign, a renaissance of strategic thinking has been taking place within the U.S. Navy. This renaissance has been marked by a series of internal and external discussions and debates in which naval strategy has received more attention than in any peacetime period since the days of Alfred Thayer Mahan.[1] One important result has been to weave traditional naval thinking into a coherent concept for using early, forceful, global, forward deployment of maritime power both to deter war with the Soviet Union and to achieve U.S. war aims should deterrence fail. The concept, which has come to be called "The Maritime Strategy," was initially codified in classified internal Navy documents in 1982

The author wishes to thank Rear Admiral William Pendley, Captains Thomas Daly, Michael Hughes, and Peter Swartz, Lieutenant Commander Joseph Benkert, and Mr. Bradford Dismukes for their assistance. All have contributed to the development of the Maritime Strategy as well as to this paper; none are responsible for the use I have made of their thoughts and insights.

Linton F. Brooks is a Navy Captain now serving as Director of Defense Programs on the staff of the National Security Council.

---

1. For a bibliographic summary of the professional debate, see Captain Peter M. Swartz, "Contemporary U.S. Naval Strategy: A Bibliography," Supplement to U.S. Naval Institute *Proceedings*, January 1986, pp. 41–47, and his "1986 Addendum" to the bibliography, U.S. Naval Institute *Proceedings* (hereinafter cited as *Proceedings*), forthcoming. Both the volume of the literature and the seniority of the military authors are significant. See, for example, Admiral James D. Watkins, "The Maritime Strategy," Supplement to U.S. Naval Institute *Proceedings*, January 1986, pp. 2–17; Admiral Sylvester R. Foley, Jr., "Strategic Factors in the Pacific," *Proceedings*, Vol. 111, No. 8 (August 1985), pp. 34–38; Admiral Wesley McDonald, "Mine Warfare: A Pillar of Maritime Strategy," *Proceedings*, Vol. 111, No. 10 (October 1985), pp. 46–53; and Vice Admiral H.C. Mustin, "The Role of the Navy and Marines in the Norwegian Sea," *Naval War College Review* (hereinafter cited as *NWC Review*), Vol. 39, No. 2 (March–April 1986), pp. 2–7. Contrast these articles by the Chief of Naval Operations, Commanders-in-Chief of the Atlantic and Pacific Fleets, and Commander of NATO's Striking Fleet Atlantic, all appearing within eight months, with the historical paucity of articles on strategy in Navy professional literature documented in Linton F. Brooks, "An Examination of Professional Concerns of Naval Officers as Reflected in Their Professional Journal," *NWC Review*, Vol. 33, No. 1 (January–February 1980), pp. 46–56. Only three articles by flag officers on *any* aspect of strategy appeared in a typical five-year period in the 1960s. Of 719 articles in *Proceedings* during 1964–1968, only two were directly concerned with overall naval strategy.

International Security, Fall 1986 (Vol. 11, No. 2)
© 1986 by the President and Fellows of Harvard College and of the Massachusetts Institute of Technology

and was gradually revealed to public scrutiny through Congressional testimony and public statements culminating in a January 1986 Supplement to the *Proceedings* of the United States Naval Institute, the professional journal of the Navy and Marine Corps.[2] This supplement, jointly authored by the Secretary of the Navy, Chief of Naval Operations, and Commandant of the Marine Corps, has been called "the nearest thing to a British 'White Paper' . . . that we are likely to encounter in the American political system."[3]

Both uniformed and civilian experts agree that the days of separate land, sea, and air strategies are long gone. No meaningful single-service strategy is possible in the modern era, a fact that the Navy has recognized both in its increasingly frequent references to the Maritime Strategy as "the maritime component of the National Military Strategy"[4] and in its explicit inclusion of the contribution of Army and Air Force operations in describing the strategy.[5] Given this, why have a concept or document like the Maritime Strategy at all? Navy leaders give several reasons. First and foremost, the strategy embodies the professional consensus of the leadership of the Navy and the Marine Corps on how to deter or, if necessary, fight, a future war. As such, it "offers a global perspective to operational commanders" and "provides a foundation for advice to the National Command Authorities."[6] In addition, the strategy gives "coherence and direction to the process of allocating money among competing types of ships and aircraft and different accounts for spare parts, missile systems, defense planning, and the training of forces."[7] Civilian observers quickly add another purpose, to ensure that those programs are properly funded;[8] the Maritime Strategy has unquestionably contributed to the Navy's success in articulating and justifying programs before Congress.

---

2. The term "Maritime Strategy" has two meanings. It refers physically to a series of briefings at various levels of classification maintained by the Strategic Concepts Group of the Office of the Chief of Naval Operations. A scripted version of this briefing is extensively distributed throughout the Navy over the signature of the Chief of Naval Operations and is frequently updated. More generally, "Maritime Strategy" refers to the overall professional consensus of Navy and Marine Corps leaders on the proper use of seapower; it is this second sense in which the term is used in this essay.

3. James A. Barber, Jr., "From the Executive Director," Supplement to U.S. Naval Institute *Proceedings*, January 1986, p. 1.

4. Watkins, "Maritime Strategy," p. 4.

5. Ibid., p. 5

6. Ibid., p. 4.

7. John F. Lehman, Jr., "The 600-Ship Navy," Supplement to U.S. Naval Institute *Proceedings*, January 1986, p. 36.

8. Norman Friedman, "US Maritime Strategy," *International Defense Review*, Vol. 18, No. 7 (1985), p. 1072.

To these reasons should be added one more, seldom explicitly expressed but important nonetheless. The strategy provides a common frame of reference for Navy and Marine Corps officers, a way of considering the purpose of their profession, and a catalyst for strategic thought. While it has not been noted in the public debate, a significant effort is under way within the Navy to ensure that its officers understand the strategy and their role in it and are active in its continued refinement. Thus, the Maritime Strategy not only is the professional consensus of senior leaders as to how the Navy should fight today, but also is a vehicle for transmitting that vision to the future through programs, plans, and people.

While the strategy represents the consensus of the Navy's military and civilian leadership on the best employment of maritime forces in war, it has been greeted with anything but consensus outside the Navy. Critics assert that "a primarily maritime strategy cannot adequately protect our vital interests in Eurasia because it cannot adequately deter a great land-based power like the U.S.S.R.,"[9] or that "even the appearance of such a campaign could trigger dire consequences,"[10] or that "a classic strategic error has been made in devoting so much money to the aircraft carriers and all that goes with them."[11] Some of this criticism represents honest differences of opinion among reasonable men. Other critics misunderstand the strategy. Still others are simply wrong. Finally, some criticism reflects the fact that the strategy does not support critics' preconceived force structure preferences. This article seeks to demonstrate that, far from being irrelevant or dangerous, the on-going renaissance in Navy strategic thinking offers a method of keeping the national strategy of the United States—which includes both global commitments and a commitment to the continental defense of Europe—viable in an era of nuclear parity and substantial imbalance in European land forces. Critics should welcome the new emphasis on forward maritime options as strengthening deterrence and aiding the nation in continuing the historic American guarantee of Western Europe's security while still denying the Soviets the initiative in other areas of the world.

---

9. Robert W. Komer, *Maritime Strategy or Coalition Defense?* (Cambridge, Mass.: Abt Books, 1984), p. 67.
10. Barry R. Posen, "Inadvertent Nuclear War? Escalation and NATO's Northern Flank," in Steven E. Miller, ed., *Strategy and Nuclear Deterrence*, (Princeton: Princeton University Press, 1984), p. 100.
11. Edward N. Luttwak, *The Pentagon and the Art of War: The Question of Military Reform* (New York: Simon and Schuster, 1984), p. 264.

The Strategic Environment

The Maritime Strategy cannot be considered unless we first understand the national military strategy it is intended to implement, the Soviet military strategy it is designed to counter, or the forces with which it would be undertaken.

NATIONAL MILITARY STRATEGY

Self-styled military reformers often assert that the United States has no national strategy.[12] Such a declaration is understandable since no public document sets forth an overall military strategy except in generalities. The 99-page annual report of the Organization of the Joint Chiefs of Staff, for example, devotes only a page and a half to U.S. military strategy, describing its fundamental elements as "nuclear deterrence supported by negotiated arms reductions . . . strong alliances; forward-deployed forces; a strong central reserve; freedom of the seas, air, and space; effective command and control; and good intelligence."[13] Statements by civilian leaders are similarly general.[14] Indeed, early in the Reagan Administration, the Secretary of Defense specifically rejected "early elaboration of some elaborate 'conceptual structure,' a full-fledged Reagan strategy."[15]

There is, however, a national security strategy, promulgated by the President on May 20, 1982 in National Security Decision Document (NSDD)-32. It designates the Soviets as the main military threat, rather than more common but less significant adversaries such as Libya. To counter this threat, the strategy calls for balanced conventional forces, expects a war to be global, envisions sequential operations during that global conflict, places increasing importance on allied contributions, and directs the forward basing of U.S. forces in peacetime.[16] "In what was probably its most significant strategy

---

12. Ibid., *passim*, for example.
13. Organization of the Joint Chiefs of Staff, *United States Military Posture for FY 1987*, p. 8. The discussion of strategy is found on pp. 7–9.
14. Samuel P. Huntington, "The Defense Policy of the Reagan Administration, 1981–1982," in Fred I. Greenstein, ed., *The Reagan Presidency: An Early Assessment* (Baltimore and London: Johns Hopkins University Press, 1983), pp. 82–116.
15. Caspar W. Weinberger, "The Defense Policy of the Reagan Administration," Address, Council on Foreign Relations, New York, June 17, 1981, quoted in Huntington, "Defense Policy of the Reagan Administration," p. 89.
16. Senate Armed Services Committee, 98th Congress, 2nd Session, *Hearings on the Department of Defense Authorization for FY 85: Part 8* (Washington, D.C.: U.S. Government Printing Office, 1985), p. 3854, describes key elements of NSDD-32 as seen by the Navy. These significant

innovation, the Reagan administration consciously and formally substituted the threat of escalation in space and time for the threat of escalation in weapons," thus leading to an emphasis both on prolonged conventional conflict and on denying the Soviets the ability to choose the geographic limits of that conflict.[17]

In the development and refinement of the Maritime Strategy, Navy leaders took into account this NSDD, other Administration documents, the war plans of the Unified and Specified Commanders,[18] and the treaties and other agreements the United States has with 43 nations, the most important of which is the North Atlantic Treaty. From this series of documents, many unavailable to the public, Navy leaders concluded that "our national strategy is built on three pillars: deterrence, forward defense, and alliance solidarity."[19] This view of U.S. and NATO military strategy is the first factor shaping the Maritime Strategy.

SOVIET MILITARY STRATEGY

The second factor shaping the Maritime Strategy is Soviet military strategy and the role of the Soviet navy within that strategy. The Maritime Strategy is based on a Soviet strategy that assumes that any future war with the West "would be a decisive clash on a global scale . . . a coalition war"[20]—a war that the Soviets would prefer to fight with conventional weapons, but one

---

hearings, joint testimony of the Secretary of the Navy and Chief of Naval Operations, occurred March 14, 1984. Hereinafter cited as *FY 85 SASC Hearings.* See also Huntington, "Defense Policy of the Reagan Administration," pp. 92–102.

17. Huntington, "Defense Policy of the Reagan Administration," p. 101.

18. The commanders of the Unified and Specified Commands, under the direction of the President and the Secretary of Defense, are by law responsible for planning for actual combat operations. Unified commands include the Atlantic, Pacific, European, Southern (Central and South America), Central (Middle East), Space, and Readiness Commands. The Atlantic, European (which includes the Mediterranean), and Pacific (which includes the Indian Ocean) are most relevant to naval forces. Specified commands include those with forces of a single service only, presently the Strategic Air Command and Military Airlift Command.

19. Watkins, "Maritime Strategy," p. 4.

20. *Soviet Military Power 1986* (Washington, D.C.: U.S. Government Printing Office, 1986), p. 10. Discussion of Soviet strategy and doctrine is drawn from this publication; Watkins, "Maritime Strategy"; Office of the Chief of Naval Operations, Department of the Navy, *Understanding Soviet Naval Developments (Fifth Edition)* (Washington, D.C.: U.S. Government Printing Office, 1985); and the testimony of Rear Admiral John Butts, Director of Naval Intelligence, Senate Armed Services Committee, 99th Congress, 1st Session, *Hearings on the Department of Defense Authorizations for FY 86: Part 8* (Washington, D.C.: U.S. Government Printing Office, 1986), pp. 4344–4370. While excellent and accurate non-government publications are available on these topics, these references are more representative of the official viewpoint and thus of actual influences on the Maritime Strategy.

that is "still a 'nuclear' war in the sense that the nuclear balance is constantly examined and evaluated in anticipation of possible escalation" and in which the Soviets place "high priority on changing the nuclear balance, or as they term it, the nuclear correlation of forces, during conventional operations."[21] Soviet war aims would be to defeat and occupy NATO, to neutralize the power of the United States and China, and to dominate the postwar world.[22] The probable centerpiece of Soviet strategy in such a global war would be a "combined-arms assault against Europe, where they would seek a quick and decisive victory. . . . the Soviets would, of course, prefer to be able to concentrate on a single theater. . . ."[23]

The most important Soviet navy roles in global war would be protecting (in Soviet terms, "ensuring the combat stability of") Soviet ballistic missile submarines (SSBNs) and protecting the approaches to the Soviet homeland. Consistent with the Soviet stress on the nuclear correlation of forces, the Soviet navy must give high priority to destroying Western sea-based nuclear assets, including aircraft carriers, SSBNs, or ships equipped with land-attack cruise missiles, although for the foreseeable future the Soviets can expect to have essentially no capability to locate U.S. ballistic missile submarines. Other traditional naval roles, such as attacking reinforcement and resupply shipping or supporting the Soviet army, are clearly secondary, at least at the start of a war.[24]

To implement this strategy, the bulk of the Soviet navy must be used to protect defensive bastions near the Soviet Union, with only limited forces deployed into the broad ocean areas. This essentially defensive initial role for the Soviet navy is confirmed by the overwhelming majority of Soviet naval exercises.[25]

---

21. Watkins, "Maritime Strategy," p. 7. The Soviet emphasis is based on the view that a war "would probably escalate to nuclear conflict," although they would prefer that it remain conventional. Testimony of Rear Admiral Butts, p. 4367. This view may be changing. For a detailed discussion of evidence that Soviet doctrine contemplates conventional war, see James M. McConnell, "Shifts in Soviet Views on the Proper Focus of Military Development," *World Politics*, Vol. 37, No. 3 (April 1985), pp. 317–343. See also fn. 65.
22. *Soviet Military Power 1986*, pp. 13–14.
23. Watkins, "Maritime Strategy," p. 7.
24. Ibid.; Soviet navy missions are described in *Understanding Soviet Naval Developments*, pp. 11–18. On the vulnerability of U.S. SSBNs, although occasional allegations are made in the press that such SSBNs will be vulnerable in the near future, CIA officials testified on June 26, 1985 that "we do not believe there is a realistic possibility that the Soviets will be able to deploy in the 1990s a system that could pose any significant threat to U.S. SSBNs on patrol." Cited in Jonathan E. Medelia, "Trident Program," Congressional Research Service, *Issue Brief*, February 6, 1986.
25. See the chart of Soviet observed exercises in Watkins, "Maritime Strategy," p. 7.

AVAILABLE FORCES

The final factor shaping the Maritime Strategy is the structure of the military forces available to carry it out. As will be discussed more fully below, much public debate over strategy is really a debate over what forces the nation should procure for the future. While in theory strategy should determine forces, in practice the relationship is a reciprocal one, with available forces determining the limits of achievable strategy. The first duty of the professional military is to determine how to deter, or if necessary fight, a war *today*, and today's wars cannot be fought with future budgets. Thus, while Navy leaders have repeatedly used the Maritime Strategy to justify the ongoing buildup (actually a "buildback") to the so-called 600-ship navy,[26] they are equally quick to stress that the strategy is for "today's forces, today's capabilities, and today's threat," with "today's forces" invariably including those of allies, especially NATO.[27]

The Maritime Strategy Described

The strategy derived from these three factors is deliberately broad and general. Since it provides global guidance, rather than a detailed timetable, the strategy has no timelines attached. Based on the premise that "Sea power is relevant across the spectrum of conflict, from routine operations in peacetime to the provision of the most survivable component of our forces for deterring strategic nuclear war,"[28] it includes port visits and peacetime exercises to support alliances and short-notice response in time of crisis to deter escalation. While recognizing the importance of these aspects of the strategy, the Navy has devoted most of its attention—and critics have devoted almost all of theirs—to those aspects of the strategy dealing with global conventional war.

---

26. The term "600-ship navy" refers not only to a specific size but also to a specific composition, including 15 carrier battle groups, 4 battleship battle groups, 100 attack submarines, amphibious shipping for the assault echelons of a Marine amphibious force and a Marine amphibious brigade, and "an adequate number" of ballistic missile submarines. See, among many others, Lehman, "The 600-Ship Navy," p. 35.
27. Watkins, "Maritime Strategy," p. 4.
28. Ibid., p. 7. See also Figure 3, p. 8, for a curve invariably used in Navy briefings showing a "Spectrum of Conflict" from peacetime presence to strategic nuclear war. Despite this, there is no public explication of the Navy's role in strategic nuclear war beyond simple assertions of the importance of SSBN survivability. Unless otherwise noted, the remainder of the description of the strategy is from Watkins, "Maritime Strategy."

The Maritime Strategy foresees a global war as unfolding in three phases. In the first, *Deterrence or the Transition to War,* recognition that a specific international situation could lead to hostilities requires rapid, worldwide forward deployment of the Navy and Marine Corps (along with similar deployments by other services).[29] Actions taken in this phase will include the surge deployment of anti-submarine warfare (ASW) forces (particularly submarines), forcing Soviet submarines to retreat into defensive bastions; the assembly and forward movement of multi-carrier battle forces; and embarkation of Marine amphibious forces. At the same time, execution of Presidential authority to call up reserves and to place the Coast Guard under Navy control will help prepare for the implementation of plans for sealift to Europe. The massive nature of the forward movement (indicating national will) and its global nature (indicating an unwillingness to cede any area to the Soviets, to defend only some U.S. allies, or to allow the Soviets their preferred strategy of concentrating on a single theater) are both designed to reinforce deterrence while being easily reversible if deterrence prevails.

Should deterrence fail, a second phase, *Seizing the Initiative,* comes into play. The object is establishment of sea control in key maritime areas as far forward and as rapidly as possible. U.S. and allied ASW forces will wage an aggressive campaign against all Soviet submarines, including ballistic missile submarines. Carrier battle forces will fight their way into the Norwegian Sea, Eastern Mediterranean, and Pacific approaches to the Soviet Union, depending on their location when hostilities begin. The integrated nature of modern naval warfare is especially relevant to this phase. For example, in the Norwegian Sea, air superiority is needed to permit operation of Maritime Patrol Air ASW aircraft, which in turn are needed to destroy Soviet cruise missile-carrying submarines, which in turn are necessary to protect aircraft carriers and missile ships required to ensure air superiority, which is essential for offensive operations. The strategy assumes that the use of amphibious forces or the strike power of carrier battle forces against targets ashore may also be appropriate in this phase, and Navy leaders specifically note that "the main threats to our fleet during this phase are . . . missile-carrying aircraft of Soviet Naval Aviation [land-based in the Soviet Union]. The United States cannot allow our adversary to assume he will be able to attack the fleet with impun-

---

29. The emphasis on early movement results from the distances involved, the importance of attack submarines reaching forward areas before conflict begins, and the belief that early forward operations have deterrent value. U.S. attack submarines have been surge-developed to verify their ability to conduct such early forward movements. *FY 85 SASC Hearings,* p. 3888.

ity, from inviolable sanctuaries."[30] The last essential aspect of this phase is the establishment of a logistics structure to support sustained forward operations, including advanced bases, sealift, and mobile logistics support forces.[31]

The final phase, *Carrying the Fight to the Enemy*, begins once sea control has been established and involves using carrier air power and Marine amphibious forces directly against targets ashore. At the same time, the vigorous ASW campaign, including the campaign against Soviet SSBNs, would continue. Direct conventional attacks on the Soviet homeland, while not ruled out (or required) in earlier phases, would be more likely in this phase in order to threaten the bases and support structure of the Soviet navy.

Throughout all phases of the strategy, close cooperation with allied navies and with other services, particularly the U.S. Air Force, is mandatory. Naval operations in the Baltic and Black seas, for example, would be almost entirely allied responsibilities. Allied contributions to Mediterranean and Norwegian sea ASW and to worldwide control and protection of shipping would be significant. Consistent with the direction of NSDD-32, the strategy envisions global operations, with the Pacific equal in importance to other theaters.[32] Operations within a particular theater may be conducted sequentially to permit adequate massing of forces; such an approach is consistent with the sequential guidance of NSDD-32.

The overall objectives of this strategy, in the words of Admiral Watkins, are to:

—Deny the Soviets their kind of war by exerting global pressure, indicating that the conflict will be neither short nor localized.
—Destroy the Soviet Navy: both important in itself and a necessary step for us to realize our objectives.
—Influence the land battle by limiting redeployment of forces, by ensuring reinforcement and resupply, and by direct application of carrier air and amphibious power.

---

30. Watkins, "Maritime Strategy," p. 12.
31. There is an old military saying that "amateurs discuss strategy; professionals discuss logistics." One of the important and little noticed results of the growing prominence of the strategy within the Navy is a renewed interest in wartime logistics support.
32. On the Pacific roots of the strategy, see Harlan K. Ullman, "The Pacific and U.S. Naval Policy," *Naval Forces*, Vol. 6, No. 6 (1985), pp. 36–48. On the current situation, including why a swing strategy is wrong, see Foley, "Strategic Factors in the Pacific," pp. 34–38.

—Terminate the war on terms acceptable to us and to our allies through measures such as threatening direct attack against the homeland or changing the nuclear correlation of forces.[33]

Assumptions Explicit and Implied

In addition to embodying a specific view of U.S. and Soviet strategy, the Maritime Strategy incorporates a number of inherent assumptions seldom made explicit by Navy spokesmen. Among the more important of these are:

AS A GLOBAL POWER, THE UNITED STATES HAS MILITARILY IMPORTANT INTERESTS BEYOND NATO. Europe and NATO are vital interests of the United States. The United States is committed to the defense of Europe by treaty, by self-interest, and by the simple fact that over three hundred thousand American troops are stationed on European soil. In addition, however, the United States has formal defense agreements of varying types with a number of nations outside of Europe. Demonstrating American readiness to honor commitments to, for example, Japan and Korea is also vital. An important function of any peacetime military strategy is declaratory; announcing a willingness to abandon Pacific allies in time of global war is unlikely to contribute to either deterrence or the furthering of peacetime foreign policy goals.

THERE WILL BE NO IMMEDIATE COLLAPSE IN CENTRAL EUROPE. Maritime power inherently requires time to take effect. If land and air forces in Germany are overrun in days, neither the Maritime Strategy nor any alternative use of seapower is likely to be able to prevent that event (although, as will be argued below, there is an implicit assumption that loss of Central Europe is the loss of a campaign, not a war).

THE BEST USE OF SEA-BASED AIRPOWER IS NOT DIRECTLY IN CENTRAL EUROPE. In time of war, the Navy will, of course, fight wherever it is told to do so by the President, the Joint Chiefs of Staff, and the Unified Commanders. The Maritime Strategy, among its other purposes, provides the prewar professional consensus of the Navy's leadership about where it *should* be told to fight. While Navy leaders stress that the flexibility of seapower allows many options, there is an implicit assumption that Germany is not the optimum location for employing carrier air power. The relatively small increase in air

33. Watkins, "Maritime Strategy," p. 14.

power from early arriving carriers, the command and control complexity of adding sea-based forces to a complex air war, and, above all, the fact that it is not in NATO's interest to allow Soviets the luxury of their preferred strategy, all argue for using sea-based air power to tie down and divert Soviet forces elsewhere rather than using such air power directly on the Central Front.[34]

NUCLEAR WEAPONS USE IS NOT INEVITABLE. By altering the nuclear correlation of forces, the Maritime Strategy seeks to make nuclear escalation "a less attractive option to the Soviets with the passing of every day."[35] Lacking the ability to directly influence the nuclear decision ashore, however, the strategy must assume that early use of nuclear weapons in Europe is not required, since nuclear use at sea will almost certainly follow.[36]

THE APPROACH TO PROTECTING THE SEA LANES MUST BE DIFFERENT FROM THE PAST. It is arguable that the allies won the World War II Battle of the Atlantic, not by sinking submarines, but by building ships faster than they could be sunk. In that conflict, in both theaters, interdiction of strategic raw materials, the resources of war, was an important mission. It is tempting to use that experience as a model for the future. The strategy, however, recognizes that "we will neither be able to tolerate attrition typical of World War II nor provide adequate dedicated sealift to transport the strategic raw materials we will require."[37] The most frequently discussed economic shipping is oil from the Persian Gulf through the Straits of Hormuz; Navy "analysis shows the straits could be closed for about 2 months without . . . impacting . . . warfighting capability. . . ."[38]

NATO IS NOT THE SAME AS THE CENTRAL FRONT. Unlike other assumptions, this premise is frequently articulated by Navy leaders. It deserves mention

34. For an alternate view, see Captain Andrew C. Jampoler, "A Central Role for Naval Forces? . . . to Support the Land Battle," *NWC Review*, Vol. 37, No. 6 (November–December 1984), pp. 4–17.
35. Watkins, "Maritime Strategy," p. 14.
36. Given the stated policy of NATO to use nuclear weapons if necessary, this is a significant assumption. In a September 1983 interview, for example, General Bernard Rogers, NATO's Supreme Allied Commander in Europe, said: "if we are attacked conventionally, we can only sustain ourselves conventionally for a relatively short time. I then will be forced to . . . ask for the authorization . . . to use nuclear weapons." Quoted in *Armed Forces Journal International*, September 1983, p. 74. For an argument that an immediate collapse requiring nuclear use will not occur, see John J. Mearsheimer, *Conventional Deterrence* (Ithaca and London: Cornell University Press, 1983), pp. 165–188. The unlikelihood of initial nuclear use at sea is discussed below; see fn. 64 and accompanying text.
37. Watkins, "Maritime Strategy," p. 11.
38. *FY 85 SASC Hearings*, p. 3854.

nonetheless because of the erroneous charge that the Maritime Strategy is somehow an alternative to NATO support. In contrast, the framers of the strategy clearly assume that defense of those allies on the flanks (Norway, Denmark, Iceland, Italy, Turkey, Greece) must have equal priority with defense against a Soviet thrust in Germany since "No coalition of free nations can survive a strategy which begins by sacrificing its more exposed allies to a dubious military expediency."[39]

THE EUROPEAN CAMPAIGN IS NOT THE SAME AS THE WAR. The Maritime Strategy, like the overall military strategy it supports, recognizes that the defense of Europe is vital to the United States. But as strategists outside government have pointed out, military leaders "need to consider unpleasant as well as satisfactory futures."[40] Destruction of the Soviet fleet and establishment of maritime superiority are necessary to prevent defeat in Europe and provide leverage for war termination; they also enable the United States to ensure its own security and to enhance deterrence by demonstrating the ability to continue the conflict regardless of the outcome in Europe.

Contributions and Limitations

CONTAINING CRISES AND ADVANCING U.S. POLICY GOALS
While the attention and the debate in recent years have focused on those aspects of the strategy dealing with global war, the most common use of the U.S. Navy is to advance U.S. foreign policy aims in circumstances short of global war. An important aspect of this role is responding to crises, both to contain specific crises and to demonstrate the national will that is an inseparable component of deterrence. Success in such Third World efforts limits the expansion of Soviet influence and thus preserves the status quo.

Given innate Soviet conservatism and the lack of any plausible incentive for the Soviet Union to risk destruction in order to change the status quo, war in Europe is not likely to occur as a long-planned act of Soviet policy. Instead, a modern equivalent of the downward spiral of August 1914 during a European crisis, perhaps one growing out of an extra-European situation, offers the most plausible path to war. Containing extra-European crises is,

---

39. Lehman, "The 600-Ship Navy," p. 37.
40. Colin S. Gray, "Maritime Strategy," *Proceedings*, Vol. 112, No. 2 (February 1986), p. 41. A similar point is made in F.J. West, Jr., "Maritime Strategy and NATO Deterrence," *NWC Review*, Vol. 38, No. 5 (September–October 1985), pp. 5–19.

therefore, of obvious importance to the West. As the chief agent for dealing with Third World crises, the Navy has a unique role to play in preventing escalation of a local problem to the point where war in Europe might seem necessary or inevitable to the Soviets. Navy spokesmen routinely emphasize that Navy and Marine Corps forces were used in 80 percent of the 250 instances of American military force employment since World War II, a point no less valid for being frequently stated.[41]

DETERRENCE

The most important contribution any strategy can make is to deter, or help deter, major war. Contemporary concepts of deterrence require a capability not only to inflict punishment but also to deny war aims by holding at risk the military capability of a potential aggressor. In addition, since deterrence exists in the mind of an adversary, it depends on perceptions of national will as well as of national capabilities. How are navies and the Maritime Strategy relevant to deterrence?

First, maritime forces help demonstrate national will. Easy to move both physically and politically, the fleet can be deployed as a unilateral U.S. action, nonprovocative because reversible. Fleet movements in time of crisis can both demonstrate U.S. commitment to *all* its allies and, by their global nature, demonstrate that any war will not unfold along preferred Soviet lines. Such movements are an important *political* deterrent, precluding the Soviets from believing that they can coerce individual allies and fragment the alliance through what might be called a "political blitzkrieg."

Second, early forward fleet movements demonstrate that the Soviets will be able neither to cut off Europe from the United States nor to draw down remote theaters to reinforce Europe. Thus, they must win quickly in Europe or not at all. Since, as John Mearsheimer has argued, the Soviets probably cannot win quickly, their remaining rational choice is to not attack at all.[42]

Finally, the Maritime Strategy recognizes that the United States must deter, not a collection of American theoreticians and scholars, but a Soviet leadership that constantly calculates the nuclear correlation of forces and uses those calculations in the decision-making process. By making it clear at the outset

---

41. Watkins, "Maritime Strategy," p. 8.
42. Mearsheimer, *Conventional Deterrence*, pp. 165–188. Since Mearsheimer argues that denial of a quick victory is, in itself, sufficient for deterrence, he would of course reject the argument that early fleet movements have utility. Even in terms of his own theory, this conclusion ignores the risk of a political blitzkrieg.

that Soviet SSBNs will be at risk in a conventional war, the strategy alters Soviet correlation of forces calculations, and thus enhances deterrence.[43]

WAR-WAGING

War is the ultimate test of any strategy; a strategy useless in war cannot deter. Since Soviet strategy views Europe as central, a counter-strategy must contribute directly or indirectly to the European battle. Maritime power can make four significant contributions:

PROTECTING THE SEA LINES OF COMMUNICATION (SLOCS). Any strategy for war in Europe must ensure the unimpeded flow of supplies to Europe, the overwhelming majority of which must go by sea. The potential threat to these lines of communication represented by the Soviet submarine force is immense. Sea lane interdiction, however, is a lower priority Soviet navy task than protecting sea-based strategic forces or homeland defense.[44] Thus, early aggressive forward ASW operations directly protect the reinforcement and resupply of Europe, initially by tying down the Soviet submarine force in a pro-SSBN protective role and ultimately by destroying Soviet general purpose submarines.

DIRECTLY SUPPORTING THE LAND BATTLE ON THE EUROPEAN FLANKS. While Soviet literature suggests the Western TVD (NATO's Central Front) would be the scene of the decisive conflict, the Soviets also envision operations in the Northwestern TVD to seize northern Norway and in the Southwestern TVD to attack Thrace and to seize both sides of the Turkish straits.[45] A considerable fraction of the available combat power to thwart these thrusts is in the form of carrier-based aircraft, with three or four aircraft carriers potentially available in each theater.[46] Success in such defense serves both military and political aims by interrupting Soviet strategy and ensuring that alliance cohesion is not sacrificed by an apparent unwillingness to defend *all* members of an alliance that extends from Norway to Turkey.

TYING DOWN FORCES, ESPECIALLY AIR FORCES, THAT MIGHT OTHERWISE BE AVAILABLE FOR THE EUROPEAN BATTLE. Some 1700 Soviet tactical aircraft are

43. This point and the resultant escalation risks are discussed at greater length below.
44. *Soviet Military Power 1986*, pp. 88–89.
45. Ibid., pp. 61–62. TVD stands for "Teatr Voennykh Deistvii" or "Theater of Military Operations." TVDs are the basic Soviet organizational structure for planning strategic operations. See ibid., pp. 11–14, for additional details.
46. Lehman, "The 600-Ship Navy," p. 36, gives somewhat larger numbers; these relate to the future when all of the planned 15 carrier battle groups are available.

deployed in the Far East TVD, compared with 2300 Soviet and 1600 other Warsaw Pact tactical aircraft available in the Western TVD.[47] While the more capable forces are in the European area, Far East tactical air forces could be made available to augment them if not pinned down by the necessity to defend against aggressive forward employment of American carrier and amphibious forces in the Pacific. There is no suggestion here that the Soviets necessarily *plan* on such a shift; rather, the result of maritime operations in the Pacific will be to limit Soviet flexibility to respond to changing conditions.[48] Maritime forces are less likely to have an effect on a possible Soviet decision to shift ground forces from the Far Eastern to the Western TVD. Such forces are more difficult to transfer and are held in place by the constant Soviet need to consider the position of China. While it is exceptionally unlikely that China would find it in its interest to become an active belligerent, the active use of naval power, by demonstrating a clear U.S. intent to remain a Pacific power, may encourage a Chinese posture of armed neutrality rather than cooperation, thus complicating Soviet decision-making.

DENYING THE SOVIETS THEIR PREFERRED STRATEGY. Two-front wars are difficult. While the United States and its allies cannot open a "second front" through maritime power, the global use of such power can have a similar strategic effect on the minds of Soviet leaders, and it is in the minds of leaders that decisions to continue or terminate wars are made. Denying the Soviets their preferred strategy of a short, single theater war is an important, although intangible, contribution of maritime forces.

In addition to contributions directly relevant to Europe, maritime power and the Maritime Strategy offer a means for *defending other allies*. NATO, while America's most important defense commitment, is not its only one. It is not possible to predict which nations will become involved in a global war with the Soviets; Soviet paranoia may cause them to see enemies all around them and to preemptively attack those enemies. Thus, for example, Japan might well be the target of Soviet aggression in the context of a global war.[49] If such allies are attacked, it will be important to defend them; if they are

---

47. *Soviet Military Power 1986*, p. 13.
48. For the effect of Far East operations—or lack thereof—on Central/East Europe in World War II, see John Erickson, *The Road to Stalingrad* (Vol.I) and *The Road to Berlin* (Vol. II) (Boulder, Colo.: Westview Press, 1983), especially Vol. I, pp. 14, 55–57, 218, 237–240, 271–272, 295, and Vol. II, pp. 43, 132, 156.
49. *Soviet Military Power, 1986*, p. 63.

not, it may still be important to the United States' postwar position for them to realize they would have been defended.

WAR TERMINATION

Should deterrence fail, the most important contribution of the Maritime Strategy, as well as the least understood, will be in war termination. While deterrence is the chief component of U.S. policy, deterrence can fail. In that case, maritime power may offer a unique form of war termination leverage through its ability to dramatically alter the nuclear balance, or, in Soviet terms, the nuclear correlation of forces.[50] While Soviet doctrine may be shifting towards a conventional war option, Soviet leaders still assume a war between the two superpowers has a high probability of escalating and therefore place great importance on the constant calculation and evaluation of the nuclear correlation of forces.

The Navy can alter those Soviet calculations, most obviously by attacking SSBNs which, as a principal component of Soviet strategic reserves, are central to correlation of forces calculations. Such attacks offer significant war termination leverage. In addition to altering the Soviet estimate of their own nuclear capabilities, maritime forces can increase the magnitude of the nuclear threat that the Soviets must face. As the Soviet fleet is eliminated, both carrier strike aircraft (which the Soviets view as a significant nuclear threat) and nuclear Tomahawk missiles will be in a position to threaten the Soviet homeland. Objectively, the incremental increase in allied nuclear capability that these forces offer is small. Similarly, the destruction of even a large fraction of the Soviet SSBN force will result in only a limited decrease in total Soviet nuclear strike capability. But the Soviets, with their military conservatism and penchant for constant algebraic calculation of the correlation of forces, will not ignore either factor. They will evaluate the correlation of forces as growing constantly less favorable.

The fear that the war may escalate and the fact that such escalation is less and less attractive every day provide a powerful incentive for war termination. It is particularly powerful if combined with a prospect of stalemate on

---

50. The discussion of both war termination and the escalation considerations associated with it is drawn from Linton F. Brooks, "War Termination Through Maritime Leverage," in Keith Dunn and Stephen Cimbala, eds., *Conflict Termination and Military Strategy: Coercion, Persuasion and War* (Boulder, Colo.: Westview Press, forthcoming).

the Central Front (i.e., if NATO, while not winning there, is at least not losing) and with war termination aims that seek neither Soviet dismemberment nor the destruction of the socialist system in the Soviet Union, but rather the restoration of the status quo ante bellum.

LIMITATIONS

By its very nature, the Maritime Strategy considers only one aspect of U.S. military capability and only some of the military tasks U.S. armed forces might be required to undertake. Thus, there are inherent limitations in its applicability and in the contribution maritime forces can make. Among the more important are:

THE MARITIME STRATEGY (LIKE ANY OTHER USE OF MARITIME POWER) CANNOT ALONE BRING VICTORY IN A WAR AGAINST A MAJOR LAND POWER. This point is self-evident. Nonetheless, it needs to be stated because, in conversation if not in print, maritime critics often act as though "maritimists" were claiming that nothing else but a navy is needed. They should (and probably do) know better. As the Secretary of the Navy has noted, "Maritime superiority alone may not assure victory but the loss of it will certainly assure defeat."[51]

THE U.S. NAVY ALONE CANNOT IMPLEMENT THE MARITIME STRATEGY OR ANY CONCEIVABLE ALTERNATIVE TO IT. Again the point is self-evident; again it requires restatement. Just as a land-based defense of Europe is inconceivable without allies, so too the Maritime Strategy cannot be implemented without major contributions from allied navies, at a minimum those of NATO and, depending on circumstances, those of Japan and other non-NATO nations as well.

THERE ARE INHERENT UNCERTAINTIES IN THIS OR ANY OTHER STRATEGY THAT, BY THEIR NATURE, CAN NEVER BE RESOLVED SHORT OF WAR. One element of the strategy, for example, is early forward deployment of forces in time of grave crisis. From the coastal areas of the United States, it takes at least a week to reach the Norwegian Sea, at least nine days to reach the area off Japan, and at least ten days to reach the Mediterranean.[52] Thus, timely forward movement depends upon receiving and reacting to warning. It is,

---

51. Lehman, "The 600-Ship Navy," p. 36.
52. These figures assume a 20 knot speed of advance which may well be optimistic under some scenarios.

by definition, impossible to be certain in advance that such warning will be available, be recognized, and, most importantly, be acted upon.[53]

In a broader vein, the strategy seeks to counter a specific Soviet strategy and a specific Soviet navy role within that strategy and thus might prove inappropriate to counter a different Soviet approach. If, for example, the Soviets were to deploy their entire submarine force to the open ocean before the outbreak of war, a very different U.S. approach to ASW might be required. By adopting a declaratory strategy of threatening SSBNs, the United States limits the Soviets' options to change their strategy before hostilities begin; by conducting a successful forward ASW campaign, NATO limits Soviet ability to adopt an alternate strategy during hostilities. While the strategy thus seeks to limit Soviet options, no one can be certain that Soviet wartime strategy (or Soviet decisions on the use or nonuse of nuclear weapons) will be what their prewar doctrine suggests. Finally, should war come, the Maritime Strategy, like the national military strategy, seeks war termination on favorable terms and shares the inherent uncertainties embodied in such a goal.[54]

THERE ARE PLAUSIBLE CONFLICTS FOR WHICH THE MARITIME STRATEGY (BUT NOT NECESSARILY MARITIME POWER) IS INAPPLICABLE. The first is a major conflict (Vietnam War-scale) not involving the Soviet Union. Drawing on NSDD-32 guidance that the Soviets are the main threat, the Maritime Strategy does not deal directly with such conflicts except in terms of crisis response and containment. Such a limitation is acceptable since only the Soviet Union offers a sufficient challenge to require a global strategic response. Similarly, the strategy does not deal with a regionally limited conflict outside Europe in which the United States and the Soviet Union are directly involved. This

---

53. Early forward fleet movement, which can be unilaterally undertaken by the United States, not only can be done in advance of NATO as a body recognizing the extent of a crisis, but also may serve a "pump priming" function in convincing allies to act.
54. Fred Charles Iklé, *Every War Must End* (New York: Columbia University Press, 1971), documents the inherent difficulty of war termination and the tendency of states to continue fighting well past the point where they "should" stop. These problems apply to *any* attempt at war termination, including the approach inherent in the Maritime Strategy. A particular problem in a war with the Soviet Union is ensuring that U.S. war aims are not perceived as extending to the destruction of the Soviet state or the replacement of Communist party control. In commenting on an early draft of Brooks, "War Termination," Stephen Cimbala observed that the Soviets may not perceive United States war aims as limited if their homeland and strategic forces are at risk and second echelon forces are being attacked on Warsaw Pact (or even Soviet) territory. This underscores the need for some method of communicating the nature of U.S. war aims during hostilities.

latter point is significant in light of the occasional criticism that the Maritime Strategy is inconsistent with the national policy to limit the scope and intensity of war.[55] The strategy deals with the case in which war has spread to Europe; in such a case, war is, in essence, already global and a global maritime response is appropriate. Little in Soviet doctrine suggests that any stakes other than Europe are important enough for war with the United States; however and wherever a future war may start, it will ultimately be for Europe.[56] If, however, a regional war does occur and does remain limited, a variation of the Maritime Strategy would be required. Although a major concern of the Navy in the 1950s and 1960s, this point has not been addressed publicly by today's Navy leaders; presumably, such a strategy would involve global operations on the high seas against Soviet general purpose forces and direct projection of power ashore within the region of conflict.

Criticisms Valid and Otherwise

Critics of the Maritime Strategy focus on three major issues: risk, relevance, and resources. They argue that the strategy will not accomplish anything important, is too dangerous, or costs too much, thus diverting resources from where they are really needed. In each case, a legitimate issue is buried in the criticism, but in each case the critics are, on balance, wrong.

RISK

Those who focus on risk have two concerns. The first is that the Navy is incapable of implementing such an aggressive and ambitious strategy. While the strategy deals with more than aircraft carrier operations, critics often frame their arguments as assertions about carrier vulnerability when operating in close proximity to the Soviet homeland.[57] Retired Admiral Stansfield Turner's name is frequently invoked to buttress the judgment that such operations are excessively risky. Admiral Turner finds it difficult "to believe

---

55. See, for example, John Collins, "Comment and Discussion," *Proceedings*, Vol. 112, No. 2 (March 1986), p. 20.
56. A key consideration in responding to such wars is the prospect of U.S. forces being maldeployed should the conflict spread. This problem is somewhat less severe for naval forces because of their mobility.
57. "To venture U.S. carrier battle groups close enough to the Soviet Union to launch air strikes on the Soviet navy's home ports is to venture into the jaws of defeat." Jeffrey Record, "Jousting with Unreality: Reagan's Military Strategy," *International Security*, Vol. 8, No. 3 (Winter 1983/1984), p. 13.

that thoughtful military planners would actually do this," is certain that no President "could possibly permit the Navy to attempt such a high risk effort,"[58] and suggests that he has "yet to find one Admiral who believes that the U.S. Navy would even attempt it."[59]

Without actual combat, estimates of the Navy's ability to operate carrier battle groups in so-called "high-threat" areas are professional military judgments, best made not by those who must rely on experience with past fleet conditions, but by the men who would have to carry out such operations today. Chief among this latter group is the Commander of NATO's Striking Fleet Atlantic. Writing recently, the current Commander, Vice Admiral Henry Mustin, said:

concern over our forward strategy is frequently couched in terms of whether U.S. aircraft carriers . . . can survive in the Norwegian Sea in a conflict with the Soviet Union. No one has ever said that war with the Soviet Union would be easy. In war, ships get sunk, aircraft get shot down and people get killed. The Soviet Union and the Warsaw Pact would be very formidable. . . . they would not be invincible. The Striking Fleet can get . . . assistance in beating down Soviet air attacks through joint operations with NATO AWACS and Norwegian air defenses—including the U.S. Air Force—and we have demonstrated this capability in exercises. . . . The Soviets . . . acknowledge that a moving target ranging over thousands of square miles of blue water is much more survivable than a fixed airfield ashore. No one suggests that we should abandon all airfields in Norway at the start of hostilities, and yet some quake at the notion of less vulnerable carriers operating hundreds of miles at sea.[60]

The second risk issue concerns escalation. Critics see the strategy as provocative and likely to result in war rather than deterring nuclear war. They

---

58. Admiral Stansfield Turner and Captain George Thibault, "Preparing for the Unexpected: The Need for a New Military Strategy," *Foreign Affairs*, Vol. 61, No. 1 (Fall 1982), pp. 126–127.
59. Admiral Stansfield Turner, "Comment and Correspondence: Maritime Strategies," *Foreign Affairs*, Vol. 61, No. 2 (Winter 1982/1983), p. 457.
60. Mustin, "The Role of the Navy and Marines in the Norwegian Sea," p. 3. A related argument is the extent to which the use of nuclear weapons might make the Navy unable to carry out its strategy. See, for example, the questions of Senator Sam Nunn in *FY 85 SASC Hearings*, pp. 3878–3879. See fn. 64 and accompanying text for reasons why such use is unlikely. There is no open-source evidence that the Navy has a coherent plan for continuing or modifying the Maritime Strategy should nuclear weapons be used. See Captain Linton F. Brooks, "Tactical Nuclear Weapons: Forgotten Facet of Naval Warfare," *Proceedings*, Vol. 106, No. 1 (January 1980), pp. 28–33; Lieutenant Commander T. Wood Parker, "Theater Nuclear Warfare and the U.S. Navy," *NWC Review*, Vol. 35, No. 1 (January–February 1982), pp. 3–16; and Robert C. Powers, "The Impact of Nuclear Weapons on Naval Strategy," *Military Science and Technology*, Vol. 1, No. 5 (October 1981), on the impact of tactical nuclear weapons on war at sea. Despite the title of the last, none deal with strategy.

condemn the Pacific aspects of the strategy because it "threatens to fuel the arms race there," makes Japan "partner to a provocative strategy over which the Japanese have no influence," and "it seems virtually certain that such a conflict would 'go nuclear.'"[61] Others see dual-capable Navy ships (i.e., those capable of employing both nuclear and conventional weapons) coupled with offensive forward operations as eroding "the time-honored firebreak between nuclear and nonnuclear combat, raising the likelihood of nuclear war."[62] Operating aircraft carriers in the Norwegian Sea to defend Norway or to threaten Soviet military installations on the Kola Peninsula is also viewed by critics as risking escalation. The most serious criticisms, however, deal with the risks associated with attacking Soviet SSBNs.

To many, deliberate attacks on SSBNs seem dangerously escalatory and destabilizing, and must be avoided. One critic claims that, of "all the possible Navy strategies, this one is the most likely to cause the other side to reach for nuclear weapons."[63] Basing their logic on traditional theories of arms control, in which secure second strike strategic forces are indispensable to stability, such critics conclude that attacks on SSBNs raise the specter of a "use or lose" situation. While critics of the first sort fear the strategy cannot succeed, critics focusing on escalation fear it will, and in doing so lead to nuclear war, perhaps at sea, perhaps involving a strategic nuclear exchange.

The facts are not that clear. With regard to Pacific operations, critics and advocates are seeing opposite sides of the same coin. Forcing the Soviets to divert resources and attention from Europe is a strength of the strategy, not a weakness. Japan's central role arises not because the United States seeks to involve its allies in war, but because the United States has treaty obligations to defend Japan, and there is good reason to believe the Soviets will threaten that nation regardless of what action the United States takes. Little in history suggests that removing U.S. naval forces will reduce the chance of attack on Japan or, for that matter, of a North Korean attack on U.S. and South Korean forces under cover of a more general war.

Concern that U.S. actions at sea could force the Soviets to use tactical nuclear weapons to counter American naval superiority, especially aircraft

---

61. William M. Arkin and David Chappell, "Forward Offensive Strategy: Raising the Stakes in the Pacific," *World Policy Journal*, Vol. 2, No. 3 (Summer 1985), pp. 482, 488, 492.
62. Michael T. Klare, "Securing the Firebreak," *World Policy Journal*, Vol. 2, No. 2 (Spring 1985), p. 229. See also Posen, "Inadvertent Nuclear War?"; and Desmond Ball, "Nuclear War at Sea," *International Security*, Vol. 10, No. 3 (Winter 1985/1986), pp. 3–31 for similar points.
63. Barry Posen, quoted in Michael R. Gordon, "Navy Says in a Nonnuclear War It Might Attack Soviet A-Arms," *The New York Times*, January 7, 1986, p. A14.

carriers, is based on a misreading of Soviet doctrine. The Soviets place nuclear weapons under the same tight political control as does the United States. An extensive study of Soviet military literature found:

no literature evidence to support the view that release authority for tactical nuclear weapons is a Navy matter nor that a tactical nuclear war at sea alone would be initiated by the Soviets. The decision to initiate tactical nuclear war at sea appears neither a Navy decision nor one that will hinge on Navy matters.[64]

Simply put, a nation with a military dominated by artillerymen, a strategy focused on land, and a doctrine that suggests nuclear war cannot be limited[65] is not going to cross the nuclear threshold based on at-sea tactical considerations.

This leaves the most difficult question: attacking SSBNs. The disagreement between those who see the risk of escalation in such attacks and those who see war termination leverage is based on very different models of escalation.[66] Those with intellectual roots in traditional arms control theory view threats to SSBNs, by general agreement the most secure component of strategic forces, as escalatory by definition. They further assume that this conclusion is universally valid, based on an objective reality that does not depend on the particular characteristics of the decision-makers involved. Even viewing escalation through this lens, it is not clear that the stability model is valid. The loss to conventional attack of one SSBN at a time over a period of days or weeks provides no single event sufficient to warrant the catastrophic decision to escalate to the strategic level.[67]

---

64. James John Tritten, *Declaratory Policy for the Strategic Employment of the Soviet Navy*, P-7005 (Santa Monica, Calif.: Rand Corporation, 1984), p. 210. Tritten's conclusions are based on a content analysis of over 260 documents authored by or in the name of the Commander in Chief of the Soviet navy or the Soviet Minister of Defense. See also Donald C. Daniel, *ASW and Superpower Stability* (London: International Institute for Strategic Studies, forthcoming), especially chapter 6. Daniel argues that, since it is impossible to prevent some retaliation, escalation will not occur unless the Soviets believe the United States is about to launch a first strike.

65. "Any so-called limited use of nuclear facilities will inevitably lead to the immediate use of the whole of the sides' nuclear arsenal." Interview with Marshal of the Soviet Union N.V. Ogharkov, "The Defense of Socialism: Experience of History and the Present Day," *Krasnaya Zvezda* (Moscow), p. 3.

66. I am indebted to Barry Posen for this point.

67. Much of the discussion of Soviet response to SSBN losses is based on insights provided by Bradford Dismukes of the Center for Naval Analyses (CNA), both in personal discussion and in his unpublished CNA paper, "Pros and Cons of the Pro-SSBN Mission; What will the future bring?," June 1980.

Advocates of the anti-SSBN facet of the strategy, however, reject the conclusion that traditional arms control theory offers the proper escalation model. They base their assessment of escalation risks not on arms control theory, but on Soviet military doctrine. Soviet navy acceptance of attacks on SSBNs as an integral component of conventional war has been made clear by such authoritative spokesmen as the former commander of the Soviet navy, Sergei Gorshkov, who noted several years ago that, among the "main efforts of a fleet," the "most important of them has become the use of the forces of the fleet against the naval strategic nuclear systems of the enemy with the aim of disrupting . . . their strikes. . . ."[68] Such an approach is no more than the at-sea analogue of the priority, long recognized in the West, that the Soviets give to the destruction of nuclear weapons during the conventional phase of the land war.

Not only have the Soviets long accepted anti-SSBN operations as a legitimate military task (and one they would undertake were they able to do so); they have also long assumed that the United States will conduct such operations in time of war. Such an assumption is reasonable from the Soviet standpoint, both because of doctrinal mirror-imaging and because senior naval officers giving Congressional testimony have consistently stressed the practical difficulties of distinguishing between types of submarines and have indicated that all types of submarines would be legitimate wartime targets.[69] Thus, while in the West the explicit acknowledgment that attacking SSBNs was a component of the overall Maritime Strategy was news,[70] in the Soviet Union it was not.[71]

Even if Soviet doctrine did not recognize the prospect of attacks on SSBNs as legitimate, escalation serves no useful Soviet purpose. A nuclear strike on the United States would result in immensely destructive retaliation. It is

---

68. Admiral of the Fleet of the Soviet Union Sergeï G. Gorshkov, *The Sea Power of the State* (Annapolis, Md.: Naval Institute Press, 1979), p. 221.
69. John Perse, *U.S. Declaratory Policy on Soviet SSBN Security* (Arlington, Va.: Center for Naval Analyses, forthcoming).
70. Gordon, "Navy Says in a Nonnuclear War It Might Attack Soviet A-Arms," ignores all aspects of the strategy except anti-SSBN operations. A similar exclusive anti-SSBN focus is found in other press accounts. Swartz, "1986 Addendum."
71. Valentin Falin, "Back to the Stone Age," *Izvestiya*, January 23, 1986, p. 5, and January 24, 1986, p. 5, provides a strident Soviet attack on the Watkins article, replete with quotes such as "It is hardly possible to imagine anything worse." Of the 26 paragraphs in the two articles, only three deal with anti-SSBN operations, primarily as a vehicle to attack general U.S. nuclear policy and the Strategic Defense Initiative. There is no suggestion that such operations are either new or impermissible.

difficult to see why the Soviets would elect the physical destruction of their country unless the only alternative were its political destruction. If, therefore, allied war termination aims do not extend to the breakup of the Soviet state or the replacement of the Soviet leadership, but rather to some form of restoration of the status quo ante bellum, a Soviet nuclear strike is exceptionally unlikely.[72]

Once again, critics and advocates are seeing two sides of the same coin. Almost by definition, any U.S. action important enough to exert war termination leverage carries some risk of escalation. But no war with the Soviet Union is without immense risk, and the escalatory risk associated with conventional attacks on SSBN forces at sea should be acceptable as a unique means of gaining war termination leverage. Threatening SSBNs by conventional means carries far *less* risk of escalation than does the use of tactical nuclear weapons to restore a declining battlefield situation, a risk that NATO has accepted for years.

RELEVANCE

Quite apart from any notion of risk, some critics question the relevance of the strategy even if it should work and even if it were risk-free.[73] As one of the more prolific critics asserts, the "basic flaw in any maritime strategy is that, even if we swept the other superpower from the seas and pulverized all its naval bases, this would not suffice to prevent it from dominating the Eurasian landmass."[74] The short answer to this criticism is that no one ever claimed it would. Even the most vigorous advocates of maritime power do not suggest that it is a substitute for ground forces and air power in Europe. Wars are won on land, but they can be lost at sea.

In discussing the issue of relevance, the proper issue is whether some alternate employment of maritime forces is *more* relevant. The first alternative that critics propose deals with protecting the sea lanes to Europe. Critics and supporters alike agree with the need to protect U.S. resupply shipping. They differ over how such a mission should be accomplished. The Maritime Strategy seeks to discourage early Soviet forward deployments by adopting a declaratory strategy of threatening SSBNs, thus forcing the Soviets to with-

---

72. Absence of escalation, of course, is not the same as successful termination. See fn. 54 for war termination difficulties.
73. Risk and relevance are obviously related. Those who believe that forward maritime operations contribute little will see no point in accepting even minimal risk.
74. Komer, *Maritime Strategy or Coalition Defense?*, p. 106.

hold general purpose naval forces to protect those SSBNs, and to preclude later deployment by conducting a successful forward ASW campaign. In contrast, critics typically advocate what they term "defensive sea control," asking "why wouldn't a passive defense line across the Greenland–Iceland–United Kingdom gap protect our sea-lanes? Why are U.S. maritime strategists concerned . . . if the Soviet Navy stays home?"[75]

Advocates of such a Maginot Line strategy miss several points. First, of course, there is no reason to assume that the Soviet navy, particularly its attack submarine component, will "stay home" once it is clear that the United States has no plans to challenge the bastions. Second, such a passive strategy cedes control of the Norwegian Sea (and of the coast of Norway) to the Soviets, violating the obligation of the United States to defend *all* its allies. Finally, such a strategy forgoes both the advantages to the ASW campaign of early deployments and the war termination and deterrent leverage attained by holding Soviet SSBNs at risk.

Critics recognize the relevance of SLOC protection; they simply disagree about how to do it. In contrast, skeptics question whether it matters if other aspects of the strategy are accomplished at all. Specific doubts have been raised about the warfighting relevance of forward power projection and of Pacific operations and about the relevance to deterrence of the strategy as a whole.

While they take issue with using amphibious forces and carrier battle groups in forward power projection operations, critics have yet to come up with attractive alternatives. In theory, alternatives are available. Aircraft carriers could be deployed to engage Soviet clients and surrogates (Cuba, Vietnam), used to augment the air battle over the Central Front, or maintained as some form of strategic reserve. Amphibious forces, instead of being employed as envisioned in the Maritime Strategy,[76] could be committed early in the war to augment defenses in Europe or used against Third World Soviet surrogates. On examination, none of these alternatives appears attractive.

Use of carriers against surrogates may be useful and necessary during the strategy's second phase (Seize the Initiative). One advantage of mobile,

---

75. Collins, "Comment and Discussion," p. 20. A similar argument is found in Turner and Thibault, "Preparing for the Unexpected." Even recognizing that Turner and Thibault are discussing future forces more than current strategy, their article is curious. While discussing sea control to ensure resupply, they appear to ignore both ASW and Soviet submarines.
76. General P.X. Kelly, Commandant of the Marine Corps, and Major Hugh K. O'Donnell, Jr., "The Amphibious Warfare Strategy," Supplement to United States Naval Institute *Proceedings*, January 1986, pp. 18–19.

flexible forces is that they can be diverted to alternate tasks. But no operations against any surrogate appear to have as much prospect for direct influence over the Soviets as does the existing strategy. It is difficult to see how surrogates can be more relevant to a NATO battle than the NATO flanks themselves.[77] Using carriers as a strategic reserve denies the early benefits of mobile forces and fails to recognize that Soviet strategy increasingly assumes that "the role and significance of the initial period of the war and its initial operations [have] become incomparably greater."[78]

The remaining option, using Marine Corps or carrier air power (perhaps without the carriers) in Central Europe, is therefore the alternative that critics presumably prefer. Several problems arise. First, because carriers are the chief U.S. tool for responding to crises and because crisis control is an important aspect of war prevention, one cannot be certain where carriers would be at a war's start, thus making integrated European planning difficult. Second, the 50–60 fighter and attack aircraft per carrier are a relatively small addition to the 2100 ground attack and 900 interceptor aircraft already in place in Europe,[79] but, by virtue of their ability to threaten different areas, they can tie down far more resources on the flanks and in the Pacific. Since Soviet aircraft in these areas are distributed across a wide battle area, carrier aircraft can be concentrated in numbers that *do* make a difference. Finally, while direct use of carrier air power in Germany may or may not be more relevant to the Central Front, it is less relevant to NATO as a whole. NATO's effectiveness depends on its solidarity, which in turn requires the defense of *all* its members.

Similar arguments apply to amphibious forces. They may be required to act against surrogates or, if the European ground war goes badly, to "support a NATO defense which is *in extremis* on the English Channel coast."[80] Given the option, however, their use as mobile reserves capable of forcible insertion and flexible operation seems preferable to some alternate plan for early integration of these comparatively light forces into the direct defense of Germany.

---

77. The sole exception is Cuba. "It will be a real problem to get merchantmen moving through this area . . . until Cuba is taken care of either politically or militarily." Lehman, *FY 85 SASC Hearings*, p. 3870. To avoid diverting forces, Cuba must be induced or coerced to remain neutral.
78. Ogharkov, "The Defense of Socialism," p. 3.
79. *Soviet Military Power 1986*, p. 89, for European figures; *FY 85 SASC Hearings*, p. 3857, for composition of carrier air wings.
80. Kelly and O'Donnell, "The Amphibious Warfare Strategy," p. 26.

Maritime critics reserve a special form of criticism for Pacific forward operations. Accustomed to thinking in theater terms, critics doubt the relevance of global operations and urge against any form of horizontal escalation.[81] Since Pacific operations are unlikely to draw ground forces from other theaters, they are deemed useless, even though the critics acknowledge that the Soviets "might reinforce their Pacific Fleet air forces."[82] But air power is exactly what will be crucial in a European war. The result the critics denigrate would be a clear gain for NATO forces; indeed, even if the Soviets do not reinforce the Far Eastern TVD but simply fail to draw on its resources to augment the West, maritime power will have made a significant contribution. The argument that a return to the swing strategy of the 1970s would be more relevant than aggressive forward operations in the Pacific is fallacious even in terms of European defense; it is even more so when the political impact on America's Asian allies is taken into account.

The most important criticism to analyze is the allegation that the Maritime Strategy is irrelevant to deterrence. The first question is what it is one seeks to deter. Mearsheimer, for example, in focusing on deterrence of large-scale conventional attack in Europe, concludes that denying the Soviets the ability to conduct a blitzkrieg is both necessary and sufficient for deterrence and that maritime forces are irrelevant to such denial.[83] This approach may be flawed. First, it ignores what was termed earlier a "political blitzkrieg," a fragmenting of the alliance in a crisis if the United States appears to be setting priorities among its allies. Second, conservative Soviet planners must consider unfavorable outcomes as well as favorable ones. Real decisions are always based on difficult judgments; by demonstrating an ability to deny the Soviets their preferred strategy and to adversely alter the nuclear correlation of forces, an announced Maritime Strategy can make failure of a blitzkrieg even more unattractive and thus enhance deterrence.

A more general problem with the assertion that maritime forces are irrelevant to deterrence is that it considers only the deterrence of large-scale

---

81. Joshua M. Epstein, "Horizontal Escalation: Sour Notes of a Recurrent Theme," *International Security*, Vol. 8, No. 3 (Winter 1983/1984), pp. 19–31.

82. Ibid., p. 23. Epstein suggests that the Soviets will draw down forces facing China. Forcing the Soviets to face such a decision would complicate their planning, further denying them their preferred strategy. To make the maritime case, however, it is not necessary that the Soviets transfer *any* forces to the Far East, simply that they be inhibited in transferring forces from the area.

83. John Mearsheimer, Remarks at the Naval War College, May Conference on "Maritime Strategy: Issues and Perspectives," May 15–17, 1985. Mearsheimer also doubts that the strategy can be executed or that the President will permit it to be attempted.

conventional attack in Europe. Such an attack is only likely to be considered after serious deterioration in the international situation. By responding to Soviet global encroachment, containing extra-European crises, demonstrating U.S. support for allies, and serving as a well-understood symbol of national will, maritime forces can deter the Soviets from the type of adventurism that could escalate into a grave crisis warranting Soviet consideration of war. This form of deterrence complements rather than substitutes for that provided by the ability to deny a blitzkrieg, just as wartime global maritime operations complement direct defense in Europe.

The most important problem with those who argue against the Maritime Strategy on grounds of its relevance is that they ignore the entire question of war termination. It often seems that, for all their stress on innovation at the theater level, at the strategic level critics are espousing a strategy whose components are: reinforce Europe, pray for a miracle, and be ready to use nuclear weapons if no miracle occurs.[84] Ensuring alliance solidarity and providing war termination leverage appear to supporters of the Maritime Strategy to be eminently relevant.

RESOURCES

Arguments about the relevance of the Maritime Strategy are often really arguments about resources, focusing not so much on how existing forces will be used as on what will be procured for the future.[85] Much debate about any strategy is really a debate about money. This is as it should be; strategy should guide resource allocation. Thus, critics who recognize this fact attack the strategy precisely *because* it has helped justify the ongoing naval buildup. They assert either that there is no strategy, only a budget document, or that the United States can not afford to buy the Navy required to implement such a strategy. Neither point is valid.

THE NAVY DOES NOT HAVE A REAL STRATEGY, ONLY A SPEECH IT USES FOR BUDGET PURPOSES. Professional Navy officers are frankly puzzled by this criticism. In contrast to the fuzzy descriptions of national strategy indicated

---

84. This specifically does not apply to Mearsheimer, who believes ground and air forces on the Central Front can restore NATO's territorial integrity. Private discussions with military officers reveal relatively few who accept this view.
85. Despite the title *Maritime Strategy or Coalition Defense?*, this is Robert Komer's real issue. Komer does not actually examine the strategy as a way to use the Navy today; his concern is with spending for the future.

above, the Maritime Strategy has been set forth in considerable detail.[86] Assertions that the Navy has no strategy are based on a misunderstanding of what strategy is. The official Joint Chiefs of Staff definition of military strategy is the "art and science of employing the armed forces of a nation to secure the objectives of national policy by the application of force or the threat of force."[87] In this sense, the Maritime Strategy, whether one endorses or condemns it, clearly qualifies as a global strategy. Operational details— whether aircraft carriers will hide in the Norwegian fjords or fight through the center of the Norwegian sea, whether the Sixth Fleet will engage Soviet forces in the Central Mediterranean before moving to the Aegean, how attack submarines and other ASW forces will balance their two important missions of attacking Soviet SSBNs in bastions and clearing Soviet anti-carrier forces from the approaches to the Soviet Union, the sequence in which the Pacific Fleet will operate over one-half the surface of the globe, who will lay mines and when—all these belong not in a global strategy, but in the theater campaign plans shaped by that strategy. Those who expect such details in the Maritime Strategy are confusing the operational or theater level of war with the global or strategic level.

EVEN IF THE MARITIME STRATEGY IS A GOOD IDEA, THE UNITED STATES CAN NOT AFFORD IT SINCE BUILDING A CAPABLE FLEET DIVERTS RESOURCES THAT ARE NEEDED TO IMPROVE EUROPEAN DEFENSE. Some critics who might be prepared to accept the validity of the strategy for current forces still conclude that the United States cannot afford the type of future navy such a strategy implies. They argue that, if only the funds devoted to new carrier battle groups were shifted to Central Front defense, NATO would be capable of successful direct defense, preserving the territorial integrity of the Alliance without the need for risky attacks on Soviet strategic forces. The argument is superficially plausible. On closer examination, however, it is flawed.

The first flaw is the implicit assumption that funds "saved" from the Navy would be devoted to European defense in sufficient quantities to dramatically alter the situation. Little in recent history suggests that such a proposition is valid. NATO's unwillingness to devote sufficient resources to direct defense

---

86. In addition to Watkins, "Maritime Strategy" (the only statement to discuss the rationale for anti-SSBN operations), Lehman, "The 600-Ship Navy," and the *FY 85 SASC Hearings* cited herein, see the sixteen authoritative descriptions by senior Navy officials cited in Swartz, "Contemporary U.S. Naval Strategy: A Bibliography," and his forthcoming "1986 Addendum."
87. Joint Chiefs of Staff, *Dictionary of Military and Associated Terms* (JCS Pub 1) (Washington, D.C.: U.S. Government Printing Office, January 1, 1986), p. 228.

in Europe is a long-standing problem; there is no logical reason to assume that the Alliance would become more willing if the United States reduced spending on its Navy.

The second flaw in this argument is a blurring of time frames. If increased conventional capability is a solution at all, it is a solution for the future. But strategists have an obligation to decide how to fight a war *today*. While the Maritime Strategy logically requires the naval buildup that has come to be called the 600-ship navy, the strategy is valid today, before all of that navy is at sea. Alternatives are not.

Finally, those who would shift resources away from maritime capabilities have elected to compete with the Soviets almost entirely on their terms. Such an approach—opposing one of the largest land armies in history in a high intensity conflict on the territory of U.S. allies—carries with it the twin possibilities of the political collapse or the devastation of NATO, possibilities equally as grave as the escalation risks that critics deplore.

Conclusion

The Maritime Strategy—the early, forceful, global use of naval power in a future war with the Soviet Union—offers unique benefits, benefits that more than justify continuing to use the strategy today and procurement of the type of navy that will allow its use in the future. While it is not without risk (a risk-free war with a nuclear superpower is a contradiction in terms), its risks are not as great as critics assert. While maritime forces cannot alone prevent war or guarantee victory, they are directly relevant to deterrence, to a war in Europe, and to war termination. While there will always be arguments over resources in the American budgetary process, the strategy is clearly a prudent use of resources. Most of all, the Maritime Strategy offers a vehicle for the professional military, especially the Navy and Marine Corps, to apply sound strategic thinking to the solution of national problems. It is particularly ironic that military reformers, who applaud, correctly, the flexible operational and theater level concepts embodied in the AirLand Battle doctrine, fail to see that these same concepts on a strategic and global scale underlie the Maritime Strategy.[88]

---

88. For a discussion of these conceptual similarities, see Rear Admiral William T. Pendley, "The Navy, Forward Defense, and the Airland Battle," prepared for Tufts Fifteenth Annual Conference on "Emerging Doctrines and Technologies: Implications for Global and Regional Political–

The challenge for critics is not to bemoan Navy successes in budget battles but rather to look for serious alternatives to those aspects of the Maritime Strategy that they find unpalatable. As this article demonstrates, there are serious objections to the alternatives most frequently set forth, objections that, to maritime advocates, are far more persuasive than any drawbacks the critics have yet discovered with the Maritime Strategy.

In *Billy Budd*, Herman Melville noted that "Everything is for a time remarkable in navies." What has been remarkable about the U.S. Navy recently is its attention to strategy and to a reexamination of the fundamental purposes of maritime power in both deterrence and war. Like all intellectual trends, this one will not last forever. Its immediate legacy is a maritime component of national strategy that can contribute to deterrence, promote alliance solidarity, ensure unimpeded reinforcement of Europe, divert Soviet resources and attention from the Central Front, and provide unique war termination leverage. Its mid-term legacy is a larger and more capable Navy and an increased understanding within that Navy of the need to plan, train, and operate with other services and with allies. Its long-term legacy, perhaps the most important of all, is the forging of a new professional consensus on the purpose of the Navy and the importance of systematic thought and study. It is ironic that those outside the professional military who have called for more strategic thinking by those in uniform have failed to recognize that fact.

Military Balances Toward the Year 2000 and Beyond," April 16–18, 1986. Despite these conceptual similarities, many Army officers are skeptical of the Maritime Strategy, primarily out of fear that it has or will divert resources from Army forces to the Navy.

A Strategic Misstep | *John J. Mearsheimer*

The Maritime Strategy and Deterrence in Europe

$\mathbf{A}$ core element of the Reagan Administration's defense buildup lies in its plan to increase the size of the U.S. Navy to 600 ships.[1] This 600-ship force is purportedly required to implement "The Maritime Strategy," which is the Navy's blueprint for fighting a global conventional war against the Soviet Union. It is being built at the expense of American air and ground forces in Central Europe, which have not been significantly strengthened during the Reagan Administration's tenure, even though the Administration has expressed the view that the NATO–Warsaw Pact conventional balance in Europe clearly favors the Pact.[2]

Serious controversy has surrounded both the naval buildup and its attendant Maritime Strategy. Critics have charged that the Maritime Strategy is not coherent or complete, and does not provide an adequate rationale for

This article was originally prepared for the Naval War College's May 1985 conference on the Maritime Strategy. I would like to thank James Kurth, who was then Director of the Strategy and Campaign Department at the War College, for suggesting the topic to me, as well as the many conference participants who offered comments on my original draft. I also am deeply indebted to the following individuals for comments on later drafts of this article: Robert Art, Richard Betts, Daniel Bolger, Michael Brown, Owen Cote, Michael Desch, Benjamin Frankel, Charles Glaser, Karl Lautenschläger, Robert Pape, Barry Posen, George Quester, Rade Radovich, Jack Snyder, Peter Swartz, and Andrew Twomey. None bears any responsibility for the arguments offered here. Finally, I would like to thank the MacArthur Foundation for providing support.

John Mearsheimer, an associate professor in the political science department at the University of Chicago, is completing a book on B.H. Liddell Hart.

1. See Richard Halloran, "Reagan Selling Navy Budget As Heart of Military Mission," *The New York Times*, April 11, 1982, p. 24; James Meacham, "The United States Navy," *The Economist*, April 19–25, 1986, pp. 57–69; Peter T. Tarpgaard, *Building a 600-Ship Navy: Costs, Timing, and Alternative Approaches* (Washington, D.C.: U.S. Congressional Budget Office, March 1982); and Caspar W. Weinberger, *Annual Report to the Congress, Fiscal Year 1983* (Washington, D.C.: U.S. Government Printing Office, 1982), pp. I-30, II-12–II-17, III-19–III-36.

2. President Reagan, for example, told a press conference in his first year in office that NATO is "vastly outdistanced" by the Warsaw Pact in terms of forces on the Central Front. See *The New York Times*, October 3, 1981, p. 12. Admiral Watkins, former Chief of Naval Operations, recently wrote that the Soviets have a "massive ground force advantage" over NATO. Admiral James D. Watkins, "The Maritime Strategy," in *The Maritime Strategy*, Supplement to U.S. Naval Institute *Proceedings*, January 1986, p. 8. The supplement is hereinafter cited as *The Maritime Strategy*. On the Reagan Administration's defense allocations, see Michael R. Gordon, "The Pentagon Under Weinberger May Be Biting Off More Than Even It Can Chew," *National Journal*, February 4, 1984, pp. 204–209.

International Security, Fall 1986 (Vol. 11, No. 2)
© 1986 by the President and Fellows of Harvard College and of the Massachusetts Institute of Technology.

increasing the size of the Navy. They also charge that no sound rationale can in fact be offered for investing so heavily in the Navy.[3] The Navy has answered that the Maritime Strategy and its attendant buildup are vital to the protection of American interests and the preservation of peace.

This article explores the wisdom of the Reagan Administration's naval buildup by assessing the overall effect of the Maritime Strategy on deterrence in Europe. America's central military objective, aside from deterring a direct attack on the United States, is to deter the Soviet Union from starting a European war. Strategically, Europe is the most important area of the world for the United States, and is the place where the Soviet Union has concentrated its most formidable military assets.[4] A European war would therefore directly threaten America's vital interests. Such a war also could jeopardize the survival of the United States if it escalated to a nuclear exchange. Deterring the Soviet threat to NATO, especially the Soviet conventional threat, is therefore the baseline case against which the Maritime Strategy should be measured.[5] This is not to deny the importance of other contingencies, but simply to point out that NATO is the most important and most demanding contingency confronting the American military. Indeed, the Navy itself has

---

3. See, for example, Keith A. Dunn and William O. Staudenmaier, *Strategic Implications of the Continental–Maritime Debate*, Washington Paper No. 107 (New York: Praeger, 1984); William W. Kaufmann, "The Defense Budget," in Joseph A. Pechman, ed., *Setting National Priorities: The 1983 Budget* (Washington, D.C.: Brookings, 1982), pp. 51–99; Joshua M. Epstein, *The 1987 Defense Budget* (Washington, D.C.: Brookings, 1986); Robert W. Komer, *Maritime Strategy or Coalition Defense?* (Cambridge, Mass.: Abt Books, 1984); Edward N. Luttwak, *The Pentagon and the Art of War: The Question of Military Reform* (New York: Simon and Schuster, 1985); and Jeffrey Record, *Revising U.S. Military Strategy: Tailoring Means to Ends* (New York: Pergamon Brassey's, 1984). Also see Michael Gordon, "Lehman's Navy Riding High, But Critics Question Its Strategy and Rapid Growth," *National Journal*, September 21, 1985, pp. 2120–2125.

4. Classic statements on the subject are: Walter Lippmann, *U.S. Foreign Policy: Shield of the Republic* (Boston: Little, Brown, 1943); Hans J. Morgenthau, "The United States and Europe in a Decade of Detente," in Wolfram F. Hanrieder, ed., *The United States and Western Europe: Political, Economic and Strategic Perspectives* (Cambridge, Mass.: Winthrop, 1974), pp. 1–7; Nicholas J. Spykman, *America's Strategy in World Politics: The United States and the Balance of Power* (1942; reprint ed., Hamden, Conn.: Archon, 1970); and Alfred Vagts, "The United States and the Balance of Power," *The Journal of Politics*, Vol. 3, No. 4 (November 1941), pp. 401–449. A recent analysis of this matter is Stephen Van Evera, "Why Europe Should Remain the Focus of American Strategy: A Geopolitical Assessment," Testimony prepared for the European Affairs Subcommittee of the Senate Foreign Relations Committee, October 3, 1985.

5. Secretary of the Navy John Lehman succinctly expressed this point when he noted: "Every dollar has to be justified by what it can do to defeat the Soviet maritime threat in time of war, and that is it and it only." Michael R. Gordon, "John Lehman: The Hard Liner Behind Reagan's Navy Buildup," *National Journal*, October 3, 1981, p. 1765. At the same time, Lehman made clear his strong opposition to justifying the Navy's budget on the grounds that it facilitated peacetime presence. Ibid.

long recognized that its chief role lies in its contribution to a global conventional war against the Soviet Union and that such a war would be, above all, a war for the control of Europe. Accordingly, the Navy has argued that the Maritime Strategy would indeed contribute to NATO's European deterrent posture. In this article, I explore whether this is true.

I will offer four principal conclusions. First, the Navy has not defined the Maritime Strategy clearly and, moreover, has defined it in different ways at different times. The strategy therefore tends to have an amorphous and elastic quality about it. Nevertheless, it seems apparent that the strategy is a package of four different offensive postures. The Navy has occasionally shifted its rhetorical emphasis from one to another of these four postures, but all four have remained elements of the Maritime Strategy since it was formulated in 1981.

Second, the four offensive concepts encompassed by the Maritime Strategy contribute little to deterrence in Europe and may actually detract from it. Importantly, the latest variant of the strategy, which emphasizes using American attack submarines (SSNs) to strike at Soviet ballistic missile submarines (SSBNs) so as to shift the strategic nuclear balance at the start of a conventional war, is destabilizing in a crisis and potentially escalatory in a conflict, and therefore is a dangerous strategy.

Third, the Navy's main value for deterrence lies in the realm of sea control, where protection of NATO's sea lines of communication (SLOCs) might matter to Soviet decision-makers contemplating war in Europe. The Navy can counter this threat with a defensive sea control strategy. It is not necessary or desirable to adopt an offensive strategy to protect the SLOCs.

Finally, as a result of the Reagan Administration's policy of favoring the Navy over NATO's ground and tactical air forces, a significant opportunity to improve NATO's deterrent posture has been missed. Moreover, the seeds of future defense policy crises have been sown. In fact, if (as is widely forecast) future defense budgets do not grow significantly and if the Reagan Administration continues to favor the Navy at the expense of the forces in Europe, NATO's deterrent posture may actually be weakened.

Conventional Deterrence

When two large armies directly face each other in a crisis, under what conditions is deterrence likely to fail?[6] What are the circumstances, in other

---

6. For an elaboration of the ideas presented in this section, see John J. Mearsheimer, *Conventional*

words, that would lead one side to go to war? This question lies at the heart of conventional deterrence. It is especially relevant because of the situation in Central Europe, where NATO and Warsaw Pact armies stand opposite one another.

Four general points are in order. First, war is most likely to start when a potential attacker believes that he can score a quick and decisive victory. Deterrence is best served when decision-makers conclude that war would be a ghastly and destructive experience, which happens when the war is protracted. The objective of the deterrer is therefore to decrease his opponent's ability to gain a quick and decisive victory and instead to increase the likelihood that a long and costly war would result. In short, the threat of a war of attrition is the bedrock of conventional deterrence. It is a particular military strategy, the blitzkrieg, that provides the means to win rapidly and decisively on the modern battlefield. This point has been resoundingly demonstrated by the Israelis on at least two occasions and by the Germans in the early years of World War II. Thus, whether or not deterrence obtains in a future superpower crisis would depend in large part on whether the Soviets think they could launch a successful blitzkrieg. This point is clearly reflected in Soviet military literature as well as in the organization of Soviet ground forces. They have no desire to engage in a lengthy war of attrition.

Second, conventional deterrence is ultimately a function of both these military considerations and a broader set of political considerations. Nation-states, after all, go to war to gain political objectives. In a crisis, political incentives may place significant pressure on decision-makers to go to war. Generally, deterrence is most likely to hold when the risks and costs of military action are very high. In certain cases, however, decision-makers might still opt for war even though the risks of military action are very high—simply because the political pressures for war are so great that pursuing a risk-laden military policy may be preferable to the status quo. The risks of doing nothing in such situations may seem greater than the risks of military action.

Third, the discussion up to now has portrayed deterrence as a function of the relationship between the prospects of an easy victory on the battlefield and the political considerations underlying the movement toward war. It is

Deterrence (Ithaca, N.Y.: Cornell University Press, 1983); John J. Mearsheimer, "Nuclear Weapons and Deterrence in Europe," *International Security*, Vol. 9, No. 3 (Winter 1984-85), pp. 19–46; and John J. Mearsheimer, "Offense and Deterrence," unpublished manuscript, August 1985.

assumed here that the opponent is bent on aggression and that deterrence would hold only if the potential aggressor was confronted with formidable military power that could deny him his objectives or punish him for his transgressions, or both. These calculations comprise the degree of *deterrence stability*. Deterrence, however, is not only affected by military calculations about what might happen on the battlefield, but might also be affected by calculations about *crisis stability*. This kind of stability, which applies in cases where neither side is firmly committed to aggression, is a function of the structure of the rival deterrent postures. They may be configured so as to cause fears and provide incentives to strike first in a crisis.

Three factors determine the degree of crisis stability: each side's perceptions of the other's aggressiveness; the degree of military advantage accruing to the side striking first; and the tendency of peacetime military operations to activate the opponent's rules of engagement—the standing orders under which local commanders are permitted to fire their weapons. Certain strategies, especially offensive ones, are likely to signal aggressive intentions to an opponent, and thus give that opponent cause to think about launching a preemptive strike. The rationale for such a strike would be that war is inevitable and that there are military advantages to striking the first blow. This dynamic is not a problem if the side with the provocative strategy is intent on aggression; war is inevitable anyway. It is a major problem, however, if there is no intention to attack, but the strategy, because it appears offensive to the adversary, creates a perception of aggressive intentions. Some strategies also can cause forces to intermingle in a crisis in a manner that produces a tactical or strategic first-strike advantage, creating an incentive to preempt. Finally, some strategies can raise the risk that forces will collide with one another in a manner that activates one side's rules of engagement, leading it to commence firing. In each instance crisis stability is undermined, and crises are more likely to erupt into war.[7]

To determine how a military strategy affects the probability of war, it is necessary to consider its impact on the adversary's calculations about crisis stability as well as its effect on deterrence stability. It is important to empha-

---

7. There is no universally accepted definition of crisis stability. It is usually taken to refer to the absence of a first-strike advantage. However, since other factors affect the stability of crises in addition to the size of first-strike advantages, I have used the concept with a broader meaning, to include the three factors discussed above. The classic discussions of this subject are Thomas Schelling, *Arms and Influence* (New Haven: Yale University Press, 1966), pp. 221–259; Thomas Schelling, *The Strategy of Conflict* (New York: Oxford University Press, 1960), pp. 207–254; and Glenn Snyder, *Deterrence and Defense* (Princeton: Princeton University Press, 1961), pp. 104–110.

size that there is sometimes a real tension between these two forms of stability. A particular military strategy might very well enhance deterrence stability while serving to undermine crisis stability, or vice versa. There is no simple way to resolve this dilemma.

Finally, the historical record suggests that navies should not be expected to play an important role in deterring major conventional wars. Consider the German decision to launch the Schlieffen Plan in July 1914 and the German decision-making process between October 1939 and May 1940 which led to the fall of France. These two cases are analogous to the present situation in Europe since they involved a decision by a land power (Germany) with a significant navy to attack westward against a coalition that included a formidable naval power (Great Britain) that was heavily dependent on SLOCs. The evidence from these cases shows that calculations regarding the naval balance did not play an important role in the decision for war.[8] German decision-makers paid little attention to the consequences of the naval war that was sure to ensue and focused instead almost exclusively on what would happen in the land war.

Competing Views of the Navy's Role in Deterring War in Europe

The U.S. Navy has four broad missions.[9] The first is nuclear deterrence, for which the Navy relies mainly on SSBNs. The second mission is peacetime presence, which calls for maintaining naval forces around the world on a day-to-day basis. The aim is to use them to influence both allies and potential adversaries. The third mission is direct military intervention in Third World conflicts, while the fourth is deterring and fighting a large-scale conventional war with the Soviet Union. The first three missions, although certainly essential, are not directly relevant to the Maritime Strategy. Rather, the Mari-

---

8. For the 1914 decision, see Gerhard Ritter, *The Sword and the Scepter: The Problem of Militarism in Germany*, Vol. II, trans. Heinz Norden (Coral Gables: University of Miami Press, 1970), chapter 10; and Gerhard Ritter, *The Schlieffen Plan: Critique of a Myth* (New York: Praeger, 1958), especially pp. 69–72. For the 1939–1940 decision, see F.H. Hinsley, *Hitler's Strategy* (Cambridge: Cambridge University Press, 1951), chapters 2–3; Hans–Adolf Jacobsen, "Dunkirk 1940," in Hans–Adolf Jacobsen and Jurgen Rohwer, eds., *Decisive Battles of World War II: The German View*, trans. Edward Fitzgerald (New York: Putnam, 1965), pp. 29–68; and Mearsheimer, *Conventional Deterrence*, chapter 4.

9. There is not much literature on this important subject. For two exceptions, see Stansfield Turner, "Missions of the U.S. Navy," *Naval War College Review*, Vol. 26, No. 5 (March–April 1974), pp. 2–17; and John A. Williams, "U.S. Navy Missions and Force Structure: A Critical Reappraisal," *Armed Forces and Society*, Vol. 7, No. 4 (Summer 1981), pp. 499–528.

time Strategy is concerned primarily with the fourth mission, which is largely synonymous with deterring a war in Europe. This fourth mission is the most demanding of the lot and therefore the baseline case against which the Maritime Strategy must be assessed.

The public debate about the Navy's role in deterring a European war is a confusing one. Basic concepts are not well defined, and arguments are often not clearly articulated. To impose order on this subject, it is helpful to think in terms of five alternative views about the role that the Navy might play.[10]

THE "NAVAL IRRELEVANCE" POSITION

This first position holds that the Navy contributes very little to deterrence in Europe. The fundamental assumption here is that the Soviet Union is essentially a land power and, in deciding whether to initiate war with NATO, would pay little attention to naval considerations.[11] The Soviets would instead focus almost wholly on calculations regarding those ground and air forces that would be directly engaged in the land battle. The Soviet decision-making process would, in effect, be very similar to that of the Germans in 1914 and 1939–1940. The case for naval irrelevance is buttressed by the fact that blockade, a traditional naval weapon against land powers, has virtually no utility against a largely autarkic power like the Soviet Union. The principal implications of this position for NATO are that maintaining SLOCs should not be accorded a high priority and that the United States does not need a large navy.

THE "SEA CONTROL" POSITION

A second position holds that sea control *might* matter for deterrence. The Soviets, so the argument goes, would certainly focus on the balance of forces on the Central Front, asking themselves whether a blitzkrieg is feasible. A situation might arise, however, in which they conclude that their prospects for success in the ground war are slim, but that there is a good chance that they can sever the Atlantic SLOCs and thereby bring NATO to its knees. This would be a defensible option only if: there was tremendous political

---

10. Although these views are treated largely as ideal types in this section, they are clearly reflected in the current debate about the Maritime Strategy, as will become apparent below.
11. See James A. Barry, Jr., "Soviet Naval Policy: The Institutional Setting," in Michael MccGwire and John McDonnell, eds., *Soviet Naval Influence: Domestic and Foreign Dimensions* (New York: Praeger, 1977), pp. 107–122; and Bryan Ranft and Geoffrey Till, *The Sea in Soviet Strategy* (Annapolis: Naval Institute Press, 1983), chapter 4.

pressure on the Soviets to go to war and as a result they were willing to pursue a risk-laden military strategy; and it appeared that the SLOCs could be cut in some reasonably short period of time—say six to nine months.[12] Although it is difficult to imagine the Soviets turning to their navy to provide the margin of victory in a European land war, this might happen. In such a case, the U.S. Navy would matter for deterrence. Therefore, NATO must ensure that the Soviets are never in a position where they might conclude that although a war of attrition on the Central Front is likely, they could win that war in some reasonably short time frame by cutting NATO's SLOCs.

To succeed, a sea control strategy must neutralize the Soviets' Northern Fleet, which poses the principal threat to NATO's SLOCs. This fleet is based on the Kola Peninsula, in the Arctic, east of northern Finland.[13] Its principal elements are surface ships, attack submarines, ballistic missile submarines, and land-based aircraft. Of these forces, Soviet attack submarines comprise the main element of the threat to the SLOCs, although NATO must also be concerned about Soviet long-range bombers, Backfires in particular, that could reach the North Atlantic.

NATO could deal with this threat by either defensive or offensive sea control.[14] Defensive sea control would involve three tasks: sealing off the Soviet attack submarines with a formidable barrier defense in the Greenland–

---

12. It is difficult to define with any precision what the Soviets might consider a reasonably short time frame. It is difficult to imagine them banking on a victory that takes a year or more to achieve. It seems much more likely that they would insist on securing a victory in a period of a few months. Six to nine months seems like a reasonable outer limit since, if it takes longer than that to cut the SLOCs, the Soviet advantage in the SLOC battle could not be very great and a strategy resting on SLOC cutting therefore would not provide the Soviets with a high-confidence option.

13. See Jean L. Couhat and A. David Baker III, *Combat Fleets of the World 1986/87* (Annapolis: Naval Institute Press, 1986), p. 481; Department of Defense, *Soviet Military Power*, 5th ed. (Washington, D.C.: U.S. Government Printing Office, March 1986), pp. 11–12; *The Military Balance, 1984–1985* (London: International Institute for Strategic Studies, 1984), pp. 17–22; *The Military Balance, 1985–1986* (London: International Institute for Strategic Studies, 1985), pp. 26–27; Major Hugh K. O'Donnell, Jr., "Northern Flank Maritime Offensive," U.S. Naval Institute *Proceedings*, Vol. 111, No. 9 (September 1985), pp. 42–57; Tomas Ries, "Defending the Far North," *International Defense Review*, Vol. 17, No. 7 (1984), pp. 873–880; and the subsequent correspondence about this article in "A New Strategy for the North–East Atlantic?," *International Defense Review*, Vol. 17, No. 12 (1984), pp. 1803–1804.

14. Good primers on these two kinds of sea control are Dov Zakheim, *Planning U.S. General Purpose Forces: The Navy* (Washington, D.C.: U.S. Congressional Budget Office, December 1976); Dov Zakheim, *Shaping the General Purpose Navy of the Eighties: Issues for Fiscal Years 1981–1985* (Washington, D.C.: U.S. Congressional Budget Office, January 1980); and Dov Zakheim, *The U.S. Sea Control Mission: Forces, Capabilities, and Requirements* (Washington, D.C.: U.S. Congressional Budget Office, June 1977).

Iceland–Norway (GIN) gap, a major choke point through which those submarines must pass to reach NATO's SLOCs; conducting open-ocean antisubmarine warfare (ASW) operations in the area of the Atlantic directly below the GIN gap to neutralize those attack submarines that penetrate the barrier; and providing troop and supply convoys with ASW assets.[15] The Backfire threat would be met with interceptor aircraft based in Britain, Iceland, Norway, and possibly Greenland. An offensive sea control strategy would include those same tasks but would add the task of moving north of the GIN gap to strike directly at Soviet SSNs, Soviet surface ships, naval bases, air bases, and aircraft located on the Kola Peninsula. This added task would be emphasized over the others.

Proponents of offensive sea control make their case on three grounds. First, they argue that the threat to move north forces the Soviets to hold their attack submarines in their home waters, thus keeping them away from the SLOCs. Second, they believe that it is militarily more efficient to defeat the Soviet SSN and Backfire threats in the far north than at the GIN barrier. The best way to deal with the threat to the SLOCs, in other words, is not to wait for the adversary to attempt to surge attack submarines into the North Atlantic, as mandated by defensive sea control, but to go directly to the root of the problem and eliminate it. Third, proponents suggest that an offensive strategy is essential to protect northern Norway from a Soviet attack. In this view, carrier operations in the Norwegian Sea are essential to fight off Backfire attacks against northern Norway and to provide air cover for NATO forces operating in that region. The Soviets would gain, without doubt, an important strategic advantage by capturing this area, since it would allow them to project power more easily into the lower reaches of the Norwegian Sea as well as against the SLOCs. There is, however, disagreement about whether offensive sea control is necessary to prevent that outcome.

Offensive sea control is the more demanding of the two strategies since it alone calls for taking offensive action against a powerful naval force that would have considerable assistance from land-based forces. The offensive, in its most ambitious form, would be comprised of two major operations.

---

15. The GIN gap, which covers about 1000 miles of water and runs from Greenland to Iceland to the United Kingdom to Norway, is sometimes referred to as the GIUK gap. This is a misnomer, however, since the GIUK gap actually includes only the western part of the GIN gap—covering as it does only the 750 miles of water between Greenland, Iceland, and the United Kingdom, while excluding the water between the U.K. and Norway. The other key barrier in this area is the Bear Island–North Cape line which is north of the GIN gap and which essentially separates the Barents and Norwegian seas.

First, American SSNs would destroy the Soviet SSN force in the Norwegian and Barents seas. In what is called a "rollback campaign," attacking American SSNs would form themselves into lines in the Norwegian Sea and then move north to engage Soviet SSNs. The northernmost line might be located as far north as the Bear Island–North Cape gap, and even possibly in the Barents Sea. Second, after completion of the first operation, carrier battle groups would move into the upper reaches of the Norwegian Sea and launch air strikes or cruise missile strikes against naval and air bases on the Kola Peninsula. This operation would presumably eliminate the air threat to NATO's SLOCs. The Soviet Northern Fleet's surface navy, which would be involved in defense of the Barents Sea, would surely be sought out and attacked in both of these operations. Soviet SSBNs are another matter. They are not essential targets in an operation concerned with sea control and, in all likelihood, many of them would not be in harm's way. Nevertheless, as will be discussed, a successful offensive sea control strategy would pose a significant threat to these strategic nuclear forces.

Two further points about offensive sea control bear mentioning. First, it is possible to pursue a truncated version of this strategy. One might, for example, abandon the second operation and concentrate on the anti-SSBN rollback operation. One might even limit the rollback operation to the Norwegian Sea, not going beyond the Bear Island–North Cape line. Second, offensive sea control is not a time-urgent strategy. It does not matter how long it takes the Navy to destroy the Northern Fleet, since the adversary will not be wreaking havoc in the SLOCs while this massive naval war is taking place in his home waters. It should be emphasized, however, that the strategy mandates that carriers move into the Norwegian Sea early in a conflict to provide air cover over northern Norway.

The force posture requirements of defensive and offensive sea control are quite different.[16] Defensive sea control does not require aircraft carrier battle groups and cruise missile platforms. The emphasis instead is on attack submarines, land-based patrol aircraft, destroyers, and frigates. These weapons are ideally suited for barrier defense, convoying, and wide-area ASW operations in the North Atlantic. By contrast, an offensive sea control strategy requires aircraft carrier battle groups, cruise missile platforms, and a very

---

16. See Peter T. Tarpgaard, *Naval Surface Combatants in the 1990s: Prospects and Possibilities* (Washington, D.C.: U.S. Congressional Budget Office, April 1981); Williams, "U.S. Navy Missions and Force Structure"; and the sources cited in note 14.

robust SSN force, while it places less emphasis on destroyers and especially frigates.

THE "DIRECT NAVAL IMPACT" POSITION

Sea control is all about denying the Soviets the capability to interrupt the flow of men and materials from the United States to Europe. Regardless of whether a defensive or an offensive strategy is employed to achieve that end, NATO's aim of protecting itself from Soviet attacks is fundamentally defensive in nature. Deterrence, in other words, is enhanced because NATO's ability to protect itself has been increased. A third position on the role of the Navy—direct naval impact—reverses this situation by suggesting that the U.S. Navy strike Soviet targets and thereby directly lessen Soviet prospects of winning a conventional war in Europe. In essence, this is the classic task of power projection.

Such a strategy might include three components, which need not be accorded equal weight. First, the Navy and the Marines might launch amphibious attacks onto the European continent—specifically, the Soviet or Eastern European coasts. The aim would be to create a significant threat in the rear of the Soviet forces fighting on the Central Front. This would presumably divert ground and air forces from the battles along the intra-German border, thus decreasing the Soviets' prospects of winning the critical land and air battles in that area. Second, the Navy could introduce its carrier-based aircraft into the air war on the Central Front. Third, the Navy could mount carrier-based air attacks directly against the Kola Peninsula, which, if successful, might force the Soviets to shift air assets from the Central Front to the Northern Flank. The key assumption underlying the latter two components is that, in a closely contested air war over Europe, these actions by the Navy might provide the margin of victory for NATO.

Striking directly at the Kola Peninsula, the third component, is obviously congruent with offensive sea control. However, critical time constraints are involved when attacking the Kola for the purpose of drawing forces away from the continent. Specifically, the attacks must come early enough in the war to influence events on the Central Front. An air offensive against the Kola coming after NATO had lost the air war in Central Europe would matter little. The same principle applies, of course, to amphibious operations. In short, direct military impact, unlike sea control, involves time-urgent operations.

The force posture implications of direct military impact are similar to those of offensive sea control, since both are concerned with projecting power against the Soviet homeland. Both strategies require a very large and powerful navy with a substantial carrier battle group component. The most important difference is that direct military impact could call for a truly robust amphibious capability, which is not necessary for offensive sea control.

THE "HORIZONTAL ESCALATION" POSITION

A fourth position calls for using the Navy to threaten Soviet vital interests *outside* of Europe—for instance, Soviet Third World allies such as Cuba or Vietnam, or Soviet naval bases in East Asia. The Soviets, it is assumed, are particularly vulnerable in the Third World or on their own periphery. When struck there, they would be forced either to draw units away from Europe or simply to make concessions in Europe because of the grave threat to these other areas. Horizontal escalation would obviously require a navy built around a large number of carrier battle groups and with a substantial amphibious assault capability. In this regard, it would have much in common with the forces needed for offensive sea control and direct military impact. Horizontal escalation would differ from these two positions, however, in that it would place greater emphasis on strategic mobility assets than would the other two. This reflects the fact that horizontal escalation is a truly global strategy while the other two are focused mainly on Europe. Horizontal escalation, in other words, involves simultaneous operations around the world, while the other two call for sequential operations (which is shorthand for concentrating first on the European theater).[17]

THE "COUNTERFORCE COERCION" POSITION

A fifth position holds that the Navy can deter the Soviets from moving against NATO, or could persuade them to terminate the war, by threatening to use American SSNs to eliminate significant numbers of Soviet SSBNs. This policy would represent a counterforce attack against the Soviets' strategic retaliatory forces—although conducted with conventional weapons.

Advocates of this position suggest that the threat of such a counter-SSBN campaign could enhance deterrence in two ways. First, the United States could threaten to sink enough Soviet SSBNs to shift the strategic balance

---

17. See note 41.

against the Soviets. The key assumption here is that they place so much importance on the nuclear balance that the threat to shift it significantly would deter them or compel them to halt their attack and withdraw.

Secondly, the Navy's counterforce campaign could produce deterrence simply by generating the risk of nuclear war, even if it did not necessarily change the strategic balance. Such a campaign would exemplify Thomas Schelling's notions of "manipulation of risk" or "rocking the boat"—actions taken that endanger both sides in order to deter one side.[18] There is widespread agreement that striking directly at the Soviets' strategic forces in a conventional war would be a risky strategy, simply because of the threat of nuclear escalation. By threatening to pursue such a dangerous strategy— with its potential for events spinning out of control—the Soviets, so the argument goes, would be deterred from starting a war in Europe in the first place. Deterrence, it is said, is not based simply on calculations about the balance of nuclear forces, but also on the stark fear that the U.S. Navy's actions would precipitate a nuclear war. This is deterrence based on the threat of inadvertent or accidental escalation, rather than the threat of coercion based on American nuclear superiority.[19]

The main targets of a counterforce coercion strategy would be Soviet SSBNs, although it is safe to assume that in executing this strategy American SSNs would also destroy large numbers of the Soviets' SSN force as well as a large portion of their surface navy.[20] After all, these forces would be attempting to protect the SSBNs. Thus, the principal targets of a counterforce

---

18. Schelling, *Arms and Influence*, pp. 92–125; and Schelling, *Strategy of Conflict*, p. 196.

19. "Inadvertent escalation" refers to deliberate nuclear escalation, ordered by national command authorities (NCAs) on one side, which is inadvertently provoked by actions of the other side. In contrast, "accidental escalation" arises when individual commanders use nuclear weapons, in accordance with their rules of engagement, before NCAs on either side have decided to go to nuclear war. Such escalation can develop if NCAs order operations without realizing that they will thereby bring forces into contact under circumstances where standing orders authorize local commanders on one side to fire. An error of this kind could result if NCAs fail to understand the nature of the operations that they order, the nature of their opponent's operations, the rules of engagement governing their own forces, or their opponent's rules of engagement. Accidental escalation is also often taken to include escalation arising from mechanical failure, insanity, and unauthorized use of weapons. However, I believe that these risks are very small, so I am here using the term only to refer to the risk of accidental activation of rules of engagement.

20. A counterforce coercion strategy would surely involve a major offensive against the 31 SSBNs and SSBs located with the Soviets' Pacific Fleet as well as the 41 SSBNs stationed on the Kola Peninsula. This coercion strategy therefore involves operations that are similar to those required by the variant of the horizontal escalation strategy calling for attacks against Soviet forces in the Far East. Also see note 41 and the attendant text. The subsequent analysis, because of space limitations, does not address the feasibility of an American offensive against the Soviets' Pacific Fleet, but instead focuses on the U.S. Navy's prospects against the Northern Fleet.

coercion strategy (SSBNs) are different from those of offensive sea control (SSNs), although in practice there would be considerable overlap in the actual target sets. However, the actual operational strategy associated with counterforce coercion is different from the rollback strategy needed for offensive sea control.

Counterforce coercion places a high premium on mobilizing the American SSN force early in a crisis and *inserting* large numbers of those attack submarines deep into the Barents Sea as quickly as possible. This operation is necessary because in a crisis the Soviets would surge large numbers of SSBNs—the principal target—out from their ports to hide under the polar ice cap.[21] There, they would be difficult to find and destroy.[22] The Navy would have to beat the Soviets to the punch by quickly getting attack submarines in positions outside of those ports so that they could pick up and trail the SSBNs as they head toward the ice. If American attack submarines do not move into the Barents before the SSBNs are surged, finding and destroying them would be a difficult task. Such a situation would threaten to undermine the counterforce coercion strategy, since it demands that the nuclear balance be shifted rather quickly—so as to have an impact on the Central Front. It would not make much sense for purposes of coercion for changes in the strategic nuclear balance to occur after Europe had been lost. Thus, the more pessimistic one is about the conventional balance in Europe, the more necessary it becomes to execute the strategy quickly. Since a rollback operation by its very nature would allow large numbers of Soviet SSBNs to get under the ice and would concentrate anyway on pushing back and eliminating the Soviet forces defending the bastion, it is not suited for coercion. The emphasis must instead be on a large-scale insertion operation.

---

21. See Craig Covault, "Soviet Ability to Fire Through Ice Creates New SLBM Basing Mode," *Aviation Week and Space Technology*, December 10, 1984, pp. 16–17; Richard Halloran, "Navy Trains to Battle Soviet Submarines in Arctic," *The New York Times*, May 19, 1983, p. 9; Willy Ostreng, "The Strategic Balance and the Arctic Ocean: Soviet Options," *Cooperation and Conflict*, Vol. 12, No. 1 (1977), pp. 41–62; Captain Gerald E. Synhorst, "Soviet Strategic Interest in the Maritime Arctic," U.S. Naval Institute *Proceedings*, Vol. 99, No. 5 (May 1973), pp. 89–111; William T. Tow, "NATO's Out-of-Region Challenges and Extended Containment," *Orbis*, Vol. 28, No. 4 (Winter 1985), pp. 836–837; George C. Wilson, "Navy is Preparing for Submarine Warfare Beneath Coastal Ice," *The Washington Post*, May 19, 1983, p. 5; and "3 U.S. Submarines Join at Pole," *The New York Times*, May 24, 1986, p. 7.

22. See Gordon V. Brown, "Arctic ASW," U.S. Naval Institute *Proceedings*, Vol. 88, No. 3 (March 1962), pp. 53–57; Captain T.M. LeMarchand, "Under Ice Operations," *Naval War College Review*, Vol. 38, No. 3 (May–June 1985), pp. 19–27; Ostreng, "The Strategic Balance and the Arctic Ocean"; and Norman Polmar, "Sailing Under the Ice," U.S. Naval Institute *Proceedings*, Vol. 110, No. 6 (June 1984), pp. 121–123.

This need to insert SSNs into the Barents Sea to destroy SSBNs notwithstanding, the American Navy would still have to strike at the Soviet SSNs, which would be attempting to protect the SSBNs. Thus, in the final analysis, counterforce coercion would require a large-scale insertion operation as well as a rollback operation. Obviously, this posture could only be executed with a large force of attack submarines. As defined here, counterforce coercion could be accomplished without aircraft carriers, although one could argue that destroying military targets on the Kola Peninsula would facilitate the Navy's efforts. This posture obviously requires a powerful navy with significant offensive capability.

The Maritime Strategy

The Maritime Strategy can best be understood in terms of the five positions outlined in the previous section.[23] It is important to emphasize, however, that it is not easy to describe this strategy since the Navy has often been vague in describing it. This was especially true in the first years of the Reagan Administration.[24] As a result, the public debate on this subject is often carried on without any reference to the specifics of the strategy.[25] Moreover, the core aim of the strategy appears to have changed over time. Specifically, there seems to have been a shift of emphasis during 1981–1986 from offensive sea control to counterforce coercion. This emphasis on a core aim notwithstanding, the Maritime Strategy is, in effect, an inclusive package of four offensive postures: direct naval impact, horizontal escalation, offensive sea control, and counterforce coercion. (The Navy categorically rejects the naval irrelevance and defensive sea control postures.) Although the Navy might emphasize a particular offensive posture, reference to the other three can

---

23. The subsequent discussion is based largely on the Navy's public pronouncements about the Maritime Strategy and public discussion surrounding that strategy.

24. Consider, for example, that Admiral James Watkins recently noted that in 1983 "we reviewed our extant strategy—a strategy with broad contours reasonably well understood, but one which had not been submitted to the rigor inherent in codification." Watkins, "The Maritime Strategy," p. 4. Also see Francis J. West, Jr., "Maritime Strategy and NATO Deterrence," *Naval War College Review*, Vol. 38, No. 5 (September–October 1985), p. 12.

25. For a good example of this phenomenon, see U.S. Congress, House Armed Services Committee, *Report on the 600-Ship Navy*, 99th Congress, 1st Session (Washington, D.C.: U.S. Government Printing Office, December 1985). This report, which was issued after the Seapower Subcommittee held hearings on the 600-ship navy and the Maritime Strategy, goes to some lengths to defend the Navy's strategy without ever defining the strategy or explaining why it is essential. These hearings unfortunately were not available for preparation of this article.

still be found. This fact, coupled with the often vague descriptions of the strategy offered by the Navy, lends the Maritime Strategy an ambiguous or elastic quality.

There are a number of reasons for this ambiguity. First, the Navy simply has not had or has never articulated a coherent strategy for deterring or fighting a conventional war with the Soviet Union.[26] Since the Maritime Strategy represents the Navy's first attempt to provide a coherent strategic rationale for its conventional forces, some confusion is to be expected in the early stages of such a difficult process. Second, there are almost certainly important disagreements about strategy among the different constituencies in the Navy.[27] A loosely and broadly defined strategy is an ideal device for smoothing over such differences. Third, it is hard to make the case that the Maritime Strategy significantly contributes to deterrence in Europe. It is especially difficult to justify a 600-ship navy on the grounds that it markedly enhances NATO's deterrent posture. One should therefore not expect to find an explicit or narrow rationale for the present naval buildup. Finally, and perhaps most importantly, the vagueness and hydra-headed quality of the Maritime Strategy make it difficult for observers to challenge, since evaluating it is like taking aim at a moving target that is constantly changing shape and size. This kind of ambiguity is bureaucratically advantageous, however, because it provides the Navy with multiple rationales for its forces as well as a very demanding set of military requirements, and it deflects criticism by allowing Navy spokesmen to shift the grounds of debate.

---

26. There is little evidence of a coherent strategy or even an interest in strategic arguments in the Navy's Congressional testimony and key policy statements between 1945 and 1985. See also Peter Swartz's excellent bibliography on naval strategy, the listings in which reveal little evidence of strategic thinking in the Navy during these years. Captain Peter M. Swartz, "Contemporary U.S. Naval Strategy: A Bibliography," in *The Maritime Strategy*, pp. 41–47. Also see Captain Linton F. Brooks, "Naval Power and National Security: The Case for the Maritime Strategy," *International Security*, Vol. 11, No. 2 (Fall 1986), fn. 1; and the testimony of John Lehman in U.S. Congress, Senate Armed Services Committee, *Hearings On The Nomination of John F. Lehman, Jr., To Be Secretary Of the Navy*, 97th Congress, 1st Session (Washington, D.C.: U.S. Government Printing Office, January 28, 1981), pp. 8–10. Probably the only exception to this general point is the Navy's attempt in the late 1940s to find a role for itself in a nuclear war. See David A. Rosenberg, "American Postwar Air Doctrine and Organization: The Navy Experience," in Alfred F. Hurley and Major Robert C. Ehrhart, eds., *Air Power and Warfare* (Washington, D.C.: Office of Air Force History, 1979), pp. 245–278. Also see pp. 279–282 of that volume.
27. See Richard Halloran, "The Steady Rise of the Submariners," *The New York Times*, November 21, 1985, p. 14; Morton H. Halperin, *Bureaucratic Politics and Foreign Policy* (Washington, D.C.: Brookings, 1974), chapter 3; and Admiral Elmo R. Zumwalt, *On Watch: A Memoir* (New York: Quadrangle, 1976), *passim*.

The evolution of the Maritime Strategy can be divided into roughly three periods, the first of which covers the initial two years of the Reagan Administration. The details of the strategy were obscure during this period, which is paradoxical, since that was when the key decisions were made to build a 600-ship navy, and these decisions presumably should have been informed by requirements generated by strategy. During 1981–1982, Navy spokesmen frequently advocated regaining "maritime superiority," a phrase that, however useful for public relations, says little about strategy or the precise purposes for which a 600-ship fleet is needed.[28] However, insofar as the strategy was articulated, it emphasized offensive sea control and horizontal escalation, with less emphasis on direct military impact. Counterforce coercion was not mentioned.

Horizontal escalation was one of the main tenets of the overall defense policy of the early Reagan Administration.[29] The concept proved to be controversial, especially because it was widely perceived at the time that advocates of horizontal escalation intended to substitute it for American forces in Europe, and intended to reduce greatly the American commitment to fight in Europe. There was much talk about abandoning the American military's traditional emphasis on sequential military operations in a major conventional war with the Soviet Union and stressing simultaneous operations in different theaters instead. The implication of this position was that Europe was no more nor less important than any other region of the world for the United States. There was indeed much discussion of unilateralism and the

---

28. This wording is most closely identified with John Lehman. See "Lehman Seeks Superiority," *International Defense Review*, Vol. 15, No. 5 (1982), pp. 547–548; John F. Lehman, Jr., "Thinking About Strategy," *Shipmate*, April 1982, pp. 18–20; John F. Lehman, Jr., "Rebirth of a U.S. Naval Strategy," *Strategic Review*, Vol. 9, No. 3 (Summer 1981), pp. 9–15; and "Rebuilt Navy," Transcript of "The MacNeil–Lehrer Report" broadcast on March 27, 1981, library number 1435, show number 6195.

29. See Richard Halloran, "Weinberger Tells of New Conventional-Force Strategy," *The New York Times*, May 6, 1981, p. 10; Samuel P. Huntington, "The Defense Policy of the Reagan Administration, 1981–1982," in Fred I. Greenstein, ed., *The Reagan Presidency, An Early Assessment* (Baltimore: The Johns Hopkins University Press, 1983), pp. 101–104; Fred C. Iklé, "The Reagan Defense Program: A Focus On The Strategic Imperatives," *Strategic Review*, Vol. 10, No. 2 (Spring 1982), pp. 14–16; Weinberger, *Annual Report to the Congress, Fiscal Year 1983*, pp. I-14–I-17; Dov S. Zakheim, "The Unforeseen Contingency: Reflections on Strategy," *Washington Quarterly*, Vol. 5, No. 2 (Autumn 1982), pp. 158–166. It is clear from these sources that there was a direct connection between the Administration's emphasis on the Navy and its advocacy of horizontal escalation. Also see John Lehman's comments in "Mideast Crisis: U.S. Navy at the Ready," *U.S. News and World Report*, August 2, 1982, p. 24.

adoption of a blue water strategy, although it should be emphasized that no Navy spokesman publicly identified himself with this position.[30]

There was also considerable discussion during this early period about conducting offensive operations north of the GIN gap, which, if anything, shows that the Navy was still quite interested in Europe.[31] This discussion was linked to the issue of sea control. The Navy emphasized that defensive sea control was an unsatisfactory posture and that only an offensive strategy would provide adequate sea control.[32] Navy spokesmen also maintained that only an offensive strategy would provide the means to protect northern Norway, the defense of which is important for protecting the Atlantic SLOCs.[33] The Navy's case for offensive sea control, at least in the public domain, was based largely on assertion, not on analysis. Actually, the Navy had begun making the case for offensive operations in the Norwegian and Barents seas in the late 1970s, largely in response to the perception that the Carter Administration embraced defensive sea control.[34] The specifics of the

---

30. There is a voluminous literature associated with this subject. See *inter alia* Dunn and Staudenmaier, *Strategic Implications of the Continental–Maritime Debate*; Gary Hart, "Can Congress Come to Order?," in Thomas M. Franck, ed., *The Tethered Presidency* (New York: NYU Press, 1981), pp. 229–244; Josef Joffe, "Europe's American Pacifier," *Foreign Policy*, No. 54 (Spring 1984), pp. 66–84; Komer, *Maritime Strategy or Coalition Defense?*; Christopher Layne, "Ending the Alliance," *Journal of Contemporary Studies*, Vol. 6, No. 3 (Summer 1983), pp. 5–31; Jeffrey Record and Robert J. Hanks, *U.S. Strategy at the Crossroads: Two Views* (Cambridge, Mass.: Institute for Foreign Policy Analysis, 1982); Stansfield Turner and George Thibault, "Preparing for the Unexpected: The Need for a New Military Strategy," *Foreign Affairs*, Vol. 61, No. 1 (Fall 1982), pp. 122–135; and F.J. West, Jr., "NATO II: Common Boundaries For Common Interests," *Naval War College Review*, Vol. 34, No. 1 (January–February 1981), pp. 59–67.

31. This point is clearly reflected in the sources cited in note 28. Also see Barry R. Posen, "Inadvertent Nuclear War?," *International Security*, Vol. 7, No. 2 (Fall 1982), pp. 28–54, a controversial article from this first period which focuses on the Navy's plans for an offensive north of the GIN gap; and "Battle for the Norwegian Sea," Transcript of "Frontline" Television Program broadcast on January 1, 1985.

32. The Navy emphasized the claim that the threat of an offensive strategy forced the Soviets to keep their SSNs in their home waters and away from the SLOCs. See note 104. Navy spokesmen also argued that, "If we are not able to go on the offence against the Soviet submarine and air threat, if we have to wait for them to come to us, then we can't survive." "Lehman Seeks Superiority," p. 548.

33. For a recent statement of this case, see Vice Admiral H.C. Mustin, "The Role of the Navy and Marines in the Norwegian Sea," *Naval War College Review*, Vol. 39, No. 2 (March–April 1986), pp. 2–6.

34. It is important to emphasize that the roots of the Maritime Strategy rest firmly in the late 1970s. The Navy was dissatisfied during that period over the Carter Administration's decision to emphasize defensive sea control. (It seems that the previous administration also favored defensive sea control.) In response, the Navy, largely in the person of Admiral Thomas B. Hayward (Chief of Naval Operations), began arguing forcefully that the Navy should be used offensively against the Soviet Union. See, for example, Admiral Thomas B. Hayward, "The Future of U.S. Sea Power," U.S. Naval Institute *Proceedings*, Vol. 105, No. 5 (May 1979), pp. 66–

new strategy were not clearly spelled out in the first Reagan years, and consequently considerable controversy arose about the actual conduct of offensive operations. One issue was whether aircraft carriers would be sent forward early in a war to strike at the Soviet Union. This view was identified with Secretary of the Navy John Lehman, while it was generally believed that the admirals would hold the carriers back behind the GIN gap until submarines had swept the Norwegian Sea and large portions of the Barents Sea.[35] There was also much debate about whether the Navy could strike Soviet SSNs without destroying their SSBNs.[36]

The second period in the evolution of the Maritime Strategy began in 1983 and culminated with testimony by Navy officials before the Senate Armed Services Committee in March 1984, where they presented a detailed and reasonably coherent version of the strategy.[37] During this period, offensive

---

71, which contains virtually all the principal ideas articulated by John Lehman and others in the early Reagan years. For a discussion of the bitter feud that broke out between the Navy and the Carter Administration, see Richard Burt, "U.S. Defense Debate Arises Over Whether Focus on Europe Neglects Other Areas," *The New York Times*, March 24, 1978, p. 3; "Navy Protests Limitation of Its Long-Term Mission," *The New York Times*, March 14, 1978, p. 3; Bernard Weinraub, "Brown Criticizes Navy Officers Who Oppose Changes in Strategy," *The New York Times*, February 17, 1978, p. 11; Bernard Weinraub, "Claytor Criticizes Pentagon Aides on Plans to Reduce the Navy's Role," *The New York Times*, March 28, 1978, p. 10; and Bernard Weinraub, "Dispute Over Navy Role Termed Biggest Defense Fight Since 1949," *The New York Times*, April 4, 1978, p. 16. For examples of the Ford Administration's commitment to defensive sea control, see James R. Schlesinger, *Annual Defense Department Report, FY 1975* (Washington, D.C.: U.S. Government Printing Office, March 1974), pp. 93–94; James R. Schlesinger, *Annual Defense Department Report, FY 1976 and FY 197T* (Washington, D.C.: U.S. Government Printing Office, February 1975), p. III-25; and Donald H. Rumsfeld, *Annual Defense Department Report, FY 1978* (Washington, D.C.: U.S. Government Printing Office, January 1977), pp. 96–97.
35. See Halloran, "Navy Trains to Battle Soviet Submarines in Arctic"; U.S. Congress, Senate Armed Services Committee, *Hearings on Department of Defense Authorization for Appropriations for FY 1985 (Part 8)*, 98th Congress, 2nd Session (Washington, D.C.: U.S. Government Printing Office, March–May 1984), pp. 3871–3872 [Hereinafter cited as *SASC Hearings on FY 85 Budget (Part 8)*]; and Wilson, "Navy is Preparing for Submarine Warfare Beneath Coastal Ice."
36. The centerpiece article in this debate was Posen, "Inadvertent Nuclear War?" Also see Bruce G. Blair, "Arms Control Implications of Anti-Submarine Warfare (ASW) Programs," in U.S. Congress, House Committee on International Relations, *Report on Evaluation of Fiscal Year 1979 Arms Control Impact Statements*, 95th Congress, 2nd Session (Washington, D.C.: U.S. Government Printing Office, January 3, 1979), pp. 103–119; Joel S. Wit, "Advances in Antisubmarine Warfare," *Scientific American*, February 1981, pp. 31–41; and footnote 100 and the attendant text.
37. See *SASC Hearings on FY 85 Budget (Part 8)*, pp. 3851–3900, which is the key document from this period. Also see O'Donnell, "Northern Flank Maritime Offensive"; Commodore Dudley L. Carlson's statement in U.S. Congress, House Armed Services Committee, *Hearings on Defense Department Authorization and Oversight for FY 1984 (Part 4)*, 98th Congress, 1st Session (Washington, D.C.: U.S. Government Printing Office, February–April 1983), pp. 47–51 [These hearings are hereinafter cited as *HASC Hearings on FY 84 Budget (Part 4)*.]; West, "Maritime Strategy and NATO Deterrence"; and Robert S. Wood and John T. Hanley, Jr., "The Maritime Role in the

sea control was elevated to a preeminent position, while counterforce coercion made its first appearance as a secondary mission. Although the Navy did not explicitly endorse counterforce coercion in its public statements, the first indirect evidence began to appear in Congressional hearings and elsewhere that the Navy considered this an important element of its strategy.[38] Direct military impact continued as a secondary mission, while horizontal escalation now drew little mention. The lack of attention paid to horizontal escalation was probably due to the fact that Navy spokesmen were intent on combatting the criticism that they were advocating unilateralism while emphasizing their commitment to the Alliance.[39] Not surprisingly, they also backed away from arguing for simultaneous military operations against the Soviet Union and instead identified themselves with the traditional policy of sequential military operations. John Lehman, for example, told Congress that the Navy "cannot deal simultaneously with every theater" and therefore it would be necessary in a large-scale conventional war to deal sequentially with the various threats in the different theaters.[40] The implication was that the Navy would first concentrate its efforts in Europe and turn later to the other theaters.[41]

North Atlantic," *Naval War College Review*, Vol. 38, No. 6 (November–December 1985), pp. 5–18.

38. See, for example, *SASC Hearings on FY 85 Budget (Part 8)*, pp. 3864, 3893. Although the discussion on p. 3864 does not mention attacking Soviet SSBNs, it highlights the very high premium that counterforce coercion (as opposed to offensive sea control) places on inserting large numbers of SSNs in the Barents Sea early in a crisis. Also see Melissa Healy, "Lehman: We'll Sink Their Subs," *Defense Week*, May 13, 1985, p. 18; and West, "Maritime Strategy and NATO Deterrence," pp. 6, 8, 11. Also of interest are Commander Richard T. Ackley, "No Bastions for the Bear: Round 2," U.S. Naval Institute *Proceedings*, Vol. 111, No. 4 (April 1985), pp. 42–47; Hamlin Caldwell, "The Empty Silo—Strategic ASW," *Naval War College Review*, Vol. 34, No. 5 (September–October 1981), pp. 4–14; and David B. Rivkin, "No Bastions for the Bear," U.S. Naval Institute *Proceedings*, Vol. 110, No. 4 (April 1984), pp. 36–43.

39. It is interesting to note that the Navy has now adopted the rhetorical concept of coalition defense, which was originally used by Robert Komer to attack the Navy. See, for example, *SASC Hearings on FY 85 Budget (Part 8)*, pp. 3853–3855; and Watkins, "The Maritime Strategy," p. 4.

40. *SASC Hearings on FY 85 Budget (Part 8)*, p. 3854.

41. The Navy's position on this matter is not altogether clear. Sequential military operations focusing on Europe would seem to require a "swing" strategy, under which the United States would move at least some naval forces from the Pacific to the Atlantic in a conventional war, to cope first with the war in Europe. Yet the Navy has emphasized that it will not swing forces to Europe—implying that offensive operations in the Pacific will not take a back seat to the naval offensive in Europe. See, for example, John F. Lehman, Jr., "The 600-Ship Navy," in *The Maritime Strategy*, pp. 34–36; and *SASC Hearings on FY 85 Budget (Part 8)*, p. 3893. This position is obviously not fully consistent with the notion of sequential operations that make Europe the first priority. Moreover, there is further reason to believe that the Navy does not plan to swing forces out of the Pacific, since its counterforce coercion strategy mandates that it strike at Soviet

During this period, the Navy also fleshed out some of the operational details of the offensive sea control element of the Maritime Strategy.[42] The offensive would be conducted in two phases. In the initial phase, aircraft carriers would be kept out of harm's way while attack submarines were used to roll back Soviet naval forces deployed in the Norwegian and Barents seas. Only after that task was completed would the carriers be moved forward to strike at Murmansk and other targets on the Kola Peninsula. There now appeared to be no difference between John Lehman and the admirals about the employment of carriers in a war with the Soviets.

The second period, with its emphasis on offensive sea control, came to an abrupt end in January 1986, when Admiral James Watkins, former Chief of Naval Operations, introduced a revised version of the Maritime Strategy in an important article written for a special supplement to the U.S. Naval Institute *Proceedings*.[43] Offensive sea control was de-emphasized somewhat by the Navy chief, apparently predicated on the widely agreed-upon assumption that Soviet SSNs would *not* attempt to cut NATO's SLOCs in the early stages of a war but would instead remain in the Norwegian and Barents seas to protect Soviet SSBNs from American SSNs.[44] The threat of offensive action, in other words, largely neutralized the Soviet threat to the SLOCs. Instead of offensive sea control, Watkins now made counterforce coercion the centerpiece of the Maritime Strategy. "The Soviets," he argued, "place great weight on the nuclear correlation of forces."[45] From this, he deduced that the Navy could affect Soviet behavior on the Central Front by shifting the nuclear balance against them, principally by destroying large numbers of Soviet SSBNs. Watkins also briefly noted that the Navy would place "carriers and Tomahawk platforms around the periphery of the Soviet Union," implying that such a move would further shift the nuclear balance

---

SSBNs in the Pacific as well as those located with the Northern Fleet. See note 20. Thus, it seems reasonable to conclude that, when Navy spokesmen discuss sequential operations, they really mean that major offensives would be launched simultaneously against the Soviets' Northern and Pacific fleets, while naval operations in all other theaters would be assigned secondary importance in the war's initial stages.

42. See *SASC Hearings on FY 85 Budget (Part 8), passim*; and Judy J. McCoy and Benjamin F. Schemmer, "An Exclusive *AFJ* Interview with: Admiral Wesley L. McDonald," *Armed Forces Journal International*, April 1985, pp. 68–70. Hereinafter cited as McDonald Interview.

43. Watkins, "The Maritime Strategy." Also see Michael R. Gordon, "Nonnuclear War Might Start Raid on Soviet A-Arms," *The New York Times*, January 7, 1986, pp. 1, 11; and George C. Wilson and Michael Weisskopf, "Pentagon Plan Coldly Received," *The Washington Post*, February 6, 1986, p. 14.

44. Watkins, "The Maritime Strategy," p. 7. Also see note 98 and the attendant text.

45. Ibid., p. 14.

against the Soviets.[46] In short, altering the strategic balance now appears to be the principal goal of the Maritime Strategy.

At the same time, counterforce coercion was not the only deterrent posture discussed in Watkins's article. All three of the other elements of the Maritime Strategy received roughly equal mention as matters of secondary importance. Offensive sea control, although demoted from preeminence, was nevertheless treated as a matter of concern. Horizontal escalation reappeared after its near-disappearance during 1983–1985.[47] Specific references to direct military impact on the Central Front also occurred. For example, Admiral Watkins stressed at least twice that "the full weight of the carrier battle forces could . . . contribute to the battle on the Central Front, or carry the war to the Soviets."[48] In that same special supplement, General P.X. Kelley, the Commandant of the Marine Corps, made a more elaborate case for direct military impact. He wrote:

Maritime forces offer the opportunity to avoid a long, costly, and uncertain land effort to push the Soviets back in Central Europe. Naval operations on the exposed Rimland flanks present the option of striking quickly at key Soviet pressure points in a campaign of nautical maneuver. Used in this manner, our naval forces can make the strategic difference.[49]

Not surprisingly, he placed special weight on using "massed amphibious task forces . . . to seize key objectives in the Soviet rear."[50] He went so far as to raise the possibility of landing Marines along the eastern Baltic or the Black Sea coasts.[51]

Two general conclusions emerge from this discussion. First, the Maritime Strategy has evolved through three periods, with the public emphasis changing from a somewhat vaguely defined interest in offensive sea control and

---

46. Ibid.
47. Watkins writes, for example: "Forward deployment must be global as well as early. Deployments to the Western Pacific directly enhance deterrence, including deterrence of an attack in Europe, by providing a clear indication that, should war come, the Soviets will not be able to ignore any region of the globe." Ibid., p. 10. Also see pp. 7, 10–11, 14.
48. Ibid., p. 13. Also see p. 12.
49. General P.X. Kelley and Major Hugh K. O'Donnell, Jr., "The Amphibious Warfare Strategy," in *The Maritime Strategy*, p. 26.
50. Ibid.
51. As a sidelight on this matter, the military analyst Walter Millis raised the possibility in 1951 of launching simultaneous amphibious operations in the Baltic and Black seas. "The space between the Black Sea and the Baltic is only 750 miles wide," he noted, implying that the attacking forces could link up and produce a decisive victory against the Soviet Union. Walter Millis, "Sea Power: Abstraction or Asset?," *Foreign Affairs*, Vol. 29, No. 1 (April 1951), p. 383.

horizontal escalation (1981–1982) to a relatively clear-cut preference for offensive sea control (1983–1985) to the adoption of counterforce coercion as the centerpiece of the strategy (1986). Second, although Navy spokesmen may accentuate one or more of the offensive postures, they have used all four deterrent postures to make their case. Thus, to analyze the impact of the strategy on deterrence in Europe, one cannot simply evaluate the case for counterforce coercion. It is also necessary to consider the other three offensive postures: direct military impact, horizontal escalation, and offensive sea control. As will become evident, none of these contributes much to deterring the Soviets from going to war in Europe.

Direct Military Impact and Horizontal Escalation

DIRECT MILITARY IMPACT

The Navy has not argued forcefully about the deterrent value of direct military impact, perhaps because the case is so weak.[52] All three of the principal scenarios for using the Navy in this way lack plausibility.

Consider amphibious operations against the Soviet mainland, which would entail a major landing operation along the coast of either the Baltic or Black sea.[53] The aim would be to create such a formidable threat in the Soviets' rear that they would have to pull a substantial number of forces away from the Central Front. This is a completely unrealistic scenario for three reasons. First, it would be essential that the attacking forces have control of the air and seas in and around the landing area.[54] The landing forces must be

---

52. In this section and in the following two, I have sometimes listed the number of forces on each side and drawn conclusions from those raw balances. As I have tried to show elsewhere, it is important to go beyond mere force levels and examine what is likely to happen when the opposing forces actually engage each other. [See John J. Mearsheimer, "Why the Soviets Can't Win Quickly in Central Europe," *International Security*, Vol. 7, No. 1 (Summer 1982), pp. 3–39. Also see Joshua M. Epstein, "Soviet Vulnerabilities in Iran and the RDF Deterrent," *International Security*, Vol. 6, No. 2 (Fall 1981), pp. 126–158; and Barry R. Posen, "Measuring the European Conventional Balance: Coping With Complexity in Threat Assessment," *International Security*, Vol. 9, No. 3 (Winter 1984-85), pp. 47–88. Copies of these three pieces can be found in Steven E. Miller, ed., *Conventional Forces and American Defense Policy* (Princeton: Princeton University Press, 1986), pp. 79–157, 309–341.] Given limitations of space and time, I am unable to offer such dynamic assessments here. I am nevertheless confident about the conclusions offered in the subsequent analysis.
53. The following discussion is not addressed to a small-scale landing operation in non-Soviet territory (i.e., Norway), but to a major landing on the Soviet homeland.
54. See Jeter A. Isely and Philip A. Crowl, *The U.S. Marines and Amphibious Warfare: Its Theory and Its Practice in the Pacific* (Princeton: Princeton University Press, 1951), *passim*; and Alfred Vagts, *Landing Operations: Strategy, Psychology, Tactics, Politics, From Antiquity to 1945* (Harrisburg: Military Service Publishing Company, 1946), chapter 4.

protected from enemy air and naval strikes. Moreover, a major amphibious operation against the Soviet Union would require substantial fire support from air and naval forces, support that would not be forthcoming if those forces were themselves vulnerable to enemy air and submarine attacks. Thus, it would be imperative that the assaulting forces dominate the air and sea. This would almost certainly not be the case in the Baltic and Black seas. The Soviets have ground-based air defenses in these regions as well as large numbers of land-based anti-ship missiles.[55] Furthermore, they maintain powerful fleets in both seas, and large numbers of land-based aircraft are within easy striking distance of these coasts. Also, it would be extremely difficult to achieve tactical surprise because the attacking forces would have to move through restricted routes of passage to get into either sea. Soviet surveillance would surely discover the assault forces early and move to destroy them before they reached the landing areas. The Baltic and the Black seas would be veritable hornets' nests in a war and hardly suitable for amphibious operations.

Second, the United States has a limited amphibious lift capability. The Navy, which has long considered it a low-order priority, is now in the process of procuring enough ships to increase its amphibious assault capability by approximately one Marine brigade, from one Marine division to 1½ Marine divisions.[56] These would be essentially light infantry forces which would be at a real disadvantage against Soviet heavy mechanized and armored divisions. This is hardly a formidable threat against a power like the Soviet Union, assuming, of course, that the Navy could insert that force into the Soviets' rear. To have a significant impact on the Central Front, where scores of divisions would be locked in combat, the Navy would have to insert more than a reinforced division.[57] Instead for NATO to have any chance with an amphibious operation against the Soviet mainland would require a striking force of at least five divisions, something on the scale of the Allied force that landed at Normandy in June 1944.[58] Furthermore, NATO would need rein-

---

55. For information on Soviet forces in these two areas, see Department of Defense, *Soviet Military Power*, pp. 12–14; Vice Admiral Helmut Kampe, "Defending the Baltic Approaches," U.S. Naval Institute *Proceedings*, Vol. 112, No. 3 (March 1986), pp. 88–93; *The Military Balance, 1984–1985*, pp. 17–22; and *The Military Balance, 1985–1986*, pp. 26–29.
56. See Kelley and O'Donnell, "The Amphibious Warfare Strategy," pp. 18–29.
57. For a good description of the size and scope of possible engagements on the Central Front, see Andrew Hamilton, "Redressing the Conventional Balance: NATO's Reserve Military Manpower," *International Security*, Vol. 10, No. 1 (Summer 1985), p. 116.
58. For an appreciation of the giant scale of effort that was required to launch and sustain the

forcements, most of which would have to be heavy divisions. NATO does not have and is not likely to develop such a capability, if for no other reason than because it would mean taking land forces away from the Central Front and naval forces away from their planned offensive against the Northern Fleet. No rational NATO planner would pull forces away from those main engagements to attempt a highly risky amphibious operation in the Baltic or Black sea. In fact, there would undoubtedly be great pressure to use Marine units on the Central Front, in much the same way that their predecessors were used on the Western Front in World War I.[59]

Finally, even if NATO were able to raise such a force of five or more divisions and somehow manage to land it in the Soviets' rear, it most likely would not be capable of presenting so serious a threat to the Soviets that they would have to draw forces away from the Central Front. First, it is not clear that NATO could maintain lines of communication with the attacking forces. The Soviets would bring substantial air and naval assets to bear to cut off the amphibious forces from their source of supply. Second, the Soviets should have adequate ground forces to check the amphibious forces without having to pull units from the Central Front. They have a large pool of divisions near the potential landing areas which they could draw upon to meet the attacking forces.[60] If NATO were to develop a formidable amphibious capability, the Soviets would certainly make preparations to shift those forces in the actual event. Furthermore, because the Soviets would enjoy internal lines of communication and because they have good rail and road systems, they should be able to shift forces quickly enough to contain a NATO assault force. In sum, a major amphibious operation against the Soviet mainland is *not* a serious threat and offers little promise of enhancing NATO's deterrent posture.[61]

Normandy invasion, see Susan H. Godson, *Viking of Assault: Admiral John Lesslie Hall, Jr., and Amphibious Warfare* (Washington, D.C.: University Press of America, 1982), chapters 5–6; Gordon A. Harrison, *Cross-Channel Attack* (Washington, D.C.: Office of the Chief of Military History, Department of the Army, 1951), *passim*; Roland G. Ruppenthal, "Logistic Planning for OVER-LORD in Retrospect," in The Eisenhower Foundation, *D-Day: The Normandy Invasion in Retrospect* (Lawrence: University Press of Kansas, 1971), pp. 87–103; and Martin Van Creveld, *Supplying War: Logistics from Wallenstein to Patton* (New York: Cambridge University Press, 1977), chapter 7.

59. The Marines, who have a long-standing fear of being incorporated into the Army, seldom refer to their experience in World War I; nor do they encourage discussion of their possible use on the Central Front.

60. See the sources cited in note 55.

61. It is instructive to note that before World War I some British naval officers argued for an

What about the Navy's claim that carrier-based air might play a key role in the air war over Central Europe? This argument cannot be taken too seriously simply because neither the Navy nor NATO plans to use naval tactical air for this purpose.[62] After all, the Navy's primary rationale for carrier task forces in a European conflict is that they are necessary for projecting power against the Kola Peninsula, not for helping win the air war on the Central Front. One certainly cannot rule out the possibility that carrier-based aircraft would be used in the air war over Germany, but it must again be emphasized that NATO's deterrent posture largely ignores that possibility and depends instead on Air Force tactical fighters.

There is a fundamental problem with relying on carrier-based air to provide the margin of victory in the air war over the Central Front. The cost of procuring a fixed number of naval tactical aircraft is much greater than the cost of procuring the same number of Air Force replacement aircraft.[63] It is not the aircraft that account for the difference but the fact that, when calculating the price of naval air, it is necessary to include the enormous cost of the entire carrier battle group. In short, building naval air for the Central Front is a very inefficient way to buy tactical air power. If the purpose is to improve NATO's chances in the air war over Europe, it would make much more sense to use resources allocated to the Navy to buy additional Air Force fighters.

---

amphibious assault along the German coast as a means of thwarting a German offensive against France and that the proposal was rejected as strategically unsound. See John Gooch, *The Plans of War: The General Staff and British Military Strategy c. 1900–1916* (New York: Wiley, 1974), chapter 9; Paul Haggie, "The Royal Navy and War Planning in the Fisher Era," *Journal of Contemporary History*, Vol. 8, No. 3 (July 1973), pp. 125–131; Arthur J. Marder, *From the Dreadnought to Scapa Flow: The Royal Navy in the Fisher Era, 1904–1919*, Vol. 1 (London: Oxford University Press, 1961), pp. 383–395; John McDermott, "The Revolution in British Military Thinking From the Boer War to the Moroccan Crisis," *Canadian Journal of History*, Vol. 9, No. 2 (August 1974), pp. 171–177; Nicholas d'Ombrain, *War Machinery and High Policy: Defence Administration in Peacetime Britain, 1902–1914* (London: Oxford University Press, 1973), chapter 2; and Samuel R. Williamson, Jr., *The Politics of Grand Strategy: Britain and France Prepare for War, 1904–1914* (Cambridge: Harvard University Press, 1969), *passim.* The British navy maintained interest in this scheme even after the war had begun. See Paul M. Kennedy, *The Rise and Fall of British Naval Mastery* (London: Allen Lane, 1976), pp. 255–259. Kennedy aptly concludes: "The mind recoils at the fate which would have befallen an expeditionary force landed on the Pomeranian coast, even had the fleet managed to break into the Baltic. Whilst the French political and military leaders saw such diversions as a virtual betrayal, British generals regarded them as a nonsense." Ibid., p. 258.
62. I know of no evidence in the public record that shows otherwise.
63. For a good discussion of this matter, see Arnold M. Kuzmack, *Naval Force Levels and Modernization: An Analysis of Shipbuilding Requirements* (Washington, D.C.: Brookings, 1971), pp. 27–31.

Finally, there is the issue that the threat of a successful naval offensive against the Kola Peninsula would cause the Soviets to pause when contemplating a blitzkrieg in Europe. Specifically, it has been suggested here that a naval victory on the Northern Flank might force the Soviets to transfer much-needed air units from the European heartland to the Kola—reducing their chances for success in the main land battles. Although Navy spokesmen do not often make this argument, it is important to consider.[64]

The first problem with it is lack of credibility. It is not at all obvious that the national command authorities would allow the Navy to strike at the Kola Peninsula and, even if allowed to do so, it is not clear that the Navy would achieve a major success. These matters will be discussed in greater detail in subsequent sections. Suffice it to say here that the Navy, if sent to strike north, would face a very formidable task.

Second, if the Soviets were pressed to send additional air units to the Northern Flank, they could be drawn from units in regions not directly involved with the conflict on the Central Front.[65] Third, there is the time factor. The Soviets are likely to go to war only if they believe that there is a good chance that they can win a quick and decisive victory. Should they reach such a conclusion, it would not be unrealistic for them to think in terms of effectively crippling NATO in 10–14 days.[66] After all, NATO lacks strategic depth and employs most of its forces in forward positions. An American naval offensive, on the other hand, would probably take considerable time to execute. Carrier-based strikes against the Kola Peninsula, for example, would not take place until after the Northern Fleet's SSNs and surface navy had been rolled back; and that difficult task would probably take much time (several weeks or months) to accomplish. This point is driven home by John Lehman's response when asked to specify the period of time

---

64. For examples of this argument, see Michael Getler, "Lehman Sees Norwegian Sea as a Key to Soviet Naval Strategy," *The Washington Post*, December 29, 1982, p. 4; and "Lehman Seeks Superiority," p. 547.

65. The Soviets have a very large number of tactical aircraft as well as a formidable fleet of medium and long-range bombers. Air units from regions not directly involved in the fighting on the Central Front could be easily and quickly moved to the Kola Peninsula in a conflict. For information on the size and location of those air forces, see Department of Defense, *Soviet Military Power*, pp. 12–14; *The Military Balance, 1984–1985*, pp. 17–18, 20–22; and *The Military Balance, 1985–1986*, pp. 21–30.

66. I should note that I do not believe that it is likely that the Soviets would reach this conclusion. See Mearsheimer, "Why the Soviets Can't Win Quickly in Central Europe."

needed to conduct the rollback operation. Lehman, who has great confidence in the Navy's ability to carry out difficult missions, answered:

No one has tried to put a timeframe on it because of the inherent unpredictability. . . . War is inherently unpredictable, one can't easily determine how it will break out or how long it will take, for instance, to nullify the submarine force in the Norwegian Sea. *That is a tough area to operate in.* It may take a week or it may take a month or 3 months.[67]

The key point here is that the Soviets' time frame for executing a successful blitzkrieg would, in all likelihood, be short enough that events on the Northern Flank would not upset it in any way.

Fourth, even if the Navy is capable of executing an offensive against the Kola Peninsula in a relatively short period of time, the resulting threat to the Soviet Union would not be very great. The Soviet Union is a great land power that cannot be hurt badly by naval strikes against its periphery. For the Soviets, and ultimately for NATO, Central Europe is where a major conventional war would be settled. The Navy could score a stunning victory on the Northern Flank, and it would be all for naught if NATO failed to check the Soviets in the land battle on the intra-German border. The Soviets, after consolidating their position on the continent, would then have little difficulty eliminating the threat on their northern flank. There would therefore be no compelling reason for the Soviets to pull units away from the Central Front. In short, a Soviet decision to launch a blitzkrieg would probably not be affected by the threat of naval strikes against the Kola Peninsula.

HORIZONTAL ESCALATION

The Reagan Administration was initially attracted to horizontal escalation as a deterrent posture. However, except for the Navy, there no longer appears to be much interest in this strategy.[68] This is probably because such a strategy does little to enhance deterrence—especially for NATO. One of the principal difficulties is finding an appropriate target. No area in the Third World compares in importance to Western Europe. Surely, the consequences for the Soviet Union of "losing" Angola, Cuba, or Vietnam would be nowhere

---

67. *SASC Hearings on FY 85 Budget (Part 8)*, p. 3877. Emphasis added. Also see Watkins, "The Maritime Strategy," pp. 8–9.
68. Good critiques of horizontal escalation are Joshua Epstein, "Horizontal Escalation: Sour Notes of a Recurrent Theme," *International Security*, Vol. 8, No. 3 (Winter 1983/1984), pp. 19–31; and Robert Perry, Mark A. Lorell, and Kevin N. Lewis, *Second-Area Operations: A Strategy Option*, R-2992-USDP (Santa Monica, Calif.: Rand, May 1984).

near as great as the consequences to the United States of seeing Western Europe fall into Soviet hands. As a result, the Soviets would not be deterred by the prospect of the loss of those areas and probably would not move significant forces to defend them. The net result would be that American forces, but not Soviet forces, would be diverted from the crucial battle in Europe into campaigns that held little strategic significance. Furthermore, as the United States learned in Southeast Asia, America's ability to influence the course of events in the Third World is limited, so American horizontal escalations could develop into costly enterprises. It is, in short, simply not plausible to think in terms of threatening the Soviets with a tit-for-tat strategy in which they take Europe and the United States takes an area of comparable value in the Third World.

One might argue that the threat of a major military strike on the Soviet periphery would force the Soviets to pull units away from the Central Front.[69] This kind of offensive would have to be directed at Soviet forces in the Far East, since this is the only important Soviet area *outside* of Europe that is vulnerable to attack by powerful naval forces. The logic here is analogous to the claim that direct strikes against the Kola Peninsula would weaken the Soviets' position in Central Europe. The flaws in the argument are similar. The Soviets could afford to absorb a temporary beating in the Far East while they were rolling up NATO's forces in Central Europe. A setback on the periphery would not weaken their European effort in any meaningful way and, moreover, once the Soviets had consolidated their position in Western Europe, they could move massive forces to deal with problems on their periphery. In any event, it is not clear that the Navy could inflict a significant defeat on Soviet forces in the Far East.[70] The Soviets have formidable military forces in this area, and they could transfer forces from areas other than Central Europe to this theater. The Navy likes to emphasize the ease with which it could move forces around the Soviet periphery, giving the impression that it could bring greater force to bear at the point of attack than could the Soviets. This is a dubious claim. As Fred Iklé, the Under Secretary of Defense for Policy, notes:

---

69. Navy spokesmen, when discussing horizontal escalation, focus mainly, although not exclusively, on striking at the Soviet periphery—not areas in the Third World. Also see note 20.
70. There is hardly any analysis in the public domain about the Navy's prospects against the Soviets' Pacific Fleet. It is apparent, however, that the Navy believes that the Soviets have formidable forces in that region. For information on Soviet forces in the Far East, see Department of Defense, *Soviet Military Power*, p. 13; *The Military Balance, 1984–1985*, pp. 19–21; and *The Military Balance, 1985–1986*, pp. 29–30.

The Soviets are able to exploit their interior lines of communication in order to shift rapidly the geographical pivots of their force concentrations for power projection. Thus, they are in a position to move airborne forces and air forces swiftly along their periphery, and they can shift Backfire bombers to attack our fleets more rapidly than the United States can shift its aircraft carriers between widely separated sea regions near the Soviet Union.[71]

Finally, a major non-European offensive would employ forces that could otherwise be used in Europe. It takes NATO forces to divert Soviet forces, and there is no evidence, as implied in arguments for horizontal escalation, that NATO could force the Pact to divert more forces than NATO would divert. Thus, there is no evidence that NATO could improve the force ratio in Europe by pursuing a horizontal escalation strategy.

SEA POWER IN THE INDUSTRIAL AGE
The inadequacies of direct military impact and horizontal escalation as deterrent postures extend beyond their particulars to include the general view of military power that underpins them. Advocates of both strategies tend toward a Mahanian view of military power.[72] They believe that control of the seas is the key ingredient for great power status. John Lehman, who is given to highlighting the relevance and importance of Mahan, argues that "the sea is *inevitably* the major arena of competition and conflict among nations aspiring to wealth and power."[73] Underlying this belief is the core assumption that, in the competition between land power and sea power, the latter has distinct advantages that derive mainly from the flexibility inherent in naval forces. For these neo-Mahanians, offensively oriented naval forces provide the key for gaining advantage over a land power like the Soviet Union.

---

71. Iklé, "The Reagan Defense Program," p. 15. This point notwithstanding, the author actually argues in favor of horizontal escalation.

72. For a listing of Mahan's principal writings, see Peter Paret, ed., *Makers of Modern Strategy: from Machiavelli to the Nuclear Age* (Princeton: Princeton University Press, 1986), pp. 904–905. For a discussion of Mahan's thinking about naval power, see *inter alia*: Philip A. Crowl, "Alfred Thayer Mahan: The Naval Historian," in ibid., pp. 444–477; Kennedy, *Rise and Fall of British Naval Mastery*, pp. 1–9; William E. Livezey, *Mahan on Sea Power* (Norman: University of Oklahoma Press, 1981); William Reitzel, "Mahan on the Use of the Sea," *Naval War College Review*, Vol. 25, No. 5 (May–June 1973), pp. 73–82; Donald M. Schurman, *The Education of a Navy: The Development of British Naval Strategic Thought, 1867–1914* (Chicago: University of Chicago Press, 1965), pp. 60–82; and Margaret T. Sprout, "Mahan: Evangelist of Sea Power," in Edward Mead Earle, ed., *Makers of Modern Strategy: Military Thought From Machiavelli to Hitler* (Princeton: Princeton University Press, 1943), pp. 415–445.

73. Lehman, "Rebirth of a U.S. Naval Strategy," p. 11. For a brief discussion of the enduring influence of Mahan on naval thinking, see Crowl, "Alfred Thayer Mahan," pp. 476–477.

This view of power in the international system is fundamentally flawed and has little application to the U.S.–Soviet competition. Mahan's theories, as is widely recognized by scholars, were largely outdated when they were written.[74] Furthermore, the notion that "the sea is . . . the major arena of competition" between great powers is probably an accurate description of the past conflict between Japan and the United States, but it is not an accurate assessment of the present superpower rivalry.[75] Nor is it an accurate description of the British–German competition in the first half of this century. The Soviet Union, like Germany before it, is a continental power that threatens to take control of the Eurasian heartland, an area of tremendous strategic importance. Consequently, a rival power, be it Britain or the United States, has no alternative but to treat the Eurasian heartland as the principal arena of competition.

One might accept the claim that the Soviet Union is primarily a continental power, but maintain that naval forces provide an insular power like the United States with a significant lever against a land power like the Soviet Union. This, however, is not true. The industrialization and democratization that has occurred over the past century and a half, especially the development of mass armies and of railroads to move them rapidly, has led to a significant shift in the relationship between land power and sea power in favor of the former.[76] Insular powers like the United States can do little with independent

---

74. For a superb exposition of this point, see Kennedy, *Rise and Fall of British Naval Mastery*, especially chapter 7. Also see Paul M. Kennedy, *Strategy and Diplomacy, 1870–1945* (London: Allen and Unwin, 1983), pp. 43–85.

75. The Navy, not surprisingly, has historically had a preference for the Pacific over the European theater. See Vincent Davis, *Postwar Defense Policy and the U.S. Navy, 1943–1946* (Chapel Hill: University of North Carolina Press, 1962), pp. 76–80. Former Chief of Naval Operations Admiral Arleigh Burke explained this preference in the mid-1950s: "Naval forces, because of the size of the Pacific Ocean and the geographical positions of the potential enemy, will be the primary source of U.S. strength in the Western Pacific, whereas in the European Theater, other elements of this and other nations' armed forces would most surely dominate." Quoted in Lt. Commander Joseph A. Sestak, Jr., "Righting the Atlantic Tilt," U.S. Naval Institute *Proceedings*, Vol. 112, No. 1 (January 1986), p. 66. Recently, General P.X. Kelley, the Commandant of the Marine Corps, noted: "Making a case for an offensively oriented Navy and Marine Corps is not an easy undertaking if Europe is the primary U.S. area of interest." Kelley and O'Donnell, "The Amphibious Warfare Strategy," p. 23.

76. The classic statement of this position is Halford J. Mackinder, "The Geographical Pivot of History," *The Geographical Journal*, Vol. 23, No. 4 (April 1904), pp. 421–437, with subsequent discussion on pp. 437–444. Also see Halford J. Mackinder, *Democratic Ideals and Reality* (New York: Norton, 1962); Derwent Whittlesey, "Haushofer: The Geopoliticians," in Earle, *Makers of Modern Strategy*, pp. 388–411; Martin Wight, "Sea Power and Land Power," in Hedley Bull and Carsten Holbraad, eds., *Power Politics* (New York: Holmes and Meier, 1978), pp. 68–80; and the sources cited in note 72.

naval forces to hurt a land power like the Soviet Union. This point was demonstrated in both world wars, when Britain's navy had little effect on Germany's ability to wage war.[77] To the extent that there was an impact, it involved the much over-rated naval blockade of World War I.[78] However, a blockade against an autarkic state like the Soviet Union would be fruitless.

The only suitable military lever that can bring pressure against a continental power is a strong army amply supported with tactical air forces.[79] It is worth noting here that at the time of the infamous Munich accord the British chiefs of staff concluded that Britain, because it lacked an army that could be employed on the continent, had no choice but to appease Hitler.[80] Britain's navy was simply not an effective instrument for confronting the likes of the Third Reich. The same is true with regard to the American Navy and the Soviet Union. The neo-Mahanian threats of horizontal escalation and direct military impact simply do not provide a satisfactory posture for deterring a formidable land power like the Soviet Union.

It would be a mistake to conclude from this discussion that NATO should not be concerned with the naval dimension of a conventional war with the Soviet Union. The evidence from both world wars makes it clear that a continental power with a robust submarine force can seriously threaten an insular power that is either heavily dependent on imports or has to project forces and materials across wide oceans. The Germans, in both wars, came dangerously close to knocking Britain out of the war with their U-boats, and although it is not widely recognized, American submarines greatly reduced Japan's warfighting capability by cutting its SLOCs.[81]

---

77. See Gerd Hardach, *The First World War, 1914–1918* (Los Angeles: University of California Press, 1977), chapter 2; Kennedy, *Rise and Fall of British Naval Mastery*, chapters 9, 11; and Alan S. Milward, *War, Economy and Society, 1939–1945* (Los Angeles: University of California Press, 1979), chapter 9.

78. Liddell Hart's claim that "Among the causes of Germany's surrender the blockade is seen to be the most fundamental" is often cited by other authors. See Liddell Hart, *Strategy*, 2nd rev. ed. (New York: Praeger, 1967), p. 218. This is not, however, the case. See Louis Guichard, *The Naval Blockade*, trans. Christopher R. Turner (New York: Appleton, 1930); and Hardach, *The First World War*, chapter 2.

79. This discussion leaves aside, of course, the important matter of nuclear weapons. For my views on this subject, see Mearsheimer, "Nuclear Weapons and Deterrence in Europe."

80. See Paul M. Kennedy, *The Realities Behind Diplomacy: Background Influences on British External Policy, 1865–1980* (London: Fontana, 1981), pp. 290–293; and Telford Taylor, *Munich: The Price of Peace* (Garden City, N.Y.: Doubleday, 1979), *passim*, but especially pp. 600–603, 629–633, 832–837, 995–1000.

81. Regarding the American submarine campaign against Japan, see Clay Blair, Jr., *Silent Victory* (New York: Lippincott, 1975); and U.S. Strategic Bombing Survey, *The War Against Japanese Transportation, 1941–1945*, Pacific War, Report 54 (Washington, D.C.: U.S. Government Printing

The principal lesson to be derived from the historical record is *not* that an insular power with a large surface navy can use that force to threaten a continental power but, on the contrary, that a continental power armed with submarines is a very real threat to an insular power. It is the United States, not the Soviet Union, that must concern itself with falling victim to the other side's naval power. Thus, in the final analysis, the central question is not whether the United States can hurt the Soviets with its navy, but whether NATO can protect its SLOCs from Soviet submarines. Sea control is the key issue.

Offensive Sea Control

It is not likely that Soviet calculations about the SLOC battle will have much influence on a decision to launch a war against NATO since that decision would probably be based on an assessment of their prospects in the ground war. Nevertheless, NATO must be concerned with the scenario in which the Soviets conclude that they are not likely to effect a blitzkrieg, but that there is a reasonable chance that they can defeat NATO by cutting its SLOCs in a few months' time. NATO must ensure that it has a sufficiently strong sea control capability that this situation never occurs. In this regard, the Navy matters for deterrence on the Central Front. The key issue, however, is whether offensive sea control is necessary to secure the SLOCs or whether that goal is best served by a defensive sea control strategy.

The balance of evidence suggests that offensive sea control is not an appropriate deterrent strategy for NATO. In the first place, it is not a credible strategy. Simply put, it is unlikely that the national command authorities would allow the Navy to pursue such a strategy in a conventional war. In addition, it is not necessary for the Navy to strike north to protect NATO's SLOCs. A robust defensive sea control posture would provide adequate protection.

The credibility of the Navy's threat to move north to protect the SLOCs depends on three factors. First is the matter of feasibility. It is not clear that the Navy, even a 600-ship navy, could roll back the Soviets' Northern Fleet

---

Office, 1947). Concerning the German U-boat campaigns in the two world wars, see Hardach, *The First World War*, chapter 3; Marder, *From the Dreadnought to Scapa Flow*, Vol. 5, pp. 77–120; Milward, *War, Economy and Society*, chapter 9; and Jurgen Rohwer, "The U-Boat War Against the Allied Supply Lines," in Jacobsen and Rohwer, *Decisive Battles of World War II*, pp. 259–312.

and then launch devastating attacks against the Kola Peninsula. Second are the problems of inadvertent and accidental escalation. An offensive strategy carries a real risk of nuclear escalation, and NATO decision-makers will surely want to avoid a situation in which a conventional war escalates extempore. Finally, there is the matter of necessity. Policymakers might be willing to set aside the problems of feasibility and nuclear escalation if they conclude that NATO has no choice but to strike north to guarantee protection of the SLOCs. Since this is not the case, offensive sea control is neither credible nor necessary.

THE PROSPECTS FOR SUCCESSFUL IMPLEMENTATION

The Navy's prospects for rolling back Soviet naval forces in the Norwegian and Barents seas and then eliminating the important military installations and forces located on the Kola Peninsula are not good. This would not be an easy task in general, for it has become increasingly difficult in the 20th century for naval forces to strike effectively against powerful land-based forces. One only has to recall that in the First World War the British navy would not venture near the German coast to strike at the High Seas Fleet.[82] British leaders worried that they would lose their fleet to the combined actions of mines, long-range coastal ordnance, torpedo boats, submarines, and the main units of the High Seas Fleet. The airplane further complicated the attacking naval force's problem. Although a striking force could employ carrier-based aircraft, the defending land power could always deploy many more aircraft than could a handful of carriers. The same argument applies to cruise missiles.

There are three specific reasons why it would be difficult for the Navy to roll back Soviet naval forces located in the Norwegian and especially the Barents seas. First, it is not apparent that the balance of forces would work to NATO's advantage. The Soviets have a large number of submarines in their Northern Fleet: 41 ballistic missile submarines and 140 attack and cruise missile submarines.[83] Some of these submarines are old and would be of limited utility against modern American attack submarines. Still, approximately 75 percent of the Soviets' most modern attack submarines are in these northern waters. The American Navy, once it achieves the Maritime Strategy's goal of 100 attack submarines, would normally maintain about 56 of

82. See Kennedy, *Rise and Fall of British Naval Mastery,* chapter 9; and Marder, *From the Dreadnought to Scapa Flow,* Vols. 2–5, *passim.*
83. Couhat and Baker, *Combat Fleets of the World, 1986/87,* p. 481.

those SSNs in the Atlantic and about 44 in the Pacific.[84] The Navy stresses that it would not swing forces from the Pacific to the Atlantic.[85] It seems reasonable to assume that about six of those 56 submarines would be in overhaul at any one time.[86] Furthermore, about 20 of the 50 operational attack submarines in the Atlantic would probably have to be used for defending the GIN barrier, protecting carrier battle groups, and possibly escorting high-value convoys across the Atlantic.[87] The Navy would therefore have about 30 attack submarines to send north against a total force of 181 Soviet submarines. Not counting Soviet SSBNs brings that number down to 140. There is no doubt that American submarines have a qualitative edge over their Soviet counterparts. Nevertheless, the Navy maintains that this qualitative edge has eroded markedly over the past decade and that the Navy now needs anywhere from 115 to 140 attack submarines to carry out its strategy.[88] In short, it is not easy to assess what is likely to happen in a submarine war in northern waters. After all, there has never been a submarine versus submarine war, and military history demonstrates that outnumbered forces occasionally prevail in war. Nevertheless, it is difficult to be confident about the Navy's assumption that its submarines would score greater than 3:1 exchange ratios in Soviet waters.

Second, the Soviet submarine force will be assisted by an impressive array of land-based aircraft and surface combatants assigned to the Northern Fleet. These forces are specifically designed to participate in ASW operations.

Third, the Soviets have a huge arsenal of mines, which promise to complicate American efforts to move swiftly into this northern bastion.[89] To

---

84. See the testimony of Admiral James A. Lyons, Jr., in U.S. Congress, Senate Armed Services Committee, *Hearings on Department of Defense Authorization for Appropriations for Fiscal Year 1986 (Part 8)*, 99th Congress, 1st Session (Washington, D.C.: U.S. Government Printing Office, February–March 1985), p. 4411. Hereinafter cited as *SASC Hearings on FY 86 Budget (Part 8)*.
85. See note 41.
86. As a rule of thumb, about 10 percent of the attack submarine force is in overhaul at any time. See *HASC Hearings on FY 84 Budget (Part 4)*, p. 176.
87. This number, which is admittedly a rough calculation, is based on the assumption that the Navy would send ⅔ to ¾ of its available attack submarines north to strike at the Soviet fleet. This assumption is derived from interviews.
88. Admiral N.R. Thunman, the Deputy Chief of Naval Operations for Submarine Warfare, notes that with regard to attack submarines "our force level goals [presumably 100 attack submarines] are well short of those considered necessary for a reasonable assurance of success." *HASC Hearings on FY 84 Budget (Part 4)*, p. 192. For references to the need for 115, 130, or 140 attack submarines, see ibid., pp. 177, 181, 185, 192–193, 217, 219. Regarding the erosion of the American qualitative edge in submarine technology, see McDonald Interview, p. 72; and Office of the Chief of Naval Operations, *Understanding Soviet Naval Developments*, 5th ed. (Washington, D.C.: U.S. Government Printing Office, April 1985), pp. 28–29.
89. See Ted S. Wile, "Their Mine Warfare Capability," U.S. Naval Institute *Proceedings*, Vol. 108, No. 10 (October 1982), pp. 145–151.

compound these difficulties, the NATO navies, and particularly the American, have a weak countermine capability. To quote Admiral Wesley Mc-Donald (the former Supreme Allied Commander, Atlantic), the "US countermine capability is *woefully* inadequate."[90] In sum, it will be difficult for the American Navy to roll back the Soviet Northern Fleet.

It may be the case, however, that the American Navy is such a superb fighting force that it would ultimately eliminate the majority of the adversary's fleet. Let us assume that the Navy has successfully rolled back the Northern Fleet, driving the remnants of that force into their bases on the Kola, and that the time has arrived to move the carriers forward and strike directly at the Kola Peninsula. This too would be a very difficult mission. The Soviets would undoubtedly have had considerable time to augment their already formidable forces and to fortify their defensive positions. These forces would present two different problems for the carriers. First, the Soviets would have a large number of Backfire bombers, cruise missiles, and other strike systems that could be used against the carriers.[91] Second, the Soviets would employ large numbers of fighter aircraft as well as surface-to-air missiles (SAMs) and anti-aircraft artillery (AAA) to defend the Kola Peninsula. Thus, even if the carriers survive, their strike aircraft would be flying into the teeth of a well-armed defender.

Other considerations cast doubt on the likelihood of the Navy successfully silencing the Soviet threat on the Kola Peninsula. First, the number of attacking aircraft that a handful of carriers could muster for an air offensive is not great. There are limits to how many aircraft could be placed on a carrier, and furthermore, a substantial number of them would have to be used to defend the carriers.[92] It is therefore not surprising that the Joint Chiefs of

---

90. Quoted in Giovanni de Briganti, "Europe's Navies: Can They Keep NATO Ports Open?," *Armed Forces Journal International*, April 1985, p. 56. Also see Jan S. Breemer, *U.S. Naval Developments* (Annapolis: Nautical and Aviation Publishing Company of America, 1983), pp. 56–57; and Norman Friedman, "US Mine–Countermeasures Programs," *International Defense Review*, Vol. 17, No. 9 (1984), pp. 1259–1268.

91. For information on the formidable array of assets that the Soviets have on the Kola Peninsula, see the sources cited in note 13.

92. There are generally about 90 aircraft on a carrier, and about 34 of them are designated attack aircraft. See Alan H. Shaw, *Costs of Expanding and Modernizing the Navy's Carrier-Based Air Forces* (Washington, D.C.: U.S. Congressional Budget Office, May 1982), p. 4. Thus, if three carriers were placed in the upper reaches of the Norwegian Sea, the Navy would have only about 100 attack aircraft to strike against a wide variety of heavily defended targets. This small force hardly generates confidence. The Navy, if it is serious about launching a large-scale air offensive against the Kola Peninsula, would surely have to augment its forces with *significant* numbers of land-based fighters and bombers.

Staff maintain that the Navy would need 22 carriers, not the currently programmed 15, to execute its chosen strategy.[93]

Second, even if the Navy enjoys great success with its initial air strikes, Soviet air power in that region would not be "finished off" in any meaningful sense. The Soviets would simply move air units from other areas of the Soviet Union to the Northern Flank.[94] It is not possible to inflict a knockout blow against the Soviet air forces; the air war would be a protracted one. This is one of the central lessons of the air war in World War II.[95] In that war, the United States and its allies, despite inflicting a series of major defeats on the German and Japanese air forces in the early years of the war, were not able to establish complete dominance of the air until the late stages of the war. Earlier, the Luftwaffe scored a great victory against the Soviet air forces in the opening weeks of June 1941, but the Soviets nevertheless recovered.

Third, the Navy does not have many replacements or the capability of quickly generating replacements for either carriers or air wings lost in combat.[96] The Navy, in short, is not well-suited to fight a protracted air war on NATO's Northern Flank. These considerations point up that there are substantial reasons for doubting whether the U.S. Navy could successfully execute its forward offensive strategy.[97]

---

93. See Richard Halloran, "Pentagon Draws Up First Strategy For Fighting a Long Nuclear War," *The New York Times*, May 30, 1982, pp. 1, 12; and George C. Wilson, "U.S. Defense Paper Cites Gap Between Rhetoric, Intentions," *The Washington Post*, May 27, 1982, p. 1.
94. See note 65.
95. See R.J. Overy, *The Air War, 1939–1945* (New York: Stein and Day, 1981).
96. For a good discussion of this problem, see George C. Wilson, "Navy Cites a Shortfall of Combat Aircraft," *The Washington Post*, June 5, 1983, p. 5. Also see Lane Pierrot and Bob Kornfeld, *Combat Aircraft Plans in the Department of the Navy: Key Issues* (Congressional Budget Office, Staff Working Paper, March 1985); Shaw, *Costs of Expanding and Modernizing the Navy's Carrier-Based Air Forces*; and Peter T. Tarpgaard and Robert E. Mechanic, *Future Budget Requirements for the 600-Ship Navy* (Washington, D.C.: U.S. Congressional Budget Office, September 1985), pp. 27–32.
97. Consider, for example, these two quotations. Captain (Ret.) Robert H. Smith wrote the following in response to an article in the U.S. Naval Institute *Proceedings* that called for direct attacks on the Kola Peninsula: "The author takes it for granted that sufficient forces *can* do the job of defeating all the defending Soviet naval forces and land-based air, and thence destroy the SSBNs. I don't believe it for a minute, and I don't believe any other knowledgeable naval observer believes it either." Letter, U.S. Naval Institute *Proceedings*, Vol. 110, No. 7 (July 1984), p. 17. Admiral (Ret.) Stansfield Turner wrote the following about Secretary Lehman's views on an offensive naval strategy: "Finally, the Secretary advocates a strategy for the Navy of 'maneuver, initiative and offense.' Presumably he is reaffirming his many public statements that our Navy is going to be capable of carrying the war right to the Soviets' home bases and airfields. That sounds stirring and patriotic. The only problem is that I have yet to find one Admiral who

THE THREAT OF ESCALATION

The majority of Soviet SSBNs are located with the Northern Fleet. There is widespread agreement in the Navy and the intelligence community at large that the principal mission of the Northern Fleet's other assets, its SSNs and surface forces, is to protect those SSBNs.[98] Soviet forces in the Barents Sea, in other words, would be principally concerned with preventing American naval forces—especially SSNs—from reaching their SSBN sanctuaries. An offensive sea control strategy calls for strikes into the Barents, not for the purpose of eliminating Soviet SSBNs, but to destroy Soviet SSNs, which are the main threat to NATO's SLOCs. The aim would be to ensure that the Soviets cannot stage another "Battle of the Atlantic." There is, however, a major problem with this strategy: an offensive into the Barents Sea, regardless of intentions, would seriously threaten Soviet strategic nuclear forces, thus raising the specter of nuclear escalation.[99]

The Navy, in pursuing a strategy of offensive sea control, might simply decide *not* to attempt to discriminate between Soviet SSBNs and SSNs, but to destroy all Soviet submarines. The rationale need not be linked to the counterforce coercion posture, but could instead include the following arguments: when a state goes to war, it should go all-out to defeat the adversary; because Soviet SSBNs are well equipped to destroy attack submarines, it would be dangerous to grant them immunity; the tactical situation facing the American attack submarines (i.e., the intermingling of Soviet SSBNs and SSNs) would not permit discrimination without placing the attacking SSNs in jeopardy; and the Navy simply does not have the intelligence capability to discriminate among Soviet submarines in a fast-paced conflict.[100] There is

---

believes that the U.S. Navy would even attempt it." Letter, *Foreign Affairs,* Vol. 61, No. 2 (Winter 1982/83), p. 457. Also see "Battle for the Norwegian Sea," *passim.*

98. For a discussion of the evolution of American thinking on this important matter, see Jan S. Breemer, "The Soviet Navy's SSBN Bastions: Evidence, Inference and Alternative Scenarios," *Journal of the Royal United Services Institute for Defence Studies,* Vol. 130, No. 1 (March 1985), pp. 18–21. Also see note 44 and the attendant text.

99. For a good discussion of this problem, see Posen, "Inadvertent Nuclear War?"

100. Consider the recent Congressional testimony of the Director of Naval Warfare: "I don't believe you could effectively [distinguish between attack submarines and strategic submarines when conducting ASW]. . . . I think that it would be a stricture that would be very, very onerous from the standpoint of ASW. I don't believe you could make a distinction in a combat environment—even prehostilities—with certainty. . . . It is going to get worse in the future with the quieting trends that I depicted. . . ." *SASC Hearings on FY 86 Budget (Part 8),* p. 4399. For more general discussions of this problem, see Desmond Ball, "Nuclear War at Sea," *International Security,* Vol. 10, No. 3 (Winter 1985–86), pp. 17–18; Blair, "Arms Control Implications of Anti-

little doubt that in this case an offensive sea control strategy would result in the destruction of some portion of the Soviets' strategic retaliatory forces.

There is, however, the possibility that the Navy would try to avoid striking Soviet SSBNs, but it is not clear that this would be possible in practice. Indeed, even if the Navy could technically discriminate SSNs from SSBNs, Soviet defensive strategy might call for co-mingling the two kinds of submarines. The attacking forces would then have no choice but to destroy SSBNs as well as SSNs. The best case that can be made for a discriminating strategy is that the Soviets would, in fact, accommodate it by placing their SSBNs under the polar ice cap, while locating the majority of their SSNs in the Norwegian and Barents seas to do battle with American attack submarines.[101] Thus, the rollback strategy would largely involve a battle between rival SSN forces. Assuming this proves to be the case, successful execution of the strategy would still seriously threaten the Soviet SSBN force. The shield that protects the SSBNs would be destroyed, leaving them exposed to American attack submarines. Furthermore, the Navy plans to smash all military installations on the Kola Peninsula, which means elimination of the SSBNs' ports. As John Lehman notes, "They'd lose their whole strategic submarine fleet if they lose Kola."[102] Thus, despite the best intentions, a discriminating strategy would probably not mean very much to Soviet decision-makers intent on preserving their SSBN force.

In sum, an offensive sea control strategy would seriously threaten one leg of the Soviets' strategic triad. Whether intended or not, this deterrent posture would have the markings of a strategic ASW campaign and would therefore create risks of inadvertent nuclear escalation. The Soviets probably would not stand idly by while the strategic nuclear balance shifted against them. There would undoubtedly be pressure, which would grow as Soviet strategic assets were destroyed, to strike at American nuclear forces or to use Soviet nuclear weapons against selected NATO targets. Moreover, in addition to the risk of deliberate Soviet escalation provoked inadvertently by the United States, there would be the additional risk of accidental nuclear escalation against the wishes of both sides—meaning escalation in which individual

Submarine Warfare (ASW) Programs," pp. 112–115; and Posen, "Inadvertent Nuclear War?," *passim*.
101. See Robert Pape, "Offensive Sea Control and Inadvertent Escalation," Paper prepared for Seminar on Military Affairs, University of Chicago, May 1985.
102. Quoted in Getler, "Lehman Sees Norwegian Sea as a Key to Soviet Naval Strategy," p. 4.

commanders fire nuclear weapons before national command authorities on either side have decided to go to nuclear war. This could arise because of the uncontrolled or unforeseen interactions of local forces.[103]

In a conventional war between the superpowers, American policymakers would almost surely go to great lengths to prevent nuclear escalation. Therefore, a good case can be made that they would not allow the Navy to launch an offensive against the Northern Fleet. Whether they are seriously tempted to overlook these risks and pursue such a strategy would depend on military feasibility and strategic necessity. As emphasized, there is good reason to doubt that the Navy could actually execute the strategy. Let us now consider whether offensive sea control is required for SLOC protection.

IS OFFENSIVE SEA CONTROL NECESSARY?

The Navy's case for offensive sea control rests on three beliefs: (1) an offensive strategy forces the Soviets to keep their SSNs in their home waters, where they are not a threat to the SLOCs; (2) offensive sea control is militarily more efficient than defensive sea control; and (3) an offensive strategy is essential for keeping northern Norway out of Soviet hands. These arguments are flawed. NATO does not need an offensive sea control strategy. In fact, NATO's deterrent posture would be better served by the defensive alternative.

First, regardless of which sea control strategy the Navy adopts, only a small number of Soviet attack submarines at most are going to leave Soviet home waters and attempt to move into the Atlantic. The SSNs' primary mission is to protect SSBNs, *not* to attack NATO's SLOCs. Navy spokesmen are correct when they emphasize the importance the Soviets place on protecting their strategic nuclear forces. They fear, however, that if the Navy does not have an offensive sea control strategy, Soviet SSNs would be free to roam the Atlantic since there would be no threat of the American Navy moving north.[104]

This fear is unfounded. The Soviet SSNs must remain in home waters to protect the SSBNs—regardless of American declaratory strategy—because of the threat posed by the mere presence of American attack submarines in the

---

103. See note 19.
104. See, for example, *SASC Hearings on FY 85 Budget (Part 8)*, p. 3870; and Watkins, "The Maritime Strategy," p. 9.

area around the GIN gap. They cannot risk leaving their SSBNs exposed to the formidable American SSN force.

For this reason, the Navy should always maintain a powerful attack submarine force with offensive potential. It is not necessary, however, to have an offensive sea control strategy. The best overall sea control strategy lies between pure offense and pure defense: the Navy should maintain an offensive punch but hold it in reserve to deter the Soviets from sending their SSNs to attack the SLOCs.

The validity of this argument seems to be supported by the historical record: despite the fact that the United States moved toward defensive sea control in the 1970s, it was during this period that the American intelligence community became convinced that the Soviets would keep most of their naval forces in the Barents Sea to protect their SSBNs.[105] The Soviet decision to concentrate on protecting their SSBNs, which means that a large force of SSNs would not be surged into the Atlantic, does not correlate with the Reagan Administration's much-advertised switch from defensive to offensive sea control.

Let us assume for argument's sake that NATO adopts a strategy of defensive sea control and, as a result, the Soviets attempt to cut NATO's SLOCs. Can they be confident that their navy can accomplish that end and emasculate NATO's fighting power in some reasonably short period of time?

The answer is almost surely no. First, Soviet attack submarines would confront not only the American Navy, but the not-insignificant navies of U.S. allies.[106] NATO's combined navies represent a very formidable fighting force. Second, Soviet SSNs would have to pass through the GIN gap, which NATO has turned into a strong defensive barrier, on their way to *and* from the Atlantic.[107] Third, NATO's land-based tactical aircraft can deal with the Backfire threat in and around the GIN gap.[108] Finally, NATO's dependence on reinforcement by sea in the early stages of a conflict is not great. Massive amounts of U.S. equipment and many thousands of American soldiers and airmen are already located in Europe. Moreover, much of the equipment that American reinforcements would need in the early stages of a war is pre-

---

105. See the sources cited in notes 34 and 98.
106. For a good description of the allied navies, see de Briganti, "Europe's Navies"; also see *The Military Balance, 1985–1986*, pp. 37–60.
107. See Wit, "Advances in Antisubmarine Warfare"; and Zakheim, *The U.S. Sea Control Mission*.
108. See Deborah Shapley, "New Study of Land-Based Aircraft Questions Need for Aircraft Carriers," *Science*, June 2, 1978, pp. 1024–1025.

positioned in Europe. The manpower for those units could be flown in from the United States and would not depend on sea transportation. It is not surprising, given these factors, that NATO was confident in the 1970s that it could protect its SLOCs with a strategy of defensive sea control.[109] The bottom line is that the Soviets could not be confident of winning the SLOC war—much less winning it in a reasonably short period of time—if NATO pursued a defensive sea control posture.

Even if one doubts the efficacy of a defensive sea control strategy, it still seems apparent that, on grounds of military efficiency, defensive sea control is preferable to offensive sea control. Consider dealing with the Backfire threat. It would be much easier for NATO to destroy Backfires in the area around the GIN gap than near the Kola Peninsula. The Backfires would have virtually no support in the southern part of the Norwegian Sea, while NATO would have numerous assets that it could use to target and destroy them. The situation would be reversed if the battle took place near the Kola Peninsula. There, the Backfires would be supported by fighter escorts, while the American Navy's striking forces would surely be under heavy pressure from the numerous Soviet naval and air forces in that region. The same logic applies to the submarine war. NATO could use a variety of assets at the GIN barrier that it probably would not be able to use in the submarine war in the Barents Sea. This would include land-based aircraft like the P-3, the American surface navy, and the sophisticated listening devices that NATO has deployed in the Norwegian Sea. Correspondingly, the Soviets could not use their surface navy and their land-based aircraft to support their submarines at the GIN barrier, while they could do so in the Barents Sea. The choice between defensive and offensive sea control boils down to a question of whether the Navy is best served by fighting air battles and SSN battles in the Soviets' backyard, where Soviet forces might outnumber NATO forces, or in NATO's backyard, where NATO forces would outnumber the Soviets. There is a strong case for preferring the latter location.[110]

---

109. This optimism is reflected in the Defense Secretaries' *Posture Statements* of that decade. Also see the sources cited in note 14.

110. A problem with the debate about sea control is the use of the adjectives "offensive" and "defensive" to describe alternative sea control strategies. Conversations with military officers and civilian analysts have convinced me that much opposition to a defensive sea control strategy is based on its label as "defensive." This label brings a stigma with it, since many observers believe that offensive strategies are almost always preferable to defensive ones. However, judgments based simply on the offensive or defensive nature of the two strategies are not well grounded. The real issue is: where is it best for NATO to conduct its ASW campaign—in the

Proponents of offensive sea control also argue that offensive operations are necessary to defend against the Soviet threat to northern Norway. This is not the case. Close examination of the terrain in that region and the disposition of forces reveals that NATO has adequate ground forces for thwarting a Soviet ground attack launched from the Kola Peninsula.[111] Regarding air support, which the Navy implies that only carriers can provide, NATO could rely on land-based air forces in Norway. Additional aircraft, if needed, could be flown into those bases.

One final point about sea control bears mentioning. It is not clear why the Navy is concerned about this matter since there appears to be a widespread consensus in the Navy as well as among key Reagan Administration defense officials that the conventional balance in Europe is so unfavorable that in the event of war it is almost certain that the Soviets would quickly knock out NATO. There is no point in worrying about SLOCs if the United States is going to suffer a quick and decisive defeat on the continent. Only those who believe that NATO has the wherewithal to thwart a Soviet offensive and turn the conflict into a lengthy war of attrition should worry about the SLOC battle. It would appear that this logic has not escaped the Navy. Counterforce coercion, which the Navy now emphasizes, is a strategy that says, in effect, that the Navy, acting independently, can reverse NATO's expected losses on the Central Front.[112]

Counterforce Coercion

Deterrence is a function of both crisis stability and deterrence stability. It is essential to threaten an adversary with a formidable military posture so that

---

area around the GIN gap or in the Barents Sea? This choice, as Karl Lautenschläger notes, resembles choices faced by earlier navies between close blockades and distant blockades, where the issue was whether to engage the opponent's forces close to his coast or in more distant waters. (Personal correspondence with the author, May 16, 1986.) The distant blockade was usually favored because the opponent had military advantages near his own coast that did not obtain in waters far from the coast. See Marder, *From the Dreadnought to Scapa Flow*, Vol. 1, pp. 368–373. The same logic applies to the choice between offensive and defensive sea control.

111. For an excellent discussion of this matter, see Major General Richard C. Bowman, "Soviet Options on NATO's Northern Flank," *Armed Forces Journal International*, April 1984, pp. 88–98. Also see O'Donnell, "Northern Flank Maritime Offensive," p. 46; and Ries, "Defending the Far North," p. 877.

112. This strategic concept resembles claims by air power advocates of the 1920s and 1930s that independent air forces, not ground and tactical air forces, would decide the outcome of future wars. See Edward Warner, "Douhet, Mitchell, Seversky: Theories of Air Warfare," in Earle, *Makers of Modern Strategy*, pp. 485–503; and David MacIsaac, "Voices from the Central Blue: The Air Power Theorists," in Paret, *Makers of Modern Strategy*, pp. 624–647.

he recognizes that he cannot use force to upset the status quo. At the same time, when dealing with an adversary who is not clearly bent on aggression, it is wise to avoid employing a strategy that gives him any incentive to launch a preemptive strike. The aim in such a situation should be to dampen tensions, not to exacerbate them. This matter of crisis stability was not an important issue for the previous three offensive postures, mainly because they do not require the Navy to take provocative action in a crisis. Counterforce coercion, however, could be quite destabilizing in a crisis. Of course, it is possible that counterforce coercion provides so much deterrence stability that it is worth accepting the danger of crisis instability. However, there is a strong case that this is not so. Counterforce coercion provides very little deterrence stability and is therefore a deficient naval strategy.

CRISIS STABILITY

The root of the crisis stability problem is that a counterforce coercion strategy demands that the American SSN force be mobilized early in a crisis and that large numbers of those attack submarines be inserted deep into the Barents Sea as quickly as possible.[113] If the SSNs are not moved into the Barents before the Soviets surge their SSBNs and move them under the ice, finding and destroying those SSBNs would be a difficult and time-consuming task. Inserting a large number of attack submarines into the Barents Sea during a crisis, however, would be very dangerous for several reasons. First, such a deployment would almost surely be interpreted by the Soviets as an offensive move, signalling offensive American intentions, even if the Americans meant it as a defensive measure that would buttress deterrence.[114] After all, the

---

113. Navy spokesmen make it unequivocally clear that successful execution of this strategy is dependent on early insertion of American SSNs into the Barents Sea during a crisis. Consider, for example, the following comment of Admiral Watkins: "The transition to war is perhaps the most crucial of all [phases of the Maritime Strategy]. How we position ourselves in the transitional phase, what Admiral Gorshkov calls the 'battle for the first salvo' is critical. How we handle rules of engagement and the willingness of the political authority to deal realistically with the potential threat is crucial. In every crisis we get into, we see just how determinant that can be. . . . I believe that this pattern of political inaction [observed in war games—see notes 118 and 119 and the attendant texts] prior to receiving the first blow will be devastating in the next conflict. Somehow we have to build up in this crisis period, the transition to war, in a new way. We have to think more aggressively in terms of the pre-positioning of forces. The rules of engagement must change to meet emerging circumstances." *SASC Hearings on FY 85 Budget (Part 8)*, p. 3864. Also see Part 2, p. 902 of these hearings.

114. See Robert Jervis, "Cooperation Under the Security Dilemma," *World Politics*, Vol. 30, No. 2 (January 1978), pp. 167–214; and Stephen Van Evera, "Causes of War" (Ph.D. dissertation, University of California, Berkeley, 1984), chapter 3.

United States will soon have, with Trident D-5, MX, Minuteman IIIA, nuclear-armed Tomahawk cruise missiles, and the Pershing IIs, a substantial
counterforce capability against Soviet land-based ICBMs.[115] This development, coupled with the fact that the Soviets have a small, antiquated, and
vulnerable bomber force, means that the two land-based legs of their triad
would be in good part vulnerable to an American strike. The survivability of
their SSBN force would therefore loom as a much more important matter.
Given this situation, the Soviets would almost certainly make worst case
assumptions about American intentions if U.S. attack submarines began to
position themselves to destroy the Soviet SSBN force. This would probably
intensify rather than defuse a crisis.

A second dimension to the crisis stability problem lies in the risk that the
Soviets would not stand idly by as American attack submarines moved into
their bastion. They would undoubtedly use some of their SSNs to create a
barrier defense at the Bear Island–North Cape line and maybe even at the
GIN gap. Soviet attack submarines, because they are based closer to these
two lines than their American counterparts, should be able to establish
defensive positions there before large numbers of American attack submarines reach them. The Americans would then have to decide whether to
penetrate these barriers, while the Soviets would have to face the question
of whether to attempt to destroy any American submarines that cross the
barriers. There would be several compelling reasons for the Soviets to fight
at the barriers rather than allow American SSNs to reach the Barents Sea.
First, it is important for them to keep the American SSNs far away from their
SSBNs. Second, it would not be easy for the Soviets to find the American
SSNs once they reached the Barents. They would be easier to locate and
target at the barriers. Finally, it is probable that one-on-one SSN engagements, where each submarine knows of the other's presence, would occur
at the barriers. There would be an incentive in these confrontations to fire
at the opponent, since the side that got off the first shot would stand a good
chance of destroying the adversary.[116] It is a straightforward case of the classic
gunfighter analogy. To make matters worse, the command and control of

---

115. This should not be interpreted to mean that the United States will have a splendid first-
strike capability. See note 130 and the attendant text.
116. A recent comment by a senior naval officer goes to the heart of this problem: "In submarine
warfare, the most important thing to do is shoot first." *HASC Hearings on FY 84 Budget (Part 4),*
p. 354.

submarines is generally poor.[117] It is not difficult in such a circumstance to imagine submarine commanders on either side interpreting their orders liberally and perhaps firing their weapons. This is, without a doubt, the kind of situation to avoid in a crisis.

Even if the Soviets failed to erect barriers or if the American SSNs penetrated them without prompting a naval battle, there would still be significant potential for crisis instability. The Soviets would surely send attack submarines to search the Barents for American SSNs and, moreover, Soviet SSBNs would certainly be on constant look-out for the American submarines. A deadly game of cat and mouse would ensue in which there would be the danger of one-on-one first-shooter-wins engagements as described above. There is the additional possibility that some American submarines would be lost to mines, which the Soviets would undoubtedly use at the barriers and in the Barents Sea. The United States, which would probably not know how these submarines were destroyed, might conclude that Soviet SSNs were responsible and that therefore a response in kind was in order. Finally, in a severe crisis, the Soviets might decide to declare a keep-out zone around their home ports, firing at unidentified submarines that enter. Again, these are the kinds of situations to avoid in a crisis.

DETERRENCE STABILITY

Thus, it is apparent that a counterforce coercion posture, when viewed in terms of crisis stability, weakens deterrence. Nevertheless, it might be argued that the strategy provides so much deterrence stability that it is worth accepting the danger of crisis instability. To evaluate this matter, it is necessary to answer three questions. First, how likely is it that the Navy would actually be allowed to execute this variant of the Maritime Strategy? Second, assuming that the Navy is turned loose, is successful execution of the strategy likely? Is an anti-SSBN campaign realizable? Finally, assuming the strategy is operationally effective, what is the Soviet response likely to be? In other words, is a successful anti-SSBN campaign or the threat of one likely to provide the leverage necessary for coercion? It should be emphasized that, since the focus here is on deterring the Soviets, the key issue is how *they* would answer

---

117. See Ball, "Nuclear War at Sea," pp. 10–11, 15, 18–21; Blair, "Arms Control Implications of Anti-Submarine Warfare (ASW) Programs," pp. 116–117; and Owen Wilkes, "Command and Control of the Sea-Based Nuclear Deterrent: The Possibility of a Counterforce Role," in *World Armaments and Disarmament: SIPRI Yearbook 1979* (London: Taylor and Francis, 1979), pp. 389–420.

these questions in a crisis. This is obviously impossible to determine with any precision because there is little information about Soviet thinking on these questions and also because it is difficult to predict how decision-makers will behave in an actual crisis. Still, one can make reasonable guesses about each question.

It is *not* likely that the U.S. national command authority would allow the Navy to surge submarines into the Barents Sea during a crisis with the Soviet Union, simply because American policymakers would almost surely try to dampen, not exacerbate, the crisis. This point is not lost on the Navy, which is well aware that counterforce coercion is a "strategy not without risk."[118] Admiral Watkins candidly told the Senate Armed Services Committee in 1984 that:

All of our war games, *all* of our exercises that we have run, where we have the very best people playing the roles, whether on the Soviet side or the United States side in our games, indicate that, in fact, we will *not* make the political decision to move forces early.[119]

Not allowing the Navy to move north in a crisis would cripple the strategy since, to use again the words of Admiral Watkins, it "depends on early reaction to crisis."[120] The key point, however, is that the deterrent value of the strategy would suffer if it is improbable that the Navy would be allowed to execute it. As Admiral Watkins's testimony suggests, there is at least a good chance that the SSNs would be held back.

A second reason why national command authorities are not likely to allow the Navy to execute counterforce coercion is not related to crisis stability but to the threat of nuclear escalation during a war. There is a danger that a large-scale offensive against the Soviets' northern bastion would lead to nuclear war. American policymakers, who would surely go to great lengths to keep a conventional war from escalating to the nuclear level, would certainly have serious reservations about launching a submarine offensive

---

118. Watkins, "The Maritime Strategy," p. 14. Also see *SASC Hearings on FY 85 Budget (Part 8)*, p. 3864, where Watkins notes that when playing war games that involve inserting large numbers of American SSNs in the Barents Sea during a crisis, "There is great consternation . . . on the part of the players about whether we are sending more of a deterrent signal by moving forces, or whether we are actually tearing down deterrence and encouraging adventurism."

119. *SASC Hearings on FY 85 Budget (Part 8)*, p. 3860. Emphasis added. Also see p. 3864; and Richard K. Betts, *Surprise Attack: Lessons for Defense Planning* (Washington, D.C.: Brookings, 1982), *passim*, which provides a good discussion of those factors that would make NATO leaders reluctant to mobilize military forces in a crisis.

120. Watkins, "The Maritime Strategy," p. 14. Also see note 113.

that is laden with escalatory potential. Thus, there are good reasons to believe that the Navy would not be allowed to execute a counterforce coercion strategy. The willingness of policymakers to pursue such a risky strategy would depend in large part on the likelihood of operational success and the likelihood that that success would favorably influence events in Central Europe.

It is not clear that the counter-SSBN campaign could succeed quickly if the Navy were allowed to move north. Counterforce coercion, with its emphasis on destroying large numbers of SSBNs in a short period of time, is a demanding strategy.[121] The attacking forces, as emphasized in the discussion of offensive sea control, would be outnumbered and they would be operating in a heavily defended bastion.[122] The Soviets have a variety of assets to protect their SSBNs. Furthermore, the SSBNs themselves could prove to be an elusive target. If, for example, in the very early stages of a crisis, the Soviets were able to move a large number of their SSBNs under the ice before the American attack submarines reached the Barents, those SSBNs would then be difficult to find. Finally, it should be remembered that the Navy has never conducted an operation of this kind under wartime conditions.[123]

For all these reasons, it is difficult to be confident about the military outcome in the event. This is undoubtedly why John Lehman, when asked about the timing of an offensive operation, answered that "No one has tried to put a timeframe on it because of the inherent unpredictability."[124] No evidence in the public record would lead one to view counterforce coercion as a high-confidence option.

Nevertheless, let us assume that in some future conflict the Navy is allowed to pursue a counterforce coercion strategy and, furthermore, that the Navy successfully destroys a large number of SSBNs, markedly shifting the balance

---

121. The subsequent discussion does not consider the Navy's prospects in an offensive against the Soviets' Pacific Fleet, which would surely be part of a counterforce coercion strategy. See note 20. Thus the task facing the American Navy is somewhat more difficult than the following analysis indicates. Also see note 70.

122. It is important to note that the risks to American attack submarines would increase if they were asked to destroy enemy submarines quickly, since they would more often be forced to fire in a manner that exposed themselves to answering enemy fire.

123. Two recent studies dealing with ASW provide good accounts of the difficulties associated with destroying large numbers of Soviet SSBNs quickly in a conventional war. See Donald C. Daniel, "ASW and Superpower Strategic Stability," unpublished manuscript, n.d., especially chapters 1–2; and Tom Stefanick, "Strategic Antisubmarine Warfare and Naval Strategy," unpublished manuscript, 1985, *passim*. Both manuscripts will soon be published as books. Also see Donald C. Daniel, "Antisubmarine Warfare in the Nuclear Age," *Orbis*, Vol. 28, No. 3 (Fall 1984), pp. 527–552.

124. See note 67.

of nuclear forces. What are the Soviets likely to do? In other words, is an operationally effective strategy likely to lead to coercion? The Navy assumes that the Soviets would be so disturbed by this shift that they would throw up their hands and agree, in the words of Admiral Watkins, "to end the war on our terms."[125] Presumably, this means that on the continent they would, at the very least, retreat to the prewar borders.

But this is not likely. It is more likely that the Soviets would either ignore the shifting nuclear balance, refusing to be coerced, or would lash back militarily, themselves applying nuclear coercion against NATO.[126]

If they chose to lash back, they would have three principal military options: (1) a strike against U.S. strategic nuclear forces, to redress the strategic balance; (2) a theater nuclear strike against U.S. anti-submarine forces, to cut the American noose before it closes completely on their SSBN force; or (3) a theater nuclear strike against American naval targets as a shot across the bow, or a "manipulation of risk," to introduce the threat of nuclear escalation unless NATO called off its counter-SSBN campaign.

A Soviet strategic strike against some portion of the American nuclear retaliatory force seems possible, although not likely. The Navy maintains that the Soviets would not attempt a strategic nuclear strike because the nuclear balance would be against them.[127] This argument misses the essential point that the very attraction of a counterforce strike is that it would offer the prospect of redressing that balance. If the balance of nuclear forces is as important to the Soviets as the Navy claims, then there would undoubtedly be significant pressure on them to rectify the balance with a counterforce strike. This is not a likely response; it would involve a direct attack on the American homeland and they have other options.

A more attractive response would be to mount limited nuclear attacks against American naval forces and installations, for purposes of either noose-cutting or the manipulation of risk. Targets could include American aircraft

125. Watkins, "The Maritime Strategy," p. 11.
126. Classic works on the general subject of military coercion are Morton A. Kaplan, *The Strategy of Limited Retaliation*, Center of International Studies Policy Memorandum No. 19, Princeton University, April 9, 1959; and Schelling, *Arms and Influence*, chapter 1. Also see Robert J. Art, "To What Ends Military Power?," *International Security*, Vol. 4, No. 4 (Spring 1980), pp. 3–35; Daniel Ellsberg, *The Theory and Practice of Blackmail*, P-3883 (Santa Monica, Calif.: Rand, 1968); Alexander George, "The Development of Doctrine and Strategy," in Alexander George, David Hall, and William Simons, *The Limits of Coercive Diplomacy* (Boston: Little, Brown, 1971), chapter 1; Wallace J. Thies, *When Governments Collide: Coercion and Diplomacy in the Vietnam Conflict, 1964–1968* (Berkeley: University of California Press, 1980), chapters 1, 5, 8; and Schelling, *Strategy of Conflict*, chapter 2.
127. Watkins, "The Maritime Strategy," p. 14.

carrier battle groups, which are vulnerable, and could be attacked without wide collateral damage.[128] The Soviets should have no shortage of targets since the Maritime Strategy calls for ringing the Soviet Union with carriers and other ships carrying Tomahawk missiles. The Soviets also might consider nuclear strikes against those NATO naval installations in Norway, Iceland, Britain, and Greenland that contribute to the American anti-SSBN campaign or against high value targets in continental Europe.

Such strikes might not do much to cut the American noose, because its crucial element, the SSN force, is not directly vulnerable to nuclear attack.[129] However, such attacks would signal seriousness of purpose and would make clear that the ante could be raised if the American Navy continued to destroy SSBNs. The Soviets could thereby put the last clear chance to avoid uncontrolled escalation on the United States.

As a final option, the Soviets could accept the SSBN losses and operate on the assumption that shifts in the strategic nuclear balance have no political utility. This is a viable strategy as long as the Soviet Union retains a secure assured destruction capability. The United States is unlikely to launch a first-strike against the Soviets as long as they have the capability to inflict massive damage on the American homeland. A counterforce coercion strategy will not eliminate the Soviets' assured destruction capability. The Navy, it should be emphasized, does not call for eliminating all Soviet SSBNs, but argues only for destroying enough SSBNs to shift perceptibly the nuclear balance. Even if the Navy were to eliminate this leg of the Soviet triad, the United States would still not be able to effect a splendid first strike against the triad's other two legs. The Soviets could adopt a launch-on-warning posture, and even if that failed, they could lose 95 percent of their land-based assets and

---

128. The Navy has not paid much attention to the issue of nuclear war at sea, preferring instead to concentrate on preparing for a conventional war. However, it would be very difficult to protect a carrier battle group from a nuclear strike. See Ball, "Nuclear War at Sea," pp. 8–10; Captain Linton F. Brooks, "Escalation and Naval Strategy," U.S. Naval Institute *Proceedings*, Vol. 110, No. 8 (August 1984), pp. 33–37; Captain Linton F. Brooks, "Tactical Nuclear Weapons: The Forgotten Facet of Naval Warfare," U.S. Naval Institute *Proceedings*, Vol. 106, No. 1 (January 1980), pp. 28–33; Gordon H. McCormick and Mark E. Miller, "American Seapower at Risk: Nuclear Weapons in Soviet Naval Planning," *Orbis*, Vol. 25, No. 2 (Summer 1981), pp. 351–367; and Lt. Commander T. Wood Parker, "Theater Nuclear Warfare and the U.S. Navy," *Naval War College Review*, Vol. 35, No. 1 (January–February 1982), pp. 3–16.

129. Barry Posen suggests a scenario that illustrates how the Soviets might use nuclear strikes to stop the American anti-SSBN campaign. He posits that the Soviets, after using nuclear weapons to render the GIN barrier ineffective, could then move large numbers of submarines into the North Atlantic to strike at NATO shipping. This move might force the American Navy to call off its hunt for Soviet SSBNs and move most of its SSNs into the Atlantic to counter Soviet attacks against the SLOCs. (Personal correspondence with author, May 14, 1986.)

still have a sufficient number of warheads left to wreak unacceptable damage on American society.[130] As long as the Soviets maintain this capability, they can ignore an unfavorable nuclear balance. Those who doubt the logic of this argument should be reminded that little evidence exists that Soviet behavior in the first two decades of the Cold War was affected in any meaningful way by the fact that the balance of nuclear forces clearly favored the United States.[131]

Thus three principal flaws are apparent in arguments that an anti-SSBN offensive would produce deterrence stability by shifting the strategic balance: (1) American political leaders may not allow the Navy to execute the strategy; (2) the Navy may not be capable of implementing it effectively; and (3) a successful anti-SSBN campaign may not coerce the Soviets into better behavior, since the Soviets would have options other than standing down, including escalation on their own part.

Advocates of a counter-SSBN strategy still might argue that such a campaign need not create a meaningful shift in the strategic balance in order to produce deterrence stability. Rather, in this view, such a campaign would deter the Soviets simply by generating or manipulating a shared risk of nuclear war. The Soviets, so the argument goes, would be given pause not by concern about the balance of nuclear forces, but by fear that the naval conflict would spin out of control and lead to a strategic nuclear exchange.[132]

There is no question that NATO derives some deterrence stability from this threat, although not a great deal. The principal limiting factor is that, given the triple dangers of crisis instability, inadvertent escalation, and ac-

---

130. Five percent of the existing Soviet ICBM force would represent an arsenal of more than 321 warheads delivering 208 equivalent megatons (EMTs). It is generally assumed that an American assured destruction capability of 200 EMT is enough to destroy 20 to 25 percent of Soviet population and 50 percent of Soviet industry, thereby destroying the Soviet Union as a functioning modern society. If we assume, as we probably should, that the United States is similarly vulnerable to nuclear attack, then it follows that 5 percent of the Soviet ICBM force could inflict destruction of the same scale on the United States. Data derived from *The Military Balance, 1985–1986*, p. 181. This discussion, of course, does not take into account any surviving Soviet SSBNs or bombers, a handful of which could alone inflict significant damage on the United States. See Daniel, "ASW and Superpower Strategic Stability," especially pp. 5–11, 123–133.

131. It is difficult to draw firm conclusions about this matter since so little is known about Soviet thinking during this period. See Richard K. Betts, "Elusive Equivalence: The Political and Military Meaning of the Nuclear Balance," in Samuel P. Huntington, ed., *The Strategic Imperative: New Policies for American Security* (Cambridge, Mass.: Ballinger, 1982), pp. 101–140. For a general defense of the position that shifts in the nuclear balance are not important, see Robert Jervis, "Why Nuclear Superiority Doesn't Matter," *Political Science Quarterly*, Vol. 94, No. 4 (Winter 1979–1980), pp. 617–633.

132. This argument is implied, for example, in Watkins, "The Maritime Strategy," p. 14.

cidental escalation, it is again not likely that the Navy will be allowed to execute the strategy. Therefore, the threat may not be sufficiently credible to produce deterrence.

Is it worth pursuing a counterforce coercion strategy nevertheless, in order to gain the modicum of deterrence stability that its threat of escalation produces? The answer is no, for several reasons. First, this small gain in deterrence stability is far outweighed by the danger of crisis instability inherent in the strategy. Second, it would be the height of irresponsibility for NATO to begin purposely manipulating the risk of nuclear escalation *before* determining the fate of NATO's conventional forces on the Central Front; and a counterforce coercion strategy must be launched immediately upon the outbreak of war. Finally, if it becomes necessary for NATO to manipulate the risk of nuclear escalation, NATO already has forces in Europe that can perform this function in a safer and more credible manner. The express purpose of the American Pershing IIs, ground-launched cruise missiles (GLCMs), and other theater nuclear forces in Europe is to generate the risk of nuclear escalation if the Soviets overrun Europe.[133] These forces have the advantage of producing such risks at the appropriate time (after and only after NATO conventional forces are overrun), and they can do so more credibly than can sea-based forces, since "use or lose" dynamics could operate to persuade NATO commanders to use them. In contrast, an American anti-SSBN campaign generates risk too early and with less credibility.

The bottom line is that a counterforce coercion posture promises little deterrence stability. It is a strategy built on a number of suspect assumptions, and on close scrutiny it hardly generates confidence. When one then considers that the strategy also undermines crisis stability in important ways, the net conclusion is that counterforce coercion is a badly flawed deterrent posture.

Conclusion

The Maritime Strategy, which can best be described as a loose combination of four offensive concepts (direct military impact, horizontal escalation, offensive sea control, and counterforce coercion) does not contribute much to

---

133. For an excellent discussion of NATO thinking about the employment of nuclear weapons, see J. Michael Legge, *Theater Nuclear Weapons and the NATO Strategy of Flexible Response*, R-2964-FF (Santa Monica, Calif.: Rand, April 1983), chapter 2.

deterring a war in Europe. Direct military impact and horizontal escalation, which use the Navy to project power against the Soviet Union or its allies, simply have very little deterrent value. Counterforce coercion, the posture the Navy now stresses, actually threatens to undermine deterrence, mainly because implementation of that strategy in a crisis would be highly destabilizing. This problem, coupled with the threat of nuclear escalation that attends this posture, points up that this variant of the Maritime Strategy is potentially dangerous.

Sea control is where the Navy matters for deterrence in Europe. The Soviets must not be allowed to think that they can cut NATO's SLOCs quickly. The Navy maintains that an offensive sea control posture is necessary for this purpose, but this is not so. A defensive sea control posture would satisfy NATO's needs on this count without the risks that attend offensive sea control. Very importantly, the force structure demands of defensive sea control are more modest than those of offensive sea control. Specifically, a defensive sea control strategy would allow the Navy to relinquish its requirement for 15 carrier battle groups. The Navy would still need large-deck carriers for its other missions—peacetime presence and direct intervention in Third World conflicts—but the overall number would be significantly less than 15—perhaps 10 would be enough. The Navy, however, should continue to maintain a powerful SSN force, which would not only be useful for executing a defensive sea control strategy, but would also provide the threat needed to keep Soviet SSNs concentrated in their home waters.

Moving to this smaller force structure would free up resources for the ground and air forces on the Central Front, which represent the nucleus of NATO's deterrent. This discussion points up that an assessment of the Maritime Strategy must consider the crucial issue of opportunity costs. Specifically, what are the implications for the forces *in* Europe of spending very large sums of money to procure a 600-ship navy built around 15 carrier battle groups?

The Reagan Administration has not sought to increase the size and strength of NATO's ground and air forces. Instead, it has essentially maintained the status quo on the Central Front, even though the Administration is clearly identified with the position that NATO is badly outnumbered in Europe. Perhaps this decision was made on the assumption that substantially increasing the size and strength of the Navy would markedly enhance the allies' deterrent posture in Western Europe. Unfortunately, this assumption is invalid. The best way to achieve that end is to invest more heavily in those forces that stand in the way of the Warsaw Pact armies—NATO's ground and air forces. The Administration passed up this opportunity to devote extra resources to the Navy. Thus, the Administration's defense buildup has done remarkably little to improve NATO's prospects of deterring the Soviet Union in a crisis.[134]

In sum, the Maritime Strategy is fundamentally flawed, not only because it fails to enhance the deterrent posture in Europe, but also because it has meant spending large sums of money on the Navy that might have otherwise been spent on enhancing the fighting power of those forces that matter most for deterrence. While NATO is not any worse off in 1986 than it was in 1980, the more important point is that the Administration missed an excellent opportunity to improve NATO's deterrent posture.

Even worse are the implications of the Reagan naval program for the future. The U.S. defense budget probably will not grow very much, if at all, in the next few years. Therefore, choices about defense spending will become even more painful. At the same time, however, the Navy's budget will have to keep growing simply in order to man and maintain the ships that the Administration has already begun to build.[135] In effect, early Reagan Administration decisions locked the country into future Navy budget increases that cannot be avoided without taking inefficient steps, such as mothballing new ships or failing to outfit them fully.[136] This situation is particularly worrisome because it threatens NATO's ground and air forces, which may be underfunded in order to meet the Navy's ever-growing budget demands.

Thus the Reagan naval program represents more than a missed opportunity. It also threatens to do real damage to NATO's crucial deterrent forces.

To minimize the damage from this unfortunate situation, the United States should adopt the long-range goal of moving toward a smaller navy config-

---

134. There is a growing debate about the management and direction of the overall Reagan defense buildup. See, for example, William W. Kaufmann, *A Reasonable Defense* (Washington, D.C.: Brookings, 1986), pp. 41–46; Richard Stubbing, "The Defense Program: Buildup or Binge?," *Foreign Affairs*, Vol. 63, No. 4 (Spring 1985), pp. 848–872; and R. William Thomas, *Defense Spending: What Has Been Accomplished* (Washington, D.C.: U.S. Congressional Budget Office, Staff Working Paper, April 1985).

135. See John Enns, *Manpower for a 600-Ship Navy: Costs and Policy Alternatives* (Washington, D.C.: U.S. Congressional Budget Office, August 1983); House Armed Services Committee, *Report on the 600-Ship Navy*; Bill Keller, "Strain on Budget Seen In Navy Plan," *The New York Times*, September 7, 1985, p. 8; and Tarpgaard and Mechanic, *Future Budget Requirements for the 600-Ship Navy*.

136. See Michael Getler, "'Too Late to Stop' Fleet Buildup, Says Navy Secretary," *The Washington Post*, December 2, 1982, p. 16; and George C. Wilson, "The Birth of a Spending 'Bow Wave,'" *The Washington Post*, November 28, 1982, pp. 1, 12–13.

ured for defensive sea control and away from the larger, offense-oriented navy required by the Maritime Strategy. These changes, however, will take many years to accomplish. In the meantime, the Administration should seek to slow down spending on the Navy wherever possible, while going to great lengths to avoid weakening American air and ground forces in Europe, although the Administration has limited room for maneuver at this point.

I should emphasize that this article does not argue that the Navy is irrelevant for dealing with the Soviet threat in Europe. The Navy is necessary to protect NATO's SLOCs in a war of attrition and, moreover, that mission might be important for deterrence. Nevertheless, the key to deterrence is not the Navy, but the forces that will be fighting on the Central Front. Those forces should be given first priority when deciding how to allocate defense budgets.

Horizontal Escalation | *Joshua M. Epstein*

Sour Notes of a Recurrent Theme

$\mathbf{T}$he deterrence of Soviet military aggression has been the basis of American national security policy since the Truman Administration. The means preferred to secure containment, however, have changed with each administration since. But they have all partaken of two archetypal approaches: the symmetrical and the asymmetrical. John Lewis Gaddis has characterized them succinctly:

Symmetrical response simply means reacting to threats to the balance of power at the same location, time, and level of the original provocation.

Asymmetrical response involves shifting the location or nature of one's reaction onto terrain better suited to the application of one's strength against adversary weakness.[1]

Following the invasion of Afghanistan, former President Carter committed the United States to the deterrence of further Soviet aggression in the Persian Gulf region. While that deterrent commitment was affirmed intact by the Reagan Administration, the symmetrical thrust of the Carter Doctrine[2] was not; whether one dubs it "horizontal escalation" or something else,[3] the Reagan Administration's attraction to an asymmetrical conventional strategy for the Persian Gulf was quickly evident.

This article is a slightly revised version of a chapter in Robert J. Art and Kenneth N. Waltz, eds., *The Use of Force* (Lanham, Md.: University Press of America, 1983).

Joshua M. Epstein is a Rockefeller Foundation International Relations Fellow in residence at the Brookings Institution. This article was written while he was a Postdoctoral Fellow at Harvard University's Center for International Affairs. While the author bears sole responsibility for all views herein expressed, he wishes to acknowledge the contributions of Barry R Posen, John Mearsheimer, Robert Art, Kenneth Waltz, and Melissa Healy.

1. John Lewis Gaddis, "Containment: Its Past and Future," *International Security*, Vol 5, No. 4 (Spring 1981), p. 80. For a thorough analysis of American oscillations between the two approaches, see Gaddis's *Strategies of Containment* (New York: Oxford University Press, 1982).
2. ". . . an attempt by any outside force to gain control of the Persian Gulf region—will be *repelled* by any means necessary, including military force" (emphasis mine). See 1980 State of the Union Address, U.S. Congress, Senate, Committee on Foreign Relations, "U.S Security Interests and Policies in Southwest Asia," Hearings before the Subcommittee on Near Eastern and South Asian Affairs, Ninety-Sixth Congress, Second Session, February–March 1980, p. 350.
3. "Geographical escalation" and "war-widening strategy" are other names sometimes used.

International Security, Winter 1983/84 (Vol. 8, No. 3) 0162-2889/84/030019-13 $02.50/1
© 1983 by the University Press of America

Shortly after taking office, Reagan Administration strategists reportedly issued guidelines to the military "to hit the Soviets at their remote and vulnerable outposts in retaliation for any cutoff of Persian Gulf oil." In Secretary of Defense Caspar Weinberger's view, "our deterrent capability in the Persian Gulf is linked with our ability and willingness to shift or widen the war to other areas."[4]

One possibility cited at the time was "to threaten the Soviet brigade in Cuba if Moscow or its surrogates move into the Persian Gulf." Direct conventional defense would certainly be attempted "whatever the odds," but the prospects for symmetrical response were accounted as grim "given the Soviets' inherent geographical advantages and their superior number of available ground forces."[5]

The same language was carried into the Administration's first Defense Posture Statement a year later.[6] But, in the interim, the strategy seemed to have assumed larger proportions: "If Soviet forces were to invade the Persian Gulf region, the United States should have the capability to hit back there or in Cuba, Libya, Vietnam, or the Asian land mass of the Soviet Union itself."[7] The list of remote "Soviet" vulnerabilities was longer and the Soviet homeland itself had emerged as a potential target for horizontal escalation. Similarly, the procurement of two additional large-deck nuclear aircraft carrier battle groups was advocated not merely for their capacity to lash back at Soviet weak points, but, on the contrary, for the alleged "capability of a 15 carrier 600 ship Navy to fight and win in areas of *highest* Soviet capability" (emphasis mine).[8] But, as ever more challenging horizontal options emerged, the Administration's commitment to direct (symmetrical) defense was reaffirmed: "whatever the circumstances, we should be prepared to introduce

---

4. George C. Wilson, "U.S. May Hit Soviet Outposts in Event of Oil Cutoff," *The Washington Post*, July 17, 1981, p. 1.

5. Ibid.

6. For example, ". . . even if the enemy attacked at only one place, *we* might choose not to restrict ourselves to meeting aggression on its own immediate front. . . . A wartime strategy that confronts the enemy, were he to attack, with the risk of our counteroffensive against his vulnerable points strengthens deterrence and serves the defensive peacetime strategy." Caspar W. Weinberger, Secretary of Defense, *Annual Report to the Congress for Fiscal Year 1983* (Washington, D.C.: U.S. Government Printing Office, 1982), pp. I-16, I-17 (hereafter referred to as *FY 83 Posture Statement*).

7. Leslie H. Gelb, "Reagan's Military Budget Puts Emphasis on a Buildup of U.S. Global Power," *The New York Times*, February 7, 1982, pp. Z6, Z7.

8. John Lehman, Secretary of the Navy, "America's Growing Need for Seaborne Air Bases," *The Wall Street Journal*, March 30, 1982.

American forces into the region should it appear that the security of access to Persian Gulf oil is threatened."[9]

Both direct defense *and* asymmetrical response were apparently embraced; under the latter, not just points of Soviet weakness, but points of extreme Soviet strength as well were contemplated as horizontal targets. This "do everything" quality of the articulated strategy was compounded by the Administration's pointed rejection of planning around any specific set of prototypical (or real) contingencies (e.g., the one-and-one-half or two-and-one-half war sizing devices)[10] and its exhortations to prepare for "prolonged conventional wars simultaneously in several parts of the globe."[11]

Predictably, when the military was called upon to attach a price tag to the strategy, it was whopping: hundreds of billions—by some estimates $750 billion[12]—above and beyond the $1.6 trillion Reagan five-year defense plan. The strategy's requirements were reportedly "so grandiose that the Joint Chiefs of Staff says carrying them out would require 50 percent more troops, fighter planes, and aircraft carriers than are now deployed, along with another Marine amphibious force."[13]

It was against that sobering budgetary backdrop that William Clark, in his first major speech as National Security Adviser, announced, "there is not enough money available to eliminate all the risks overnight."[14] And having thus resurfaced, the budget constraint has since imposed a slackening in the planned pace of American military growth, in turn stimulating a more animated debate on the ranking of defense priorities.

Basic questions remain unresolved, however, concerning the advisability of a "war-widening strategy" *in principle*, particularly about its horizontal counteroffensive component. And those underlying questions are not addressed merely by admitting a cutback in the pace of funding; the more basic

9. George C. Wilson, "U.S. Defense Paper Cites Gap Between Rhetoric, Intentions," *The Washington Post*, May 27, 1982, p. 1.

10. *FY 83 Posture Statement*, pp. I-15–I-16. On the one-and-one-half and two-and-one-half war concepts as peacetime planning devices, see William W. Kaufmann, "The Defense Budget," in Joseph A. Pechman, ed., *Setting National Priorities: The 1983 Budget* (Washington, D C.. The Brookings Institution, 1982), p. 81.

11. "Flood and Leak at the Pentagon," *The New York Times*, February 1, 1982, p. A-14. Richard Halloran, "Needed: A Leader for the Joint Chiefs," *The New York Times*, February 1, 1982.

12. Robert W. Komer, "Maritime Strategy vs. Coalition Defense," *Foreign Affairs*, Vol. 60, No. 5 (Summer 1982), pp. 1128–1129. See also "Flood and Leak."

13. Wilson, "U.S. Defense Paper."

14. Ibid.

issue is whether this course, this entire strategic direction, is advisable at all. Moreover, until that issue is settled, the debate on program priorities can only founder.

Recognizing that some—by no means all—within the Administration have "stepped back" to reexamine the strategy's appeal, it may therefore be constructive for us to raise some of the unanswered questions about "horizontal escalation." Five seem especially basic: First, what is horizontal escalation supposed to do; what is the goal in an operational sense? Second, given some relatively concrete goal, how is the "proper" horizontal target selected? Related to that, in what ways might horizontal escalation affect the probability of vertical escalation by the United States or the Soviet Union? Fourth, what are the risks of "counter-horizontal escalation" by the Soviets? Finally, are there otherwise-avoidable diplomatic costs associated with the strategy? Not only do these questions deserve thought, but the Administration's attraction to the strategy seems to rest on military premises that are questionable in their own right. Having discussed these issues, the most important question of all—that of credibility—will be addressed.

The Question of Wartime Operational Goals

Without some sense of the operational goal of a military action, it is not possible to select its targets. Neither, *a fortiori*, is it possible to derive the forces required or to assess the adequacy of those already in being. While "carrying the war to other arenas" and "hitting the Soviets at their vulnerable outposts" may sound clear enough at first blush, such phrases in fact provide little guidance to the military planner, charged with designing a force, or to the public, charged with paying for it. What, exactly, is the goal of horizontal escalation?

Is it *to destroy something* in order *to punish* the Soviets, perhaps holding out the prospect of further punishment unless they comply with American terms, whatever those might be? Is it *to take something hostage*, hoping thereby to bargain a return to the prewar status quo or some other political arrangement? Obviously, not both goals can be achieved: one cannot hold hostage what one has already destroyed.

Although it is hard to imagine many Soviet assets whose military acquisition would *compensate* the West for the loss of access to Persian Gulf oil, compensatory acquisition is, in principle, another possible goal, one distinct from punishment or hostage-taking for bargaining purposes.

Perhaps the most intuitively appealing goals for horizontal escalation would be to *inhibit or to induce redeployments of Soviet forces*, thereby improving Western prospects for conventional defense in the original contingency. The appeal is natural enough: attacked at point A, one counters at point B in order (i) to "fix" Soviet forces at point B, preventing their use as reinforcements at point A, or (ii) to force the Soviets to shift forces from A to B.

In the context of European war, the former has long been among the Navy's arguments for "opening up a second theater" by offensive fleet operations against Soviet naval bases in the Western Pacific.[15] And, it is true that in a protracted conventional war between NATO and the Warsaw Pact, Soviet ground forces arrayed opposite China might be redeployed to the West as a third (or fourth) echelon of the Soviet European offensive. "Opening up a second theater" in the Pacific would allegedly tie down those Soviet forces, improving NATO's chances for conventional defense.

In fact, however, the Soviets' freedom to redeploy to Europe—or to the Persian Gulf for that matter—their divisions on the Sino–Soviet border would depend primarily on the posture of the Chinese army, not the U.S. Navy. Admittedly, the Soviets might reinforce their Pacific Fleet's air forces with air power normally deployed inland opposite China. But what real role could the corresponding Soviet ground forces play in either the Pacific sea battle or in defending Soviet naval shore facilities? A role significant enough to "fix" over forty ground divisions? That seems implausible.

Rather than inhibiting Soviet redeployments by "fixing" Soviet forces, geographical escalation could, in principle, *induce* Soviet redeployments. The object in that case would be to draw Soviet forces out of the original contingency, improving Western chances for success there. One open question in this regard is simply: when?

The Reagan five-year defense plan will, by definition, take half a decade to realize. At the end of that period, horizontal escalation might succeed in *forcing* Soviet wartime reallocations, but only if, in the interim, some ceiling on Soviet military growth had been reached. Otherwise, what's to prevent the Soviets from anticipating the strategy by simply *adding forces* in each theater over that interim period? After all, when the Sino–Soviet split presented them with "a new theater" (i.e., their Chinese frontier), the Soviets

---

15. See, for example, The Congressional Budget Office, *Navy Budget Issues for Fiscal Year 1980,* March 1979, p. xviii and *Sea Plan 2000 Naval Force Planning Study, Unclassified Executive Summary* (undated), p. 15.

didn't shift many forces. True to Russian form, they essentially built. And, unless there's some "limit to growth," it may be hard to *force* Soviet reallocations, at least between theaters in the U.S.S.R.

To be sure, there may be military constraints in the Soviet offing. But if there are, the Reagan Administration has certainly not suggested them by its review of the Soviet military buildup and its projections of continuing growth as Soviet military *investment*[16] comes to fruition. In short, a reallocative goal for horizontal escalation seems to entail a Soviet military growth ceiling, while the Administration's Soviet projections seem to deny the existence of any such limits.[17]

Targets: Soviet Value, U.S. Diversion, and Vertical Escalation

Punishment (destructive retaliation), hostage-taking, compensatory acquisition, fixing Soviet forces, or inducing their redeployment—these may not be the only possible rationales for horizontal escalation.

But, whatever its operational goal (punishment, hostage-taking, etc.), the compellent effectiveness of the horizontal action will surely depend upon the value placed *by the Soviets* upon its target. After all, the mere fact that a Soviet outpost is vulnerable does not make it valuable, as Secretary Weinberger recognizes: "If it [the counteroffensive] is to offset the enemy's attack, it should be launched against territory or assets that are of an importance to him comparable to the ones he is attacking."[18]

In practice, however, it may be difficult to identify a "horizontal target" which is of sufficient Soviet value to compel the behavior sought, but at the same time is not of such great value as to stimulate rash and grossly disproportionate (e.g., nuclear) Soviet responses.

It is an open question whether offensive sea control (attacking Soviet Fleets in their home waters) would be the most efficient way to secure Western sea

---

16 Investment is stressed over spending in *FY 83 Posture Statement*, pp. II-4–II-7.

17. Some might argue that this criticism isn't fair because it presents horizontal escalation as a strategy toward which the U.S. is building for the future when, in fact, they will claim, horizontal escalation is only an interim strategy until the U.S. can build up greater conventional defenses in the Gulf. But, if horizontal escalation is an interim measure with direct conventional defense the ultimate goal, then why is the Administration spending so much on carrier battle groups for "outyear" counteroffensives and so much less on programs directly related to Gulf defense (e.g., airlift and sealift) today?

18. *FY 83 Posture Statement*, p I-16.

lines of communication.[19] But, even if it otherwise were, such offensive naval operations could well run the nuclear risk, and not simply because they would involve hitting the Soviet homeland. Beyond that, in conducting offensive operations against either the Soviet Northern or Pacific Fleets, it might be difficult for the U.S. Navy to avoid sinking Soviet strategic ballistic missile submarines (SSBNs). And if American operations were to degrade, even inadvertently and by conventional means, the Soviets' strategic nuclear retaliatory capabilities, the Soviets might not *interpret* those operations as "conventional sea control," but as conventional precursors to nuclear attack—conventionally executed nuclear damage-limiting first strikes, if you will.[20]

As for targets of too little value, were the Soviets to end up with control of oil fields in the Persian Gulf, they could certainly afford to buy a new merchant fleet,[21] had theirs been "swept from the seas" in bristling riposte to Gulf aggression. And regarding counteroffensives against Cuba, suppose, just for the sake of argument, that the Soviets were offered a straight trade: they "give up" Cuba and, in return, they get Iran's (or Saudi Arabia's) oil. Would Moscow turn it down? What assurance is there that, behind the compulsory fulminations, the Kremlin would not be willing to let Cuba take a "horizontal beating" if it meant Soviet control of Gulf oil? And, if the Soviets would accept that trade, how "offsetting" would horizontal escalation against Cuba, in fact, be?

Not only may Administration strategists have exaggerated Cuba's value to Moscow, but horizontal escalation against Cuba was reportedly envisioned "if Moscow *or its surrogates* move into the Persian Gulf" (emphasis mine).[22] Are the Iraqis Soviet "surrogates"? Does that mean they have no goals of their own? Would the threat of American retaliation against Cuba (or anyone else for that matter) deter *them*? Would it deter Communist elements in Iran? The Syrians? The PDRY (South Yemen)?

Even if Moscow wished to bridle any of its surrogates lest the U.S. respond against "Cuba, Libya, Vietnam, or the Soviet Union itself," could they do it?

---

19. See *Navy Budget Issues* and Congressional Budget Office, *The U.S. Sea Control Mission: Forces, Capabilities, and Requirements*, June 1977. Still among the most concise essays on the factors involved in evaluating offense, defense, and requirements in each case is Arnold M Kuzmack, *Where Does the Navy Go From Here?* (Washington, D C The Brookings Institution, 1972)

20. By far the most comprehensive study of this important problem is Barry R. Posen, "Inadvertent Nuclear War? Escalation and NATO's Northern Flank," *International Security*, Vol. 7, No. 2 (Fall 1982), pp. 28–54.

21. Komer, "Maritime Strategy vs. Coalition Defense," p. 1131.

22. Wilson, "U.S. May Hit Soviet Outposts," p. 1

The main point, however, is that even with clear goals for horizontal escalation in mind, the selection of an appropriate target seems to require knowledge of the *Kremlin's* valuations. An uncertain affair even in peacetime, the problem would be compounded in war when, among other things, values change.[23]

In addition to the problem of Soviet valuation, there is the problem of U.S. diversion. Although contingency-counting has been rejected under the Administration's reforms, horizontal escalation nonetheless ensures that the U.S. confronts two contingencies (main plus horizontal) where it might have faced only one (main). And, even assuming that a horizontal target of the "proper" *Soviet* value were selected, the appropriate action against that target could require a significant commitment of American forces, over and above those allocated to direct defense in the initial contingency.

The obvious question, therefore, is whether the U.S. would have to divert to horizontal offensives, forces that might otherwise have been applied (initially or as reinforcements) to the main defense, reducing the latter's prospects for success. If so, would not the strategy make it more likely that the U.S. find itself under pressure either (a) to concede defeat in the main contingency or (b) to use nuclear weapons there?

On the horizontal front, it should be noted that unless these counteroffensives are to be initiated preemptively—i.e., *before* the "provoking" Soviet attack, an option Secretary Weinberger has explicitly rejected[24]—then they will probably not achieve tactical surprise. Obviously, they might, but it would be imprudent to design forces on that assumption. And, on the prudent assumption that surprise would not be achieved, the horizontal target might enjoy a number of classical warning advantages (e.g., prepared defenses, cover, dispersal, etc.). These could well exacerbate the diversion problem and, with it, the unpleasant problems of choice between humiliation and vertical (i.e., nuclear) escalation.[25]

---

23. This, and many other problems of conflict termination, are discussed in Fred Charles Iklé, *Every War Must End* (New York: Columbia University Press, 1971).
24. *FY 83 Posture Statement*, pp. I-10–I-11.
25. Indeed, in the worst of both worlds, the U.S. would divert so much force from the Gulf as to come under enormous pressure to use nuclear weapons there, while applying the large resultant counteroffensive force in such a way as to put the Soviets under equal pressure to use nuclear weapons on the horizontal front. Professor Huntington's proposal for a prompt conventional counteroffensive into Eastern Europe seems to invite this dangerous situation. See Samuel P. Huntington, "The Renewal of Strategy," in Samuel P. Huntington, ed., *The Strategic Imperative: New Policies for American Security* (Cambridge: Ballinger, 1982).

Doubtless such considerations played a role in the Services' costing of the strategy: to avoid problems of diversion, one buys *much larger forces*.[26] But now that the Administration has essentially admitted that the forces necessary to hedge against the diversion problem are too expensive, retention of the strategy may be too risky.

Some, of course, argue just the reverse, that "because the development and acquisition of new weapons might be delayed, *more* emphasis would be placed on such flexible tactics as 'geographical escalation'" (emphasis mine).[27] Admittedly, geographical escalation conjures up attractive images of "regaining the initiative," "turning the tables," and "making the Russians play by our rules." But this ought to suggest a further problem; namely, what if the Russians *do* decide to "play by our rules," and proceed to "counter-horizontal escalate" themselves?

The Problem of Counter-Horizontal Escalation

Although the possibility was recognized by some Westerners at the time,[28] Khrushchev did not play "the Berlin card" in the Cuban missile crisis of 1962. Faced with an American "war-widening" move against Cuba, his successors might behave rather differently. One might imagine the following scenario: the Soviets invade Iran; American forces "hit" Cuba (or the Soviet brigade there); the Soviets "hit" West Berlin (or the U.S. brigade there). Then what? You got him to "play by your rules"; are you better off?

Is Cuba more valuable to the Soviets than West Berlin is to the Federal Republic? Would European leaderships dutifully "fall in" behind further American "initiatives" at that point?

Even assuming that the Soviets could attack the Berlin brigade "surgically," without killing West German citizens (the legal status of all West Berliners), would the U.S. allow it to be decimated without further response? If West Berlin had been sealed off, then at some point the brigade's survival might require airlift. Would the Soviets allow another Berlin airlift, or would they shoot down Western transports this time? Perhaps the airlift would "fight

---

26. This is just the "flip side" of the Soviet solution discussed above. To avoid dangerous diversions (in the Soviet case, forced; in the U.S. case, self-imposed), build.
27. Richard Halloran, "Reagan Aides See Pressure to Cut 1984 Military Budget," *The New York Times*, July 13, 1982, p. 1.
28. Herman Kahn, *On Escalation: Metaphors and Scenarios* (New York: Praeger, 1965), pp. 86–87.

its way in," suppressing East German air defenses at a time of great tension and high military alert.

Whether or not one finds this particular scenario to be plausible, the point is that forcing the Soviets to "play by these rules" may not be in America's interest; move and counter-move might quickly bring both sides to unforeseen and very unstable situations. And reversing the spiral might be exceedingly difficult.

Diplomacy and Military Premises

Not only crisis management but peacetime diplomacy as well may be compromised by a policy of horizontal escalation. The Soviets, Secretary Weinberger states, "can coerce by threatening—implicitly or explicitly—to apply military force."[29] But, might not a strategy of horizontal escalation, *predicated on the assumption that direct defense is infeasible,* facilitate that coercion?

In the Persian Gulf, one can only render Soviet threats the *more* credible, their local audiences the *more* compliant, by suggesting that direct conventional defense is unmanageable, "given the Soviets' inherent geographical advantages in the Persian Gulf region and their superior number of available ground forces."[30]

This, moreover, is a simplistic and misleading characterization of the Soviet–American conventional balance in the region. Geographical proximity does not per se constitute Soviet military access, and static prewar "beancounting" cannot reflect dynamic wartime effectiveness. The latter always depends on operational factors such as terrain, available avenues of advance (and their vulnerability), logistics, coordination, reconnaissance, flexibility, leadership, morale, combat technology, and troop skill. That fact is consistently bemoaned on the U.S. side, but is virtually neglected when assessing the Soviet threat. When such factors are accounted for on *both* sides, in a balanced and systematic way, the prevalent assessment is seen to be overly pessimistic.[31]

---

29. *FY 83 Posture Statement*, p. I-10.

30. Wilson, "U.S. May Hit Soviet Outposts," p. 1.

31. For an operational assessment of the U.S –Soviet conventional balance in the Persian Gulf, and a detailed argument that direct conventional defense is feasible, see Joshua M. Epstein, "Soviet Vulnerabilities in Iran and the RDF Deterrent," *International Security*, Vol. 6, No. 2 (Fall 1981), pp. 126–158. See also Dennis Ross, "Considering Soviet Threats to the Persian Gulf," *International Security*, Vol. 6, No. 2 (Fall 1981), pp. 159–180, and Keith A. Dunn, "Constraints on the USSR in Southwest Asia," *Orbis*, Vol. 25, No. 3 (Fall 1981), pp. 607–631.

Furthermore, while there are compelling reasons for the U.S. to avoid reliance on nuclear employment, the Soviets surely cannot ignore that possibility. And this, too, contributes to deterrence.

But does the attraction to horizontal escalation derive only from the unwarranted assumptions (a) that conventional defense is utterly infeasible and (b) that Soviet nuclear developments have stripped the U.S. strategic triad of deterrent value? Perhaps. But it often seems as though the true motivations run deeper—as though the advocates, at bottom, have turned away from the entire concept of limited war.[32]

Winning

MacArthur said that, "in war, there is no substitute for victory." Neither, of course, is there any substitute for national survival. While at various points, the Reagan Administration seems to have unqualifiedly adopted MacArthur's heroic dictum, at other points its strategists seem to be aware of the unheroic fact of nuclear life—"the difficulty that war invites the belligerents to regard the contest as mortal and to continue it until they are reduced to ruins."[33]

Over a decade ago, in an essay called "Peace Through Escalation?", Fred Iklé, now Under Secretary of Defense for Policy, put the enduring dilemma this way:

If a nation can overwhelm all of the enemy's forces by escalating a war, the fighting will of course be brought to an end. . . . Short of inflicting such total defeat, successful escalation would have to induce the enemy government to accept the proffered peace terms. The trouble is, the greater the enemy's effort and costs in fighting a war, the more will he become committed to his own conditions for peace. Indeed, inflicting more damage on the enemy might cause him to stiffen his peace terms. . . . It is these opposed effects of escalation that make it so hard to plan for limited wars and to terminate them.[34]

In order, as Secretary Weinberger desires, "to prevent the uncontrolled spread of hostilities,"[35] it is obviously critical that the Soviets never see an

32. In Secretary of the Navy John Lehman's view, the rapid growth of Soviet naval power has "eliminated the option of planning for a regionally limited naval war with the Soviet Union. It will be instantaneously a global naval conflict." Richard Halloran, "Reagan Selling Navy Budget as Heart of Military Mission," *The New York Times*, April 11, 1982.
33. William W. Kaufmann, "Limited Warfare," in William W. Kaufmann, ed., *Military Policy and National Security* (Princeton, N.J.: Princeton University Press, 1956), p. 112.
34. Iklé, *Every War Must End*, pp. 41–42.
35. *FY 83 Posture Statement*, p. III-101.

advantage in opening new theaters or geographically widening a local war themselves. In that sense, a strategy of limited war in fact does, and always has, required "horizontal" capabilities: "you will not succeed at the initial point of attack and we will not be drawn off by attacks elsewhere; you must fight here and you cannot win here, so stop; indeed, don't start."[36]

While multi-theater capabilities sufficient to *deny* the Soviets a war-widening strategy are thus essential to channel and limit conventional conflict should deterrence fail, it is far from clear that an American policy of *initiating* such expansions would foster military control or wartime diplomacy. It might, but having reviewed the risks, Iklé's observation is worth recalling:

It is hard to say whether treason or adventurism has brought more nations to the graveyard of history. The record is muddied, because when adventurists have destroyed a nation they usually blamed "traitors" for the calamity.[37]

Credibility

If a strategy is to deter aggression, that strategy must be *credible*. We and our potential adversaries must believe at least two things about it. First, both sides must believe that, were deterrence to fail, the United States would actually operate its forces in accord with the strategy. Second, both must believe that such operations would be likely to achieve American wartime goals.

Failure on the second count often results in failure on the first. For example, massive retaliation, America's first great experiment in asymmetrical response, failed on the second requirement. The American goal was to contain Soviet expansion at military costs acceptable to the United States. But the strategy suggested that the U.S. would run a high risk of unacceptable damage in response to even the most limited of Soviet encroachments. In the final analysis, not even the U.S. believed that it would operate its forces in such a way. And so, massive retaliation failed on the first count as well.[38]

Horizontal escalation—the United States' most recent asymmetrical experiment—also fails to meet the basic requirements of credibility. It fails on the

---

36. For a fuller discussion, see Kaufmann, "Limited Warfare."
37. Iklé, *Every War Must End*, p. 62.
38. The seminal critique is Kaufmann's "The Requirements of Deterrence," in Kaufmann, ed., *Military Policy and National Security*.

first count because it is not even clear what it would *mean* to operate American forces in accordance with the strategy: it is not clear what its targets would be or, more fundamentally, how one would go about determining them in wartime. Neither is it clear what military actions would be taken against the selected targets.

Horizontal escalation fails to meet the second requirement as well. It clearly runs daunting risks—of nuclear escalation and/or Soviet counter-horizontal attacks—but the goals of horizontal escalation remain cloudy, in both the broader strategic sense (ultimate victory vs. limitation) and in the narrow operational sense (punishment, hostage-taking, etc.).

"The threat that leaves something to chance" may be credible. But threats that leave everything to chance most assuredly are not.

Now, the Reagan Administration's early enthusiasm for the strategy seems to have cooled somewhat; it is certainly less vocal than it was. Logically, however, there are only two possibilities. Either horizontal escalation is the Administration's strategy or it isn't.

If it is, then the strategy is not credible. And if horizontal escalation is not the Administration's strategy, then *what is* the strategy? Given the economic costs allegedly required to support it, the American people surely deserve an answer.

Naval Power and Soviet Global Strategy

Michael MccGwire

The Soviet Navy has become a significant factor in the debate about intentions, detente, and arms limitations, because its submarines carry a large part of the Soviet missile inventory, and because of the navy's involvement in trouble spots around the world. The West has encouraged this Soviet emphasis on seabased strategic strike systems but is deeply suspicious of the naval activism, claiming that there is no legitimate requirement for the Soviet Union, a land power, to deploy such forces.

The Soviets have always admitted this asymmetry of interest by stressing that a major change in the international situation after World War II was that Russia's potential opponents were now the "traditional maritime powers." But we, the maritime powers, have found it hard to comprehend the naval requirements that flow from their geopolitical circumstances, or to interpret the means which the Soviet Union adopted to meet these requirements. As long as the Soviet navy remained tied to home waters and offered no direct threat to North America, this lack of comprehension was not serious. But now that the Soviet navy has become a factor in our diplomatic calculations, it is important to understand what motivates these developments and to understand the different types of political commitment that underlie various aspects of Soviet naval policy.

A further reason for understanding Soviet naval developments is that they provide a case study that illuminates the wider debate about the Soviet Union's military posture and its willingness to wage nuclear war. I have therefore approached this subject through a discussion of Soviet naval doctrine, a term which I use in its broad Western sense.[1] The evidence for such doctrine comes from what the Soviets say and write, from how they deploy, operate, and exercise their forces, and from the number and characteristics of their ships, submarines, and aircraft. Taken together, these data provide

---

1. Matthew Gallagher characterizes Soviet Military Doctrine as ambiguous and highly generalized guidelines about the nature of future war, which have been adopted by the political and military leadership as a basis for the development of the armed forces. It has little in common with the Western term and does not provide a key to Soviet strategic policy, except in the broadest sense. See "The Military Role in Soviet Decision Making" in *Soviet Naval Policy; Objectives and Constraints* (henceforth *SNP*), MccGwire, Booth and McDonnell (Eds.), Praegers, 1975, pp. 56-57.

Michael MccGwire is Professor of Maritime and Strategic Studies, Center for Foreign Policy Studies, Dalhousie University, Halifax, Nova Scotia

a reasonable body of evidence, particularly when public pronouncements can be evaluated against the more concrete types of data, although the quality of understanding will always depend on the depth of hindsight. Warship building programs are particularly important in this context since procurement decisions can often be dated with some confidence and ship characteristics give a fair idea of the then-prevailing force requirements and operational concepts.[2]

Background Factors

Doctrine is the product of an evolutionary process, and it is relevant that for the last 200 years or so the Russian Navy has generally been the third or fourth largest in the world—although its effectiveness has fluctuated widely. Russia used naval forces in the eighteenth century to gain control of her Baltic and Black Sea coasts, and four times between 1768 and 1827, she deployed sizable squadrons to the Mediterranean for a year or more. For three of these deployments, during the third, fifth, and sixth wars with Turkey, ships were drawn from the Baltic Fleet and were used in operations against the southern side of the Black Sea exits.

Increasingly thereafter, Russia found herself confronting predominantly maritime powers. In the Black Sea, Britain used her naval strength to prevent or reverse Russian gains at the expense of the failing Ottoman Empire; Britain intervened directly in the seventh Turkish war (1853-56, Crimea) and the peace treaty forbade Russia a Black Sea Fleet; in the eighth Turkist war (1876-77), British pressure ensured that Russia would not gain control of the Straits. In the Far East, Russo-Japanese rivalry culminated in a disastrous war and the loss of two Russian fleets. In 1918, the Western navies provided vital support to the forces of counter-revolution. As a consequence, Russia's naval policy was increasingly dominated by the requirement to defend four widely separated fleet areas against maritime powers who could concentrate their forces at will.

It is therefore wrong to suggest that Russia has only recently awoken to the significance of sea power. She used it in the past to her own advantage,

---

2. For a summary description of this analytical method, see my "Turning Points in Soviet Naval Policy" in *Soviet Naval Developments· Context and Capability*, (henceforth *SND*), MccGwire (Ed.), Praegers, 1973, pp. 176-209.

and has more often seen its long arm used against her. Over the years she committed very substantial resources to naval construction, and the major warship building program which was initiated in 1945 was the fourth attempt in 65 years to build up a strong Russian fleet. But national strategy involves setting priorities and balancing competing claims for scarce resources. Russia was predominantly a land power; the only threats to her territorial existence had come by land; the army was the basis of security at home and influence abroad. Naval forces were indeed required to defend against assault from the sea and to counter the capability of maritime powers to dictate the outcome of events in areas adjacent to Russia. But these forces were seen as an expensive necessity rather than a preferred instrument of policy.

This ordering of priorities and the army's domination of military thought persist today,[3] and are enshrined in the concept of a combined arms approach to military problems. This bias was, if anything, accentuated by the reorganization of the armed forces into five branches. It is characterized by the fact that out of twenty Full Members on the Central Committee, fifteen come from the Ground Forces,[4] and that the naval share of the Soviet defense budget is estimated to be only about 18 percent, and this includes the cost of the ballistic missile submarine force.

Certain tendencies in Russian naval doctrine can probably be ascribed to this persistent state of affairs. Limited resources and the relative imbalance of naval power have encouraged a spirit of technical and conceptual innovation, and a readiness to adopt new but unproven technological advances. This is exemplified by the destruction of the Turkish fleet at Sinope using highly explosive shells in 1853, the daring use of torpedo boats in the eighth Turkish war in 1876, the very early emphasis on submarines, and in the mid-1950s, the application of cruise missiles to maritime platforms. Usually start-

---

3. There was only one naval title among fifteen books in the "Officers Library" series announced in 1957, and not a single naval title in the 17 books of the new series announced in 1964. There were probably only two naval titles among 50 important military books published 1960-69. See *Soviet Military Doctrine: Its Formulation and Dissemination*. H. F. Scott, Stanford Research Institute, June, 1971, pp. 81-92. The Naval Academy is only one among fifteen arm-of-service academies at the Command and Staff level. During the period 1965-70, the navy achieved its proportionate share of Candidates of Science from the Lenin Military-Political Academy (i.e. at the Commander level) but achieved no Doctors of Science (*Ibid.*, pp. 120-124).

4. The other four branches and the Main Political Administration rate one apiece. See John McDonnell, "The Soviet Defense Industry as a Pressure Group," in *SNP*, p. 104. Those figures apply to the 1971-76 period, but there was no change in the Committee elected at the XXV Party Congress.

ing from behind in terms of conventional capabilities, and with little chance of overhauling from astern, the naval leadership has frequently sought to get ahead by taking a different (and unexpected) tack.

This tendency to innovate was reinforced by the post-revolutionary emphasis on original, "proletarian" solutions to strategic problems, and by the more enduring ideological commitment to "scientific objectivity." Current writing stresses the need to assess naval requirements from first principles and the fallacy of mirroring an opponent's capability. However, innovations are not always successful. And, during the last 20 years this combination of restricted resources, a faith in objective assessment and a belief in innovation has sometimes led to the development of task- and scenario-specific capabilities which have lacked the flexibility to cope with changes in the nature of the threat. The mid-fifties decision to place primary reliance on long-range cruise missiles (which had yet to be developed), carried by surface ship and diesel submarine, had just that result. It is also an example of the army-oriented political leadership—taking decisions against specific naval advice. Kuzenetsov's objections to this ill-founded concept cost him his job as Commander in Chief of the Navy, and 45-year-old Gorshkov was brought in to implement the new policy.

But it should not be assumed that the effect of this land forces orientation has been, or is, all bad. It can of course be argued that it has led to the hobbling of naval forces by army commanders, from the scuttling of the fleet before Sevastopol in 1854, to operations in the Baltic and Black Sea during the Second World War. But several of these examples lose their force when analyzed in terms of relative capabilities and practical objectives, rather than classical naval theory. It is also true that army dominance encouraged the centralization of command and a rather rigid approach to battle planning. But "unified command" is now the fashion, and modern warfare demands close command and control. It so happens that Russia's traditional centralized command structure is well suited to contemporary requirements, in principle, if not always in practice. Its deficiencies lie more in its style of operation than its organizational structure.

A good case can be made that the emphasis on a "combined arms" approach and the existence of an army-oriented political leadership did in fact have an invigorating effect on the development of Soviet naval doctrine, not least by saving it from the fallacy that naval strategy was a universal science, whose rules had been discovered by Colomb, Corbett, and Mahan. A pragmatic approach, combined with strictly limited resources, introduced a healthy realism to naval planning and, for example, encouraged the emergence of the "small war" doctrine in the early thirties, based on the limited types of warship that could then be built. More pertinently, when it did become possible to plan for capital ship construction at the end of the thirties, Stalin refused Kuznetsov's request for carriers, on the grounds that the Red Fleet would not be operating off distant shores. Naval procurement was to be tailored to Russia's particular requirements, and not to some idealized perception of what "a navy" should be.

The army-oriented leadership has required the navy to undertake tasks which have violated its traditional assumptions about naval operations, and forced the development of radical concepts. A particular example was the shift to foward deployment in the early sixties, whereby ill-armed Soviet naval units were required to maintain close company with U.S. forces in the Mediterranean, an area where the West enjoyed overwhelming maritime preponderance. The idea of relying on "the protection of peace" to safeguard such exposed deployments was a daring concept, given the general tenor of the Western strategic debate at that time. The decision to ignore the survivability of such units and exploit the characteristics of nuclear-missile war was a major departure from traditional naval thought.

More specifically, ground force thinking can be seen in the concept of close-shadowing Sixth Fleet carriers with a gun-armed destroyer. In artillery terms, that unit was acting as a Forward Observation Post, and in the event of war would call down fire from Medium or Intermediate Range Ballistic Missiles emplaced in South West Russia.[5] A similar concept was spelled out in Grechko's statement in December, 1972, that the wartime mission of the Strategic Rocket Forces (SRF) included the destruction of "enemy means of nuclear attack, and troops and naval groupings in theaters of military operation on land and sea,"[6] although by now, target data would be provided by satellite surveillance systems.

On balance, the injection of army concepts, the emphasis on combined

---

5. Although there was no firm evidence of such targeting, this was inferred at the time from the persistence of close shadowing, whether or not there was any sea-borne counter-carrier capability present; such a capability was not achieved on a sustained basis until the end of the sixties. Russia-based aircraft were not suitable for this strike role, since their response time was too long, and they first had to breach the NATO air defense barrier. There were, however, sufficient land-based missiles to cover this requirement, and the task was technically feasible. The existence of such a concept is supported by the subsequent claim for a more extensive capability of this type.

6. A. A. Grechko "A Socialist Multinational Army", *Krasnaya zvezda*, 17 December 1972

arms, and the common development of weapon systems like missiles and aircraft have been fruitful in terms of capabilities and doctrine. Even when a new idea was initially unsuccessful, as was the long-range surface-to-surface missile (SSM), it often served the purpose of breaking with traditional concepts and opening new avenues for development. It is true that the leadership's perception of the navy as an expensive necessity has often led to the definition of naval requirements in narrow terms of countering specific threats, instead of more general capabilities, and that this has restricted the navy's flexibility. On the other hand, the sheer preponderance of ground force opinion engenders clearly defined priorities and a readiness to apply the resources of all relevant branches of the armed forces to meet any serious threat, including those that come from the sea.

Contingency Planning and the Reality of World War

Soviet military doctrine has evolved in response to what have been seen as a series of direct threats to the state's existence. The Soviet leadership has always taken the likelihood of war very seriously, and more importantly, it has been prepared to think through the implications and to take the measures necessary to secure ultimate victory should war come. This was demonstrated by the industrial relocation policies of the thirties, and by the contingency plans for physically removing industrial plants from the path of the German advance, measures which between them enabled the post-invasion build-up of Soviet military strength. This same approach persists today, and Soviet military doctrine can only be understood within the context of contingency planning for worst case situations.

At the end of World War II, the long term threat to the Soviet Union lay with America and its atomic monopoly. Within a few years this threat had extended to include the NATO alliance, with particular reference to Britain and Germany. The danger lay in nuclear attack. Soviet assessments of the probability that, sooner or later, war would come with the West have varied over the years, but the Soviet Union has never wavered in its belief that a strong military capability is the best way of making it less likely. In the first seven years after the war, the probability was almost certainly seen as high, as Russia struggled to restore her shattered economy and restructure her armed forces. Thereafter the fear of a deliberate attack receded, but national security was seen to depend on maintaining a strong defense posture, which was also necessary to prevent the United States from exploiting its military strength to dictate the outcome of events. The probability of war was seen to rise sharply in 1961–63, as the result of Kennedy's assertive policies and the sudden build-up in American strategic weapons. At this period there appears to have been genuine Soviet concern that the United States was seeking to develop the capability for a disarming strike, which (if nothing else) would allow it to "negotiate from a position of strength." From 1969–72 the SALT process served to lower the probabilities, and it is possible that one of the more important side-effects of the negotiations was a shift in Soviet threat perceptions. It seems likely this led them to de-emphasize the danger that confrontation with the United States on non-vital issues would escalate to nuclear war.

Meanwhile, the possibility of war with China had emerged, and by 1969 this was probably seen as a more likely contingency than war with the West. It might, however, trigger such a war. In any case, the possibility of war with the West remained inherent in the opposing military postures, if only through miscalculation.[7] However unlikely, world war is possible, and as such it must serve as the basis for contingency planning, with China and the West as Russia's principal opponents.

One of the problems in understanding Soviet military doctrine (and in disentangling their intentions) is that strategic thought in the West, particularly as concerns the possibility of war with Russia, has been conditioned by the concept of "nuclear deterrence." This started from the simple idea of defending Europe against a conventional attack by demonstrating the capability and resolve to inflict "unacceptable damage" on the Soviet Union. But over the years the original concept became increasingly complex, with its own vocabulary and body of doctrine, as on the one hand Russia developed a strategic nuclear capability, and on the other the theories were spun out to their logical ends.

The Soviets do not have an equivalent concept, and they even lack a proper term for it.[8] Soviet military doctrine does not separate out the ideas of "nuclear deterrence" from the general concept of defense. Defense of the Soviet Union depends on the capability to repel (or at least absorb) any attack and then go on to win the subsequent war. The Soviets obviously hope that their defense capability will be sufficient to dissuade (or hold back—*sderzhi-*

---

7. See M. Mackintosh on this point, "Soviet Military Policy", *SND*, p. 61.
8. For a detailed discussion of this question see P. Vigor, "The Semantics of Deterrence and Defence," and G. Jukes "The Military Approach to Deterrence and Defence," in *SNP.*, pp. 471-485.

vat') an aggressor, which is of course deterrence in its traditional sense.[9] But the crucial distinction between this and the Western concept of "nuclear deterrence" is implicit in the comment that "if the deterrent has to be used, it will have failed." The Soviets do not entertain such ideas. Should war come, their defense will only have failed if their armed forces are unable to recover and go on to final victory.

This emphasis on defense through war fighting has been central to Soviet military doctrine.[10] While Western theory saw nuclear weapons as a means of threatening "unacceptable damage" to Russia, the Soviet Union saw them as adjuncts to its war-fighting capability. Where the West thought in terms of credibility, argued about the merits of counterforce and countervalue, or worried about stabilizing and de-stabilizing developments, the Soviet Union focused on achieving victory in war.[11] And even now that the reality of mutual assured destruction is increasingly acknowledged within the Soviet leadership, they will not adopt it as explicit policy, lest it undermine the fundamental doctrine of being prepared to fight and win any war which is forced upon them.[12]

The readiness to think through the implications of the nuclear arms race does not imply that the Soviet Union would willingly embark on general nuclear war with the West. Very much the reverse. Marxist-Leninist theory asserts that the *initiation*[13] of war as a deliberate act of policy can only be justified if (1) the Soviet Union is virtually certain of winning, and (2) the gains clearly outweigh the cost.[14] War with the West meets neither of these criteria. Communist theory and Soviet national interests coincide in this matter, and it is widely accepted by students of the Soviet Union that the prevention and avoidance of world war is a prime objective of Soviet foreign

---

9. I am not in disagreement with Garthoff and Ermarth, both of whom stress that deterrence has long been a central concept of Soviet defense policy (*International Security*, Summer 1978, pp. 121-123 and Fall 1978, p. 145, respectively). The point being made here is that although both sides rely on deterring the other, their concepts of how best to *achieve* such deterrence have been very different.
10. See T. W. Wolfe, "Soviet Strategic Policy" in *SND*, pp. 73-4.
11. The initial location of ABM systems illustrates the very different policies and priorities which flow from the different assumptions. The Soviet Union sought to protect Moscow, its center of government, whereas the USA sought to protect the ICBM fields, its "assured response".
12. See Raymond Garthoff, "Mutual Deterrence and Strategic Arms Limitation in Soviet Policy," *International Security*, Summer 1978, pp. 112-147.
13. This term does not cover launching a pre-emptive attack to weaken or prevent an inevitable enemy assult.
14. P. Vigor, *The Soviet View of War, Peace and Neutrality*, Routledge, Kegan Paul (1975). For a summary version see "The Soviet View of War," *SND*, p. 17-18.

policy. In fact, Malenkov and Khrushchev both expressed the opinion that there could be no winners in nuclear war, and at different times they both advocated some form of deterrence policy, as a means of reducing expenditure on defense. Neither of them was successful because the security-conscious collective leadership was unwilling to base the defense of the homeland on an unproven theoretical construct.

In his book *Military Strategy*, Sokolovskij did not discuss the Soviet strategic delivery capability in terms of "deterrent forces," but in terms of war fighting. He states that "strategic operation of a future nuclear war will comprise the coordinated operation of the branches of the armed forces, and will be conducted according to a common concept or plan . . . The main forces of such an operation will be strategic nuclear weapons . . ."[15] In discussing such nuclear missile attacks, he says that "the basic aim of this type of operation is to undermine the military power of the enemy by eliminating the nuclear means of fighting and formations of the armed forces, and eliminating the military-economic potential by destroying the economic foundations of the war, and by disrupting governmental and military control."[16] This is war-fighting with nuclear weapons.

Inevitably, such a war would be a "world war," which Marxist-Leninist theory defines as a fight to the finish between the socialist and capitalist systems; defeat would be synonymous with extinction, and victory with survival.[17] It is the catastrophic consequence of defeat which explains why, despite the admittedly low probability of such a war occurring, preparations to fight and win one are given such high priority within the Soviet Union. But, for victory in such circumstances to have any meaning, it is necessary to ensure the continued existence of (1) some kind of governmental apparatus, and (2) a social and economic base on which to rebuild a Socialist society. These minimum essential requirements, coupled with the concept of a fight to the finish, provide the framework for Soviet military doctrine.

On the evidence available, it seems likely that contingency plans to cover the possibility of world war provide for two equally important sets of objectives. The first set focuses on *extirpating the capitalist system*, the aims being to:

---

15. V D. Sokolovskij, *Voennaya Strategiya* (Moscow, 1968), pp 346-47. Explicit statements concerning the targets of strategic systems remained unchanged in the 1962, 1963, and 1968 editions.
16. *Ibid*, p. 349
17. P. Vigor, *op. cit*, SND, p. 22

1. Destroy enemy forces-in-being.
2. Destroy the system's war-making potential.
3. Destroy the system's structure of government and social control.

The second set of objectives focuses on *preserving the socialist system*, the aims being to:

1. Protect the physical structure of government and secure its capacity for effective operation throughout the state.
2. Ensure the survival of a certain proportion of the working population and of the nation's industrial base.
3. Secure an alternative economic base which can contribute to the rebuilding of society.

It is clear that the measures required to destroy certain types of enemy forces-in-being will simultaneously further the second set of objectives, by limiting the enemy's capacity to wreak destruction on the Soviet homeland. Other implications of these dual sets of objectives are perhaps less obvious.

—First, operations in NATO Europe. These come primarily under the *second* set of objectives, because of Western Europe's potential as an alternative economic base on which to help rebuild the socialist system. This has two corollaries: (1) Western Europe must be taken over,[18] and (2) battle damage must be kept to the minimum. While it will be necessary to destroy those NATO forces-in-being which can strike at Russia, and to establish effective control throughout the area, the strategy is to limit the extent of devastation through selective weapons' policies, restricting military operations to essential areas, and using the diplomatic tools of bribery, blackmail and coercion to their fullest extent. The concept of operations therefore avoids "meat-grinder" tactics except where essential to break through the main battle front. The emphasis is on high mobility and deep penetration, on the ability to seize and hold areas in the enemy's rear, and on the extensive use of chemical rather than explosive weapons.

---

18. In strict war-fighting terms, it would be simpler, more certain, and more cost-effective to ravage Europe with nuclear weapons, as is clearly planned for North America. However, the Soviet military posture in Eastern Europe, Warsaw Pact exercise scenarios, and Soviet military pronouncements, all indicate a rapid advance into NATO Europe. Planning to seize this area only makes military sense if it is intended to make subsequent use of the territory and its resources. Those who argue that the present Soviet military posture indicates an intention to seize Europe in circumstances *other* than in a world war, ignore their Marxist-Leninist theory and the importance of the cost/benefit calculus. The occupation of Europe can only be expected to yield benefits in the surreal circumstances of a world war, which the Soviet leadership has been unable to avoid.

—Second, Western sea-based strategic delivery systems. When Polaris first became operational, its most vaunted characteristic was its invulnerability which (in deterrence theory) provided for an "assured response." But from the Soviet point of view, the more important implication of this invulnerability was that these missiles could be held back from the initial nuclear exchange, with the fair certainty that they would remain available for use at a subsequent stage of the war. So, too, could carrier-based nuclear-strike aircraft.[19] What was more, the United States appeared to be shifting its emphasis from land- to sea-based systems; to quote Gorshkov, one third of the West's nuclear inventory was seaborne in 1966, and by 1971 the proportion would have reached one half.[20] These developments bore directly on the Soviet Union's strategy of using Western Europe as an alternative economic base for the rebuilding of the socialist system. If Western seabased systems were withheld from the initial exchange, they would be available to deny Russia this use of Europe.

—Third, strategic reserves. Largely ignored by nuclear deterrence theory, the requirement for strategic reserves is integral to the concept of war-fighting with nuclear weapons. No one can foretell the course of such a conflict, but Soviet strategy must assume that the availability of nuclear weapons may be critical at certain stages. It must also assume that sole possession of a substantial capability is likely to determine the final outcome of the war and the political structure of the post-war world. The West has sea-based systems which can be withheld from the initial exchange, and the Soviet Union must at least match this capability for deferred strikes. But the Soviet requirement for strategic reserves goes beyond this basic requirement. By definition, world war is a struggle between social *systems* and the Soviet Union's potential opponents are not limited to members of the NATO alliance, but extend throughout the world, at least to include the OECD countries and America's military protectorates. Although there is no direct equivalent to the targets which an occupied Europe would offer the United States,[21] Soviet strategy must allow that the conflict will become global,[22] and provide against the emergence of capitalist power-bases outside the NATO area.

---

19. This was not just Soviet paranoia. A senior NATO Commander gave this as his policy in 1961.
20. *Morskoj sbornik* (henceforth *Msb*), May, 1966, p. 9, February, 1967, p. 16.
21. Europe only becomes a U.S. target zone once it has been occupied by the Soviets. Canada (and perhaps parts of Mexico) would be included in the Soviets' initial strike plan
22. Gorshkov stresses the global character of a future world war, but his statements can be read as referring to SSBN operating areas, rather than their targets on land. C. G. Gorshkov *Morskaya moshch' gosudarstvo*, Moscow, 1976, p. 363.

—Fourth, China. Russia may hope that in a fight to the finish between capitalism and socialism, China would take its side. Nevertheless, Soviet contingency planners must assume that world war may also mean war with China. If Russia is to avoid being irrevocably committed to automatically devastating China at the outbreak of a world war[23] and thereby forfeiting the possibility of an alliance, then the nuclear weapons required to cover the contingency of war with China (or to compel its neutrality), must be of a kind that can survive nuclear strikes by the West.

—And fifth, strategic infrastructure. A world war will be fought mainly with the weapons and material that exist at the outset, and combat endurance will depend heavily on prepositioned stockpiles. Most of these need to be readily available to the main engagement zones[24], but in view of the global scale of conflict, stockpiles are also required in distant parts of the world. One way of achieving this is to supply a client state with more arms than it can absorb;[25] another is to acquire base and storage areas overseas.[26] The strategic infrastructure includes the *existence* of the physical facilities which will be required to gain access to distant areas and to sustain wartime operations there. Control of such facilities is not essential prior to the outbreak of the war, and where key pieces are missing from the strategic map (ports, airfields, roads), these can be provided in peacetime under the guise of economic aid.[27] Meanwhile, the growing possibility of war with China adds another dimension to this requirement.

The Genesis of Contemporary Military Doctrine

The essential framework of contemporary military doctrine appears to have crystalized during 1961. This of course was before the Cuban Missile Crisis, which was a by-product of the process and not its cause, although the

---

23. This could well be present policy, as the only way of ensuring that China does not emerge from the war stronger than Russia.
24. Leonard Sullivan observes that this would explain the stockpiling of superseded weapons and equipment in forward areas, after units have been re-equipped with more modern arms.
25. This idea was prompted by Avigdor Haselkorn "The Evolution of the Soviet Collective Security System." However, he sees such arms as being intended for redeployment to other client states in "peacetime" conflict situations.
26. The facilities established in Somalia were capable of supporting a much larger Soviet force than has operated in the Indian Ocean so far.
27. Development of the fishing port at Gwadar in Baluchistan, 100 miles from the Pakistan-Iranian border could be an example of such a strategy The recent coup in Afghanistan lends pertinence to this idea.

outcome of the crisis would have served to reinforce the policy decisions which had already been taken.

The most interesting aspect of this framework was that it represented a significant step back from the new direction in Soviet defense policy which appears to have been agreed by the end of 1959, and which Khrushchev announced in January, 1960.[28] This included the formation of the Strategic Rocket Force (SRF), its designation as the primary arm of the nation's defense, and the cutting back of conventional ground forces. Given Khrushchev's faith in nuclear missiles and his belief that nuclear war would be suicidal for both sides, the new policy could only indicate a shift in emphasis toward the Western concept of nuclear deterrence, and away from the traditional reliance on balanced forces and a war-fighting capability.

It is important to identify the reason for this reversal of tack, which was not just a return to traditional military values, but involved a thorough reappraisal of what was involved in war-fighting with nuclear weapons, and the development of a whole series of consequential policies. It is too facile to answer that this backtracking reflected the strength of opposition from the ground forces. The policy announced in January, 1960 was the outcome of a thorough-going defense review which appears to have taken at least two years,[29] and included a substantial reorganization of military research and development, the constitution of a new branch of the armed forces, and an overall cut in their numbers. This process could only have taken place with, and through, the armed forces, who would have been closely involved in the shape of the final package. Doubtless there were disagreements with the political leadership on matters of substance, but one way or another military acceptance would have been secured; it was, after all, only three years since Zhukov's ouster. It is most unlikely that Khrushchev would have gone public with such a fundamental policy statement unless he had the support to secure its implementation, and the timing suggests that the policy was ratified at the December Plenum of the Central Committee.[30] Moreover, what rumblings there were in the military press focused on problems caused by the premature retirement of regular officers and the lack of suitable job

---

28. *Pravda*, 14 January 1960.
29. In 1958, Admiral Gorshkov solicited the opinions of Commanding Officers of Baltic Fleet units as to the future nature of warfare.
30. See J. McDonnell, "The Organisation of Soviet Defence and Military Policy Making" in *Soviet Naval Influence; Domestic and Foreign Dimensions* (henceforth SNI), MccGwire and McDonnell (eds.) , Praegers, 1977, p. 64, where the year is misprinted as 1958.

opportunities; they were not primarily concerned with questions of high strategy.

The change of tack during 1961 represented a serious reversal of Khrushchev's efforts to emphasize a nuclear deterrent capability and reduce expenditure on the ground forces;[31] it vindicated those who had argued that professional military opinion should prevail in matters of national defense. Only a change in threat perception can plausibly explain such a major shift in the recently established balance of opinion within the leadership. We can, however, rule out as proximate causes the growing conflict with China, the U-2 incident and consequential failure of the Paris summit, and the confrontation over Berlin and the East German peace treaty, since their roots all antedate the 1960 announcement. The only *new* development which could have engendered such a reevaluation of the threat and prompted the various measures which can be dated to this period, were the policy decisions made by President Kennedy shortly after taking office.

These measures included a sharp acceleration of the Polaris program[32] and a doubling of the planned production rate of solid-fuel ICBM, which would be deployed in underground silos remote from existing centers of population. In other words, there was to be a very rapid increase in the numbers of missiles and their invulnerability. Perhaps equally important in terms of Soviet threat perceptions was the crusading rhetoric of the new Administration, with its willingness to go any place, pay any price, and the detached logic of the tough-minded academic strategists who were thinking the unthinkable, and developing theories of limited nuclear war. In the circumstances, it is perhaps not surprising that the Soviet leadership decided that they could not rely on nuclear deterrence, despite its economic attractions, and applied themselves instead to the problems of fighting and winning a nuclear war, the likelihood of which appeared to have increased.

Traditional policy was reaffirmed by the Minister of Defense, Marshal Malinovskyj, in his speech at the 22nd Party Congress in October, 1961. The 1962 edition of Sokolovski's *Military Strategy* was cleared for typesetting five months later, and the fact that a substantially revised edition of the book

---

31 It is noteworthy that Khrushchev returns to this theme in the final section of his memoirs, when he refers to the infantry as "the fat of the armed forces". *Khrushchev Remembers*, Little, Brown (Bantam), 1976, p. 612.

32. Between 1958-1960, fourteen Polaris submarines had been authorized On taking office, President Kennedy authorized a further twenty-seven, of which fifteen were to start building within six months.

appeared in just over a year[33] suggests that the earlier version was prepared in some haste. Equally significant was the unprecedented decision to make the book publicly available, thus ensuring that the West was left in no doubt that the Soviet Union was prepared to wage nuclear war if need be.[34] This served to articulate the *Soviet* concept of deterrence and gave notice that Russia would not be intimidated by the build-up of U.S. strategic forces. The build-up was remarked on in the first edition of Sokolovski, which quoted from President Kennedy's message to Congress on 28 March 1961. The second edition had a table showing the sharp jump in ICBM from 200 at the beginning of 1963 to 1190 by the end of 1966, and noted that Minutemen would be emplaced in silos.

The most clear-cut policy response was the major reorganization of civil defense which was initiated in mid-1961. Responsibility was transferred from the Ministry of the Interior to a central headquarters located in the Ministry of Defense.[35] This was headed by Marshal Chuikov, a Deputy Minister of Defense and CinC of the Ground Forces, an appointment he retained.[36] Civil defense was thus firmly established as a military problem, and part of the Soviet Union's warfighting posture. There continues to be heavy military participation in this nation-wide system, in terms of staff appointments throughout the country and of specialized military civil defense units.

A second policy response was the navy's move forward in strategic defense against the West's seabased delivery systems. The first substantial Soviet naval exercise in the Norwegian Sea took place in June, 1961 and became an annual event thereafter. In January, 1962, there appeared the first of a series of articles and statements rehabilitating the role of the much-maligned surface

---

33. See Harriet Fast Scott's invaluable source of reference, *Military Strategy (Third Edition); a translation analysis and commentary, and comparison with previous editions* (Stanford Research Institute, January, 1971). The different editions were cleared for typesetting and released to the press as follows: 1962-3/62, 4/62; 1963-4/63, 8/63; 1968-11/66, 11/67.

34. Hindsight suggests this simple explanation, which focuses on the *availability* of the book, rather than its publication. The book forms part of a substantial body of military literature which had previously been restricted to an official/professional Soviet readership, and was not publicly available, least of all to the West. Revised editions of these various books are published as required. For example, there have been three editions of *History of Naval Art* (1953, 1963, and 1969), but only the last one was made publicly available.

35. It is perhaps relevant that three months earlier (in May, 1961), responsibility for Civil Defense within the United States had been transferred to the Department of Defense.

36. Chuikov took over as Commander-in-Chief of the Ground Forces in 1960 and retained the title until it lapsed in August, 1964. He was appointed Chief of Civil Defense 17 August 1961 and remained in the job until he retired in 1972.

forces.[37] In February, 1962, Admiral Kasatanov, who had followed Gorshkov as CinC Black Sea Fleet in 1955, and would later join Gorshkov in Moscow as First Deputy CinC of the navy in 1964, took over the Northern Fleet, which was slated to play a key role in the shift to forward deployment. At about this same period, the decision would have had to have been taken to change the planned configuration-mix of the new generation of nuclear submarines (due to begin delivery at the end of 1967), to provide six SSBN a year. This represented a reversal of the decision to remove the mission of intercontinental strike from the navy.[38]

A third development which clearly can be tied to the U.S. ICBM program was the shift in Soviet targeting philosophy from one of area devastation to a primary emphasis on counter-force point targeting. The initial trend in Soviet warhead development was to ever larger sizes. The 5 MT SS-7 and -8, whose warhead was probably tested in 1958, the missile being deployed in 1961–62;[39] the 25 MT SS-9 whose warhead was tested in September, 1961, the missile being deployed in 1966; and weapons in the 25-100 MT range, whose development was claimed in 1961 but which did not go into production.[40] The trend was then broken abruptly, and the 1 MT SS-11 began deployment in 1966. At the same time, the annual production of ICBM rose from about 60 to 260 a year and ran for four years at this rate, thus matching the tempo and duration of the Minuteman build-up.[41]

The original policy of using large warheads to achieve area devastation

---

37. See R. W. Herrick *Soviet Naval Strategy* (U.S. Naval Institute, 1968) pp. 72–74

38 The decision to double nuclear submarine production from five to six to about ten per annum would have been taken in 1957/58. The exact date when the navy lost its strategic strike missions is uncertain, but would have been at about that period, by which time the inadequacies of the submarine strike force must have been apparent. Evidence of a subsequent change in the projected configuration-mix away from attack submarines in favor of SSBNs is provided by aberrations in the building programmes. See my "Current Soviet Warship Construction and Naval Weapons Development" in *SNP*, pp. 429–432.

39. There was an accelerated test schedule in the fall of 1958, immediately preceding the test ban talks, which included explosions in the MT range. A. Kramish *Atomic Energy in the Soviet Union* (Stanford University Press, 1959) p. 126.

40. Warning of the impending tests was given by *Pravda* 31 August 1961. Explosions of at least 25MT were detected by the West and these included the testing of a triggering device for what was reputed to be a 100MT warhead. See R. M Slusser *The Berlin Crisis of 1961* (Johns Hopkins Press, 1973) pp 183–203.

41. For a portrayal of this lagged matching reaction, see Table 27.1 in my "Soviet Strategic Weapons Policy 1955–70" in *SNP*, p 494. Deployment dates and numbers have been refined subsequently by John McDonnell (Doctoral thesis in progress) and Robert Berman and John Baker (*Soviet Strategic Weapon Systems*, Brookings Institution, forthcoming), but their findings support the hypothesis in my original article.

took care of the full range of targets and can be seen as the most cost-effective means of meeting Soviet requirements, as the latter would have appeared in the period 1953-58.[42] As such, the policy was a natural extension of traditional war-fighting doctrine. The decision not to persist with that targeting philosophy, but to adopt the much more costly one of matching the American ICBM build-up with an equivalent counterforce capability, clearly rejected the option of placing primary emphasis on nuclear deterrence, which the existing trend of development would have allowed.

As for other developments around this period, the escalation of the Berlin crisis can be seen as an effort to achieve Soviet objectives regarding East Germany before the balance of forces swung heavily against the Soviet Union. Similarly, Cuba can be seen as an ingenious attempt to mitigate the impending massive disparity in ICBM, by emplacing IRBM within range of the United States. In military terms, the success of both these ventures would have brought substantial gains, while nothing would have been lost if they failed. It would seem, however, that his successful orchestration of the long-drawn out Berlin crisis made Khrushchev over-confident of his ability to control the level of tension over Cuba. The nuclear testing which took place in September, 1961 cannot, however, be tied to the Kennedy initiatives. It is true that these tests became inevitable once it was clear that the United States was embarking on another round in the arms race. But they represented an essential stage in the existing Soviet weapons development program and, Kennedy or not, there would have been tremendous pressure to breach the test moratorium at this particular juncture.

The Navy as a Branch of the Armed Forces

Having established the general background we can now turn to consider the Soviet navy as a branch of the armed forces. It is fair to say that since the early twenties, the Soviet leadership has demonstrated a sustained awareness of the requirement for maritime defense. Stalin and Khrushchev both took a personal interest in naval matters, and given the geo-strategic circumstances and the scale of competing priorities, the navy appears to have had at least its fair share of scarce resources. The basic procurement and strategic policies have been well founded, if not always fully successful.

Stalin recognized the need for four operationally independent fleets and

---

42. *SNP*, pp 492-497.

established the Pacific Fleet in 1932 and the Northern Fleet in 1933;[43] he sought to ensure their self-sufficiency by providing each with its own naval construction facilities, located at some distance from the open sea to be safe from enemy seizure. These two shipyards at Komsomol'sk, on the Amur, and Severodvinsk, on the White Sea, are now the two premier nuclear submarine building yards. He also set out to link the three Western fleets by inland waterway and established a submarine building yard at Gor'kij on the Volga, which was able to continue production throughout World War II. Curvatures on the Trans-Siberian Railway were calculated to allow the shipment of submarines (whole and in sections) to the Pacific by rail.

After the war, when the Soviet Union was faced by the "traditional maritime powers" the leadership responded with mass-production warship building programs to cover the threat of invasion, and the allocation of nuclear reactors to submarines intended for strategic delivery. It is true that when the threat of seaborne invasion was downgraded in 1954, there were savage cuts in conventional warship construction, but this was partly justified by the availability of long range surface-to-surface missiles (SSM). And, when a new threat to the homeland emerged in the shape of seaborne nuclear strike systems, the leadership responded by doubling the production of nuclear submarines. Military publications acknowledge the importance of the navy's role. The 1962 edition of *Military Strategy* emphasized the great changes in maritime warfare since World War II, and stressed that in future war the navy's primary theater of operations would be the open ocean and that it must not be tied to ground force theaters of operation. Writing in 1971, Marshal Grechko noted that maritime combat was achieving a special significance and that navies could have an enormous impact on the entire course of a future war.[44]

Of course there is a difference between what the navy thinks it ought to have and what the national leadership finds "reasonable" in the face of competing priorities. We have seen that the naval establishment lost its case in 1954-55, but it appears to have been more successful in 1960-64, during the debate which stemmed from the introduction of the new defense policy in January, 1960, and its subsequent modification in the wake of the Kennedy initiatives. Disagreement seems to have focused on the navy's role in nuclear

---

43. The Pacific Fleet was called "Far East Naval Forces" until January, 1935. The Northern Fleet was called "Northern Naval Flotilla" until May, 1937
44. A. A. Grechko "The Fleet of our Homeland" *Msb*, July, 1971.

war,[45] with an extreme faction arguing that a fleet was no longer necessary, even in its traditional role of supporting the army ashore, and that land-based missiles could deal with enemy surface groups, and even with submerged submarines.[46] From its side, the navy was arguing that the threat from Polaris was being under-estimated, that the role of surface ships was being under-emphasized, and that undue reliance had been placed on ground and rocket forces to the neglect of other means of warfare.[47] However, by mid-1964 the threat from Polaris had been acknowledged as being the navy's first priority,[48] and the warship building programs provided explicit recognition of the surface ship's role, if perhaps not as generously as the navy might have wished.

In the opinion of many analysts, myself included, naval interests were again heavily engaged during 1969-73, in the wide-ranging debate about foreign and domestic policy, and about the dangers of war and the future roles of the armed services.[49] The initial evidence of naval involvement came from the series of eleven articles published during 1972-73 over Gorshkov's name in *Morskoj sbornik* under the title "Navies in War and Peace."[50] These articles were rich in information and contained a strong element of "educating the fleet," but the dominant tone was one of advocacy and justification, which extended beyond the contention that the Soviet Union needed a powerful navy for use in both peace and war, to criticisms of the formulation of naval policy and the composition of the fleet. The facts that halfway through the series *Morskoj sbornik* began to encounter unprecendented delays in being released to the press by the military censors (delays which extended

---

45. For a summary of this debate see H. Ullman "The Counter Polaris Task" in *SNP*, pp. 586-590; also D. Cox "Sea Power and Soviet Foreign Policy", United States Naval Institute Proceedings, June, 1969, p. 59.

46. S. G. Gorshkov "Razvitie Sovetskogo Voenno Morskogo Isskusstva", *Msb*, February, 1967, pp. 19-20. Gorshkov places this controversy in the middle fifties; but so too does he place the decisions which underlie the present structure of the navy, which clearly originate from a later period, although he would like to tie them to his appointment as C in C. This kind of argument is likely to have persisted throughout the later fifties, until resolved in the early sixties.

47. For a clear exposition of the navy viewpoint see V. A. Alafuzov "On the appearance of the work *Military Strategy*", *Msb*, January, 1963, pp. 88-96. Admiral Alafuzov had been Chief of the Main Naval Staff during the war.

48. V. D. Sokolovskij and M. I. Cherednichenko, *Krasnaya zvezda*, 25 and 28 August 1964.

49. For evidence of this wider debate see Marshall Shulman's "Trends in Soviet Foreign Policy," *SNP*, pp. 8-10 (also published at greater length as "Towards a Western Philosophy of Coexistence" in *Foreign Affairs*, October, 1973); also John Erickson, "Soviet Defence Policy and Naval Interests," *SNP*, p. 60.

50. *Msb*, 1972, Numbers 2-6, 8-12, 1973, No. 2.

into 1974), and that during the same period there were major turnovers of the journal's editorial board (again unprecedented), were seen as confirming the textual evidence of conflict.[51]

There are, however, other analysts who consider that the articles were authoritative. This is not an abstruse academic disputation, because the alternative assessments lead to very different conclusions concerning many important aspects of Soviet military policy. One particular school of thought considers that the articles represented a "concrete expression of doctrine,"[52] and their purpose was to announce a Soviet political decision to withhold submarine-launched missiles from the initial strikes, in order to carry out deterrence in war, conduct inter war bargaining, and influence the peace talks at the end of the war. The later analyses of this school use quotations from Gorshkov's book *Seapower and the State*[53] to support this viewpoint, making the tacit assumption that the articles served as the book's precursor.

The latter assumption is open to serious question. There was a three year hiatus between publication of the last article and the appearance of the book (32 months between typesetting dates), and while this is not the place for a detailed comparison of the two publications, it can be asserted that there are substantial differences between them in length, scope, coverage and the balance between topics.[54] Among the more obvious differences is the much

---

51. Publication anomalies were first identified by R. Weinland; see "Analysis of Admiral Gorshkov's Navies in War and Peace", *SNP*, pp. 558-565. This analysis was refined and extended by John McDonnell in his *"Bibliographic Analysis of Morskoj sbornik 1963-75"* Center for Foreign Policy Studies, Dalhousie University, September, 1977. There is additional circumstantial evidence which points to this same conclusion; see *SNI*, p. 55.

52. Doctrine is here used in its *Soviet* sense, implying authoritativeness. James McConnell is the leading exponent of this viewpoint; see his "Military-Political tasks of the Soviet Navy in War and Peace", pp. 183-209 of *"Soviet Oceans Development"*, prepared by John Hardt for the Senate "National Oceans Policy Study" (henceforth *SOD*), US GPO, October, 1967 My "Naval Power and Soviet Oceans Policy" in the same publication includes a rejoinder to the advocacy argument; See Apps. B and C, pp. 167-182 For a later exposition of McConnell's analysis see "The Gorshkov articles, the new Gorshkov book, and their relation to Policy" in *SNI*, pp. 565-620.

53. *Morskaya Moshch' Gosudarstvo*, Moscow, 1976. Cleared for typesetting 1 August 1975; released to the press 27 November 1975.

54. The book comprises some 151,000 words, the articles about 54,000 The subject matter of about 80 percent of the articles is historical and is covered by Chapter II which comprises only 35 percent of the book; there has been substantial revision of this material from the articles including rewording, additions and omissions. Only a small part of the remaining 20 percent of the articles can be compared directly with material in the book. For example, the book devotes almost 11,000 words to marine transportation and the Soviet merchant fleet, whereas the articles cover the subject in seventy-five words. Some 45 percent of the book is concerned with the development of navies after the war. (Chapter III), and the art of naval warfare (Ch. IV), compared with only 16 percent of the articles.

greater attention paid to the non-naval elements of sea power; the articles treat these cursorily, whereas the book devotes 20 percent of its length to these aspects. Emphasis on marine transportation is particularly noticeable, and extends to injecting material on the Soviet merchant fleet into the section on World War II.

Although further research is needed, there is a strong case for concluding that *Seapower of the State* is the outcome of a compromise between major interests involved in the wider ranging debate which prompted (and enabled) the publication of the Gorshkov series. The latter presented a powerful argument. The dominant theme was the importance of naval forces as an instrument of state policy in peacetime and as a means of influencing the course and outcome of wars of all kinds; furthermore, this importance was increasing. Seapower was a necessary adjunct to great power status, which could not be sustained without a powerful fleet. But naval forces had to be shaped in response to specific requirements, and this demanded a conscious policy concerning the role of seapower in each nation's plans. Gorshkov asserted that the Soviet Union lacked such a policy. In consequence, the Soviet Union had an unbalanced fleet that was deficient in surface ships, both as to numbers and range of types, and the navy had been shaped too closely to a single, restrictive and largely defensive mission. If the Soviet Union was to exploit the potential of seapower as an instrument of policy, it must have a greatly improved world-wide capability. It certainly had the economic and industrial capacity to build and sustain such a fleet.[55]

On a more specific note, Gorshkov, writing at the end of 1971, strongly opposed any weakening of Russia's position in the Eastern Mediterranean, whether through some kind of mutual agreement with the United States, or by withdrawing Soviet forces from Egypt, which would imperil the navy's access to Egyptian ports and airfields. The growing probability of such a withdrawal may well have been the precipitating cause of the series' publication.[56]

The Gorshkov series provided a valuable insight to Soviet naval require-

---

55 For a convenient version of the analysis underlying this summary, see *SOD*, pp. 79-132 An edited version of my initial but more comprehensive analysis, May, 1973, appears in *Admiral Gorshkov on "Navies in War and Peace"* (Center for Naval Analyses, Washington, CRC 252, September, 1974), which also contains analyses by R Weinland and J. McConnell.

56. Some 10 percent of the series is devoted to the Mediterranean and Black Sea. The second article (cleared for typesetting in January, 1972), includes a chapter, "The Russians in the Mediterranean," which is nominally about the period prior to 1855, but speaks clearly and at length about the strategic importance and political legitimacy of the present-day Soviet naval

ments. But the 54,000 words of sustained argument also provided a window on the wider debate and allowed certain inferences as to how opinion divided on particular issues, mainly in terms of attitudes but sometimes in terms of institutions and interests.[57] The listings (which are *not* intended to portray coalitions) are too complex to summarize here, but it is relevant to this discussion that among the inferred opponents to Gorshkov's advocacy were those who believed that the probability of world war continued to be high, that military power had limited utility as an instrument of state policy outside the Soviet Bloc, and that Soviet/U.S. confrontation risked escalation. The Merchant Fleet was inferred as an opposing institution[58] and Grechko was one of two individuals identified in the analysis by name.[59]

Grechko's opposition was inferred from his article "The Fleet of the Homeland," published in the 1971 Navy Day issue of *Morskoj sbornik,* following the 24th Party Congress.[60] Grechko does not play down the Soviet Union's very real requirement for a navy, or its vital role in the country's defense. But his initial discussion covers all branches of the armed forces, the nuclear submarines being bracketed with the SRF. The emphasis is on the navy in war and on deterring attack on Russia. An article in *Morskoj sbornik* by the Minister of Defense on this occasion was itself unusual, but more importantly, the tone and substance read very differently to what appeared subsequently in the Gorshkov series. Further evidence that Gorshkov and Grechko had divergent viewpoints was provided by the latter's booklet *On Guard for Peace and the Building of Communism,* which claims a direct link with the decisions of the 24th Party Congress.[61] It placed primary emphasis on

---

presence in the area. Gorshkov returns to discuss the Mediterranean (and its contemporary significance) throughout the series, and it is the only area to have a chapter of its own.
57. See *SOD,* pp. 127-129. This tabulation was prepared in May, 1973 (i.e., 2-3 years before the book's appearance), on the basis of the initial analysis.
58. Hence the significance of the differences in the book's structure, favoring maritime transportation. See note 54.
59. The other was Brezhnev. His opposition to certain aspects of the articles was inferred from the fact that Gorshkov not only ignored his June, 1971 proposal for mutual restrictions on naval operations, but advanced a contrary line of argument, particularly as regards the Mediterranean.
60. *Msb,* July, 1971. Exercise OKEAN '70 is described as demonstrating the navy's readiness to repel attacks on Russia and to launch its own strikes. Only submarines, naval aviation, and the landings in the arctic receive special mention, with submarines singled out for a paragraph on their own. The non-mention of surface ships, by far the most numerous component in the exercise, seems pointed. Reference to U.S. imperialism is limited to South East Asia, Soviet support being limited to "fraternal aid".
61. A. A. Grechko, *Na Strazhe Mira i Stroitel'stva Kommunizma* (Moscow, 1971). A 112 page booklet designed for a "broad range of readers" that "describes the great historical mission of the Soviet Armed Forces, and the increased tasks posed for them . . . by the 24th CPSU Congress."

combat readiness and discussed the Soviet Union's international commitments only in terms of other socialist states. This approach was poles apart from the argument in the Gorshkov series (which was written in the wake of the Party Congress but did *not* claim direct links with its decisions), for an assertive overseas policy based on military power, and for the navy's unique qualifications as an instrument of state policy in peacetime. Lying somewhere between the two viewpoints is *Military Force and International Relations* edited by Kulish. This book is linked explicitly to the Party Congress and would seem to reflect most accurately the Congress' endorsement of an expansion of the "internationalist function" of the Soviet armed forces. It discusses the increasing importance of a "Soviet military presence" in distant regions, but while it gives due attention to the navy's role (including mention of the Mediterranean squadron) it is concerned with the problem of strategic mobility and the more general requirement for "mobile and well trained and well equipped forces."[62]

It seems possible that Gorshkov's opponents intended to rebut his arguments through a series of articles on military theoretical problems in *Krasnaya zvezda*. The series was announced in April, 1973 under the general title "The Defense of Socialism. Questions of Theory," but ceased without explanation after the second article in May, 1973.[63] This coincided with the temporary lifting of censorship delays on *Morskoj sbornik* and can be seen as an imposed truce, which may have been connected with the appointment of Grechko to the Politbureau in late April.[64]

A compromise seems to have been reached by mid-1974. In an article published in May, Grechko acknowledged that the "historic function of the Soviet armed forces is not restricted merely to . . . defending the homeland and other socialist states" and that their external function had now been "expanded and enriched with new content."[65] This brought him into line

---

62. V. V. Kulish, *Voennaya Sıla ı Mezhdunarodyne Otnoshennıya* (Moscow, 1972), pp. 135–137.

63. Lt. Gen. I. Zavyalov "The Creative Nature of Soviet Military Doctrine", April 19; Maj. Gen. A. Milovidov "A Philosophical Analysis of Military Thought", May 17, 1973. Neither article addressed Gorshkov's arguments directly but both took ıssue wıth some aspect of what he said. Zavyalov stressed the primacy of political factors and the fundamental position of the political content of mılitary doctrine. Milovidov emphasızed that you cannot take examples from one historical period to support arguments in the contemporary period, which is, of course, what Gorshkov did.

64. This was the first tıme since Zhukov's ouster in 1957 that a professional military officer was co-opted onto the highest party body. Gromyko (Foreign Affairs) and Andropov (KGB) joined the Politbureau on the same date.

65. A. A. Grechko "The leading role of the CPSU in building the army of a developed Socialist Society", *Voprosy istorrı KPSS*, May, 1974, pp. 38–39. Weinland has compared this with earlier

with the policies endorsed by the 24th Party Congress.[66] In July, Gorshkov acknowledged that the main naval mission in war was coming to be operations against targets on land, rather than combatting the enemy fleet.[67] He thus accepted the formal prioritization of missions which the military leadership appears to have been trying to enforce since 1966-67.

Compromise does not of course mean the end of disagreement. With this in mind, we can turn to consider the implications of three new and important sections of Gorshkov's book, which are headed "Fleet against Fleet and Fleet against Shore," "Problems of Balancing Navies," and "Command of the Sea."

Naval Dissatisfactions

There is unlikely to have been any dispute over the absolute (as opposed to relative) importance of the fleet-against-shore mission. The Soviet navy had led the world in developing submarine launched ballistic missiles, and it was the navy which had argued within the defense establishment that the threat from Polaris was being under-estimated. Throughoit the 1960s, naval writing emphasized the fundamental nature of these technological developments and discussed their implications in terms of naval warfare.

Disagreement appears to have been (and continues to be) centered on the relevance to modern war of the navy's traditional role and of general purpose naval forces. Evidence of attempts by the military establishment to downgrade this role can be seen in the 1968 edition of *Military Strategy*, which placed the SSBN force on a par with the SRF.[68] For navies in general, it added the mission of "nuclear strikes against objects on the continents . . . and the active search for enemy naval forces, and their destruction . . . ," but *deleted* from the section on structuring the armed forces the sentence:

---

pronouncements, including a comparable article in *Kommunist*, May 1973, and concludes there was a distinct shift in emphasis. See *SNP*, p. 569.

66. J. McDonnell notes that Grechko's 1974 article followed the same general line as one by A. A. Yepishev (Chief of the Main Political Administration of the Armed Forces) in *Moguchee Oruzhie Partii*, which was released to the press at the end of 1972.

67. "Soviet National Seapower", *Pravda* July 28, 1974. One is tempted to tie Admiral Kasatanov's retirement in October, 1974 to this compromise, for the reasons given in my initial analysis (note 55 above).

68. This reflected the imminent availability of the Yankee SSBN. Additions and deletions are by comparison with the 1963 edition. See Harriet Fast Scott (*Op. Cit.*, note 33), pp. 235; 240-243; 319; 308, respectively. In the 1968 Russian edition, pp. 235; 240-243; 330; 308.

"Hence, the principal mission of our navy in modern war will be combat with enemy forces at sea and in their bases."

While the primary importance of the navy's strike and counter-strike missions is always acknowledged, this downgrading of the traditional role was not matched in contemporary naval publications.[69] Indeed, *The Combat Path of the Soviet Navy*, a book that went to typesetting the month after Gorshkov's final article, asserts that the navy must carry out "active operations" against enemy sea lines of communications, and goes on to say: "Nor will such tasks as the destruction of enemy surface groups, and cooperation with the ground forces on the maritime axes, by means of amphibious landings and other operations, be taken away from the navy."[70] The Gorshkov series only makes very brief reference to the navy's contribution to strategic strike and to its role in countering Western sea-based systems.[71] The great bulk of his descriptive analysis focuses on the use of general purpose forces in both peacetime and war. He spends a substantial part of the articles demonstrating (by historical analogy) the importance of the traditional wartime role. Some 20 percent of the series is devoted to analyzing non-Russian world-wide naval operations in the two World Wars, and concerning the second one, he concludes that although the war was won in the continental theaters (primarily on the Russian front), naval operations had a significant effect on the general course of the war, and Western operations made an important contribution to the final outcome.

In his last article, Gorshkov does list the three tasks which comprise the navy's basic wartime mission:[72] (1) contributing to strategic strike; (2) blunting strategic strikes by enemy units; and (3) "participating in the operations conducted by ground forces in the continental theatres of military operations." Apart from their order, Gorshkov does not distinguish between the tasks in importance, but he adds that the third one involves "a large number of complex and major missions." Within the context of the series as a whole, one can infer that he envisages a wide range of operations, comparable to those discharged by navies in World War II. It should also be noted that in *Combat Path of the Soviet Navy* (which was under preparation at the time of

69. For example, *Istoriya Voenno-Morskogo Iskusstva* (Moscow, 1969), p. 561.
70. *Beovoj put' Sovetskogo Veonno-Morskogo Flota* (Moscow, 1974), p. 492.
71. These are listed and analyzed in Appendix B to my "Naval Power and Soviet Oceans Policy" in *SOD*, p. 167.
72. *Msb*, February, 1973, p. 21.

the Gorshkov series), the tasks of strategic delivery and countering the enemy's naval strike forces are quite specifically assigned equal importance.[73]

These well entrenched positions, coupled with the concept of *Seapower and the State* as a compromise, help to explain why the section in that book entitled "Fleet against Fleet and Fleet against Shore" pays lip service to the newly agreed priority, but devotes most of its space to illustrating the importance of the traditional naval role. This is achieved first, by extending the definition of fleet-against-shore to include landing operations (355/1)[74] and attacks on sea lines of communications (361/1-2). Next, by establishing two categories of fleet-against-fleet operations: the "pure" form, intended to gain and maintain command of the sea (352/5, 353/4); and those operations which are "tied to the simultaneous accomplishment of other missions" (352/4). It is then shown that this second category of fleet-against-fleet (which comprises the vast bulk of naval operations—354/2) is in fact supporting operations against the shore. This allows the navy's main objective to be defined as "securing the fulfillment of all missions *related* [emphasis added] to operations against enemy land targets, and to the protection of one's own territory from the attacks of his navy" (354/3). This clearly places attacks on the shore and defense against such attacks on the same level; a later formulation is less explicit, but the full context yields a similar interpretation (360/3-4).[75]

Under the guise of exceptions to the general rule, the book smuggles in numerous examples of traditional naval operations which have had strategic significance (351/2), or have even been *more* important than the battle on land (349/3). The extensive definition of fleet-against-shore operations allows discussion of the traditional roles played by navies in World War II, including the importance of carriers as general purpose forces. But surely the nicest touch is the pointed criticism of Napoleon, who blamed his admirals for repeated failure, whereas the fault really lay with his "inability to make a timely analysis of the French navy's capabilities, and to use it in the struggle with the enemy" (356/3). Napoleon's failure to invade England was not

---

73. *Boevoj put'* . . . (*op. cit* ,), p 491. The tasks are referred to as "important" and "no less important" respectively
74. Numbers in brackets refer to page/paragraph of the Russian edition, *Op. Cit* , note 53.
75. The glaring internal contradiction in paragraph 360/4 is explained by the earlier division of fleet-against-fleet operations into two categories. Use of the term *borba* (cf. 352/5) confirms that the final sentence in this paragraph is referring to the "pure form", intended solely to gain and maintain command.

primarily due to Britain's unchallenged maritime superiority, but to his "one-sided strategy, which stemmed from his preoccupation with operations in the land theaters and his lack of understanding of the navy, his disregard for its capabilities in war, and as a result, his inability to use it in a struggle with a naval power, such as England was at that time" (355/4). The analogy with present day circumstances is striking.[76]

Apparently, Gorshkov is able (with suitable obfuscation) to maintain his advocacy of the navy's wartime role and the continued importance of its traditional mission. It would seem, however, that he has not been able (or willing) to persist in his criticisms of the navy's structural characteristics. It is true that the section entitled "Problems of Balancing Navies" provides a critical analysis of great power fleet structures since 1905 in terms of their capacity to handle the unforeseen demands of war, and it can be inferred from his invariable condemnations that Gorshkov favors maximum flexibility. It is also true that his analysis provides examples that could be used to support almost any argument, and certainly Gorshkov stresses that a future war will be fought with forces-in-being (413/4, 439/5). And, as in his articles, he points to the limiting effect of Germany's concentration on submarines, which was exacerbated by its failure to provide anti-ASW forces for their support (429/1-2). He also puts in a good word for the carrier (443/3). But the more significant aspect is Gorshkov's retreat on what is implied by the term "balanced."

In the articles, the question of balance is touched on briefly, but to some purpose. When analyzing the main types of naval operations in World War II, Gorshkov points out that the task-specific fleets were severely handicapped in comparison with those which had a broad and more balanced capability, capable of carrying out large-scale and strategic-type missions. As unfavorable examples, he cites the German navy, which was virtually limited to attacking sea communications, and the Japanese navy, which had almost no ASW capability. By contrast, the British and American navies were able to carry out "broad strategic missions."[77] However, in the book, Gorshkov castigates the British and Americans along with the Germans and Japanese, and reverts to the restrictive definition of a "balanced fleet" (413/2), which

---

76. This analogy is comparable to (but much more pertinent than) the critical remarks in the Gorshkov series about those who fail to understand the significance of sea power. *Msb*, 72/3/20/2-21/2; 72/4/9/1, 22/9.
77. *Msb*, 72/11/32/5; 73/2/20/7-8.

was publicized in 1967.[78] In that version "balance" denotes the capability to carry out assigned missions in differing circumstances, but says nothing about the mission structure. The meaning which Gorshkov developed in his articles, sees "balance" as stemming from the *choice* of mission, which must be defined in as general terms as possible, in order to exploit the navy's inherent versatility and to allow for unforeseen eventualities. Unquestionably, the definition in the book represents a retreat, but its justification may lie in a change of official attitudes toward the surface ship role. At least, such a change might be inferred from the difference in tone between what he wrote in his final article,[79] and the comparable statement in the book. The latter notes that "the priority given to the development of the submarine and aircraft . . . presupposes a matching developing of the other arms of naval forces . . . ," without which no mission can be successfully completed, and among which surface ships will play the most important role (412/6).

And finally, the section entitled "Command of the Sea." This was the first substantial discussion of the subject in recent years[80] and Gorshkov was concerned to demonstrate the continuing validity of the concept, rather than how to achieve it. He stresses its uniqueness to the maritime environment (unlike most theoretical categories such as mass, maneuver, etc., which are common to all aspects of military art), and asserts that it is the most "vital concept" in the art of naval warfare. The discussion provides powerful support to the arguments advanced in the other two sections for the continued relevance of the fleet-against-fleet role, and for the importance of general purpose forces, particularly the anti-ASW/pro-SSBN element.

The central theme of Gorshkov's discussion is that developments in nuclear missile war have increased the importance of ensuring local command of the sea in key areas for particular periods of time. The strategic significance of sea-based nuclear delivery systems makes it all the more essential to ensure a "favorable operating regime" for one's own forces. He argues that local

---

78. *Msb*, 67/2/20, n.1 ". . able to discharge assigned tasks in both nuclear-missile and conventional war, and also to secure state interests at sea in peacetime."
79. *Msb*, 73/2/20/7-8, 2/21/1. See *SOD*, pp. 116-118 for a discussion of these paragraphs, which read like advocacy, or at least justification.
80. This constitutes a sub-section of "Some Theoretical Questions of Naval Warfare" which, in other respects, reproduces verbatim the greater part of Gorshkov's December, 1974 *Msb* article "Command of the Sea" is a substantial addition, which takes almost as much space as all the other seven categories together. For a discussion of earlier references to Command of the Sea, see P. Vigor "Soviet Understanding of Command of the Sea", and my "Command of the Sea in Soviet Naval Strategy", in *SNP*, pp 601-22 and 623-36.

command of the sea is essential to the successful discharge of the navy's primary mission (379/2), and that undoubtedly, the imperialists will seek to wrest such command for themselves at the very outset of war (380/2). Furthermore, gaining command of the sea has always depended on the necessary measures having been taken in time of peace (371/1) and (by implication), the suddenness of nuclear-missile war makes these preparatory measures all the more essential.

The book brings out all the classical advantages of gaining command of the sea, including the fact that the effect of a single engagement on the balance of naval strength will endure for the whole war, and he comes close to explicitly advocating the "pure form" of fleet-against-fleet operations. But two points are of particular interest in terms of Soviet contingency plans. Drawing on Soviet experience in World War II, Gorshkov observes that the occupation of coastal regions by the ground forces greatly facilitated establishing command of adjacent sea areas (379/3). Bearing in mind the significance of the Barents and Norwegian Seas to Soviet SSBN operations, one's mind inevitably turns to the Norwegian coast, Svalbard and perhaps Iceland.

The second point concerns the necessity of "establishing the conditions for gaining command of the sea (at the outset of war) . . . while still at peace" (371/1). The measures he lists include "forming groupings of forces and so disposing them in a theater that they have local superiority over the enemy, and also providing the appropriate organization of forces in the maritime theaters of operation (sea and ocean), and a system of basing, command and control, etc., as required by their missions." These requirements could well be used to describe the pattern of Soviet activity since 1964, when the navy first began to establish significant forces in distant sea areas, a process which is still in progress today.

In summing up the source and substance of naval dissatisfaction, we are drawn back to my earlier comment that traditionally the navy has been seen as an expensive necessity, rather than a flexible instrument of state policy. It would seem that much of this attitude persists today and the nub of the navy's complaint against the military leadership is summed up by Gorshkov in his indictment of Napoleon. The Soviet leadership's preoccupation with land operations, combined with a lack of understanding of the navy and a disregard for its capabilities (which it has failed to analyze properly), has resulted in a one-sided national strategy and an inability to make effective use of the navy in the struggle with its opponent, a maritime power (355/4, 356/3).

The dissatisfaction is over the unwillingness to recognize the navy's potentialities in peace and importance in war, and the inability to grasp the complexities of maritime warfare. The failure of the military leadership to understand or even to analyze properly the navy's role, coupled with its prejudices concerning particular weapons and platforms, has meant that on the one hand the fleet is configured for a relatively narrow span of specific missions, and on the other hand that it lacks the full range of forces with which to discharge these missions effectively. The most persistent criticism (which reaches back into the fifties), is of the army-dominated leadership's inability to grasp that task-specific naval forces, however deadly in themselves, require the support of general purpose forces (particularly surface ships) if they are to be able to accomplish their tasks. The need for support to the SSBN force is the most frequently argued example. In peacetime, this lack of comprehension leads to the under-utilization of the navy as an instrument of overseas policy. In a war with maritime powers, it could lead to national disaster.

The fact that the Gorshkov series of articles could be published, and the persistence of his criticisms in the subsequent book, suggest that he does not stand alone in his advocacy of the navy's role and exploiting the flexibility of seapower. But obviously there are important interests aligned against him, including those in the military who have always argued that too much is spent on the navy; the merchant fleet, where professional jealousy sharpens the competition for ship-building resources; and the many domestic interests who give priority to building up the domestic economy over adventures abroad. The argument is likely to persist and we will have to wait on future warship deliveries to assess the final outcome.

Naval Mission Structure

The Russian navy's traditional objective has been "to defend the homeland," from which were derived its two main missions of (1) supporting army operations on land, and (2) repulsing attacks from the sea. These twin missions could be discharged most effectively by maintaining command of contiguous waters, but this was frequently not possible. In times of duress, the second mission was collapsed into the first, with operations on land having overriding priority. In more favorable circumstances, the second mission was extended to carrying the war to the enemy, and the navy has a record of daring interdiction attacks on enemy forces in their ports and home waters.

In essence, this basic mission structure persists today, but in the years following World War II, the Soviet navy was faced with three developments. First, her most likely opponents were now the traditional maritime powers, who had just demonstrated their capability to project and support continental-scale armies across vast distances at sea. Second, a series of quantum jumps in the range, accuracy, and payload of maritime weapons introduced the capability of devastating one's opponents' territory with sea-based systems. And third, China joined the ranks of the Soviet Union's likely opponents. These developments increased the relative importance of the maritime components of warfare, requiring greater differentiation within the mission structure. At the same time, the development of a new range of land- and space-based weapon and surveillance systems for use against maritime targets meant that account had to be taken of the contribution to these missions by other branches of service.

On the basis of public pronouncements and other evidence, there is a fair measure of agreement in the West on Soviet naval missions in war, although there is some dispute about their relative priority. These missions can be labelled:

—Strike against shore
—Destruction of enemy naval forces
—Interdiction of enemy sea lines of communications
—Direct support of ground force operations
—Protection of own sea lines of communications

Before discussing these missions, we must focus on the concept of area defense, which is fundamental to Soviet naval operations. The concept is based on two main zones: an inner one, where superiority of force allows local command of the sea to be secured, and the outer zone, where command is actively contested.[81] The greater part of Soviet naval policy and procurement since the 1920s can be explained in terms of their attempts to extend this maritime defense perimeter and, within it, the zone of effective command.

If the four widely separated fleets were to be ensured the superiority of force necessary to establish command in their respective areas, they would

---

81. The Soviets speak in terms of three zones, but the third one is beyond the defense perimeter. Robert Herrick discusses these in his book *Soviet Naval Strategy* (United States Naval Institute, 1967), and his current research is throwing further light on this concept.

need to deny the enemy the opportunity to concentrate his forces against any one fleet. This objective could most economically be achieved by denying him physical access to the fleet areas. In this respect, Russia was favored by her geography. Three of these areas comprised semi-enclosed seas, and access to the Northern Fleet was canalized by ice during much of the year; only Petropavlovsk, the naval base on Kamchatka, lacked any geographical advantage of this kind. Until 1961, therefore, the navy's primary concern was to extend the inner zone of effective command to the natural defensive barriers, which would be seized by Soviet forces in the event of war. The outer zones did not extend very far beyond these geographic constrictions and were primarily seen as areas for interdicting the reinforcement of the enemy who defended these natural barriers.

But after 1961, the outer zone was extended to take account of the qualitatively new threat to the Soviet Union posed by the Polaris submarine, as well as the continuing threat from carrier strike aircraft. In the Far East, the maritime defense perimeter was pushed out into the East China Sea and the Pacific. In the west, the outer zones were extended to take in the Norwegian and North Seas and the Eastern Mediterranean, and the Soviet navy progressively contested the West's erstwhile maritime domination of these sea areas.

Maintaining command of contiguous waters, and contesting command of adjacent seas, is the precondition for discharging the majority of other naval missions, including the most important ones. Area defense will therefore be treated as a mission in its own right.

STRIKE AGAINST SHORE

The mission of strategic strike was first laid on the navy in the immediate post-war years, when the submarine-torpedo was the only available means of bringing an atomic weapon to bear on the continental United States. Soviet strategic delivery submarines were given top priority in nuclear propulsion, warheads, and ballistic missiles. But despite this, advances in American antisubmarine capabilities, coupled with Soviet technological inadequacies, meant that the first generation of series production units (comprising two diesel and two nuclear classes, one of each being armed with nuclear torpedoes, the other being armed with a surface-launched ballistic missile) were unable to meet the planned operational requirements. The mission was therefore taken away from the navy towards the end of the fifties.

This decision was reversed in the sixties, to match the shift in U.S. emphasis from land- to sea-based strategic delivery systems, and the 1968 edition of *Military Strategy* placed the SSBN on a par with the SRF.[82] The SSBN force now has three overlapping roles, in that it can contribute:

1. Intercontinental strike;
2. Intra-theater strike;
3. The national strategic reserve.

The Delta SSBN can carry out *intercontinental* strikes from home waters. If Yankees are targeted in this role, they need to be within 1,500 n.m. of the U.S. coastline. Only three or four units are kept forward deployed, so the remainder would have to transit Western anti-submarine barriers to come within range.

The Yankee SSBN can cover a wide range of *intra-theater* targets from home waters, and for Northern Fleet units this includes "shore facilities which support the operations of the Western SSBN force, and those ASW forces and systems which constrain the free egress and open-ocean operations of the Soviet submarine force."[83] Six Golf II SSB were transferred to the Baltic in 1976, from where they can cover most naval facilities in NATO Europe.

In the event of world war, the ballistic-missile submarines represent an important component of the *national strategic reserve*, because of their relative invulnerability. While it is clear that the Soviets do in fact think in these terms, there are no means of knowing their exact intentions in this respect. But one must assume that their operational plans provide for the greatest flexibility in the use or withholding of these systems, as best to influence the progress and outcome of the conflict.

Endless permutations of targeting, deployment and timing in these overlapping roles are possible, but they all raise the same two requirements:

1. Until such time as the missile submarines have fired all their weapons, or deployed to the open ocean, they must be kept secure against attack. This has led to the concept of defended ocean bastions.
2. If the submarines are deployed, they must be able to transit Western anti-submarine barriers in reasonable safety, and to survive attempts to find them in the open ocean. This raises a requirement for support forces.

---

82. See note 64.
83. R. O. Welander, *et al. The Soviet Navy Declaratory Doctrine for Theatre Nuclear Warfare.* BDM Corporation, Washington, D.C.; Report No. DNA (Defense Nuclear Agency) 4434T, 30 September 1977, p. 41. This is a most useful study and I have drawn freely on the conclusions.

This mission therefore involves both the ballistic-missile submarines and the maritime forces assigned to their protection and support. This pro-SSBN mission, demanding that enemy ASW forces be countered at sea, is a force-consuming task and is an important function of the navy's submarine, air, and surface components. It is seen as one of the primary missions of the Kiev class ASW carrier and the post-1965 large anti-submarine ships.

DESTRUCTION OF ENEMY FORCES

The enemy's sea-based strategic nuclear strike systems present the primary maritime threat to the Russian homeland, and the task of countering these forces is on a par with the Soviet navy's own strategic mission.

In striving to counter this threat, the Soviets have been driven by two rather different concerns. The most immediate is damage limitation, both as it affects the Soviet homeland, and their ability to fight the land battle. But more important is the withholding of these systems as tactical and strategic reserves. In the short run they could then be used to deny Russia the use of Western Europe as an alternative economic base for rebuilding the socialist system. In the longer run, the sole possession of a residual nuclear capacity is likely to determine the final outcome of a world war, and hence the political structure of the post-war world. It was this potential as a tactical and strategic reserve which required that Soviet forces be within weapon-range contact at the onset of war, and necessitated the navy's shift to forward deployment.

COUNTERING THE BALLISTIC MISSILE SUBMARINE By 1964, the Polaris submarine had replaced the carrier as the most dangerous component of this threat and overlapping concern for damage limitation and withholding focused attention on different aspects of the weapon system. To prevent their being withheld for use at a later stage, the SSBN must themselves be disabled, and this requires one to know their location. But damage limitation only requires that (preferably) all or (at least) some of the missiles should not reach their targets. The importance of this distinction is twofold: (1) once the submarine launches a missile, it reveals its position and renders itself liable to attack; and (2) the missile itself is vulnerable in its early stages.

Although the Soviets have been addressing the SSBN problem for over fifteen years, the final shape of their response is still not certain. Ideally, they would like to deploy a system which provides continuous location of all missile submarines, backed by a rapid-response kill capability. However, they will be prepared to adopt any approach which offers to reduce the overall threat from this direction, and their search for a solution has been on a very broad front. The tracking capability required for continuous location can be provided by area surveillance or by trailing. The state of the art in the sixties meant the latter was the more realizable solution in waters closer to home, while methods of ocean surveillance had still to be developed. Russia's lack of open-ocean coastline complicates the use of fixed acoustic surveillance systems located in the water column, and combined with U.S. emphasis on submarine quietening, this directed attention to air- and space-borne systems, exploiting other forms of anomaly detection.

It seems likely that this approach distinguishes between the ocean expanses on one hand, and more confined waters like the Norwegian Sea and Eastern Mediterranean on the other, both because of their relative location to Russia, and because of the different problems of detection in these different types of areas. Confined waters lend themselves to some variant of the area defense concept, whereas the open ocean is more suitable for satellite surveillance. Surface surveillance satellites have now been in service for a decade and one must assume that a missile-launch detection capability has already been developed, if not actually deployed. Some analysts believe that non-acoustic methods of submarine detection are already operational in space.[84] Meanwhile, we must not ignore the potential attraction of direct tracking by specially designed submarines, a concept which becomes increasingly relevant when Trident enters service.

Assuming that the problem of location can be solved, the problem of attacking the submarine is comparatively simple. When Soviet forces are in direct contact, they will have available an array of long-range weapons. In the ocean expanses it seems probable that the Soviet Union intended to develop a common system for anti-surface and anti-submarine strike, using terminally-guided ballistic missiles, as described below.

If it is not possible to locate the SSBN until it launches a missile, the threat from those that are withheld for later use can be reduced if they are deprived of targeting, navigation, and command and control information by disrupting communication links and attacking the sources of such information. High

84. Especially K. J. Moore; see "Developments in Submarine Systems" and "Antisubmarine Warfare" in *SNI*, pp. 151-184, 185-200. Moore highlights the Soviet emphasis on non-acoustic methods of submarine detection, the use of satellites in this and related roles, and the concept of the extended ASW team, which includes satellites, air, submarine, surface ships and shore-based missiles. He notes discussion in Soviet journals of new methods of submarine propulsion and the development of wing-in-ground effect vehicles, suitable for high-speed area-search.

priority is therefore given to the destruction of SSBN base and operational support facilities at the onset of war.

COUNTERING THE CARRIER It appears that the Soviets feel they have developed an effective counter to the strike carrier, although it took them more than fifteen years to achieve this. The two components of this mission (target location and strike) are handled somewhat differently in each of the three main types of scenario, namely continuous company, meeting engagement, and distant targeting.

"Continuous company" describes the situation in the Eastern Mediterranean, an outer defense zone. When carriers operate within this area, primary responsibility for providing target location data lies with the surface units, which remain in close company. They also provide local command and control and a secure communications link with headquarters ashore, and are therefore likely to continue in this role, despite their vulnerability. The main strike arm is now the cruise-missile-armed submarine, probably backed by IR/MRBM. Aircraft based in Russia lack the rapid response time essential to this scenario, and while they remain a threat to Western forces they are not critical to the operational concept.

The "meeting engagement" is best exemplified by the deployment of carriers from U.S. east coast ports to launch strikes against Russia from the south Norwegian Sea. In such circumstances, target location data are provided by a mix of forward pickets (AGIs, etc.), and aircraft and satellite surveillance. The force will be harassed by air and submarine attacks en route, but the main engagement will take place in the encounter zone. Here, the force will be subjected to successive, heavy attacks by air- and submarine-launched missiles at the same time that it is transiting a torpedo-attack submarine barrier. If the carrier force is configured for nuclear strike against Russia, the encounter zone is located to the west of the probable launch area; but should it be configured for operations in the Norwegian Sea (e.g. in support of counter-SSBN operations or flank reinforcement), the main encounter is likely to be north of the Gap. SSM-armed surface ships will probably be used in the latter case but will be held back in the former. It seems likely that the large anti-submarine ships will be deployed to hinder attempts by Western ASW forces to prevent the missile-armed submarines from launching their weapons. Variations of this concept apply in the Pacific and in the Mediterranean when U.S. forces are not already deployed in the eastern basin.

The "distant-targeting" scenario covers those carriers that do not immediately threaten the Soviet Union at the outbreak of war but, if not disposed of, will contribute to the U.S. reserve. In these circumstances, target location information is provided by air and satellite reconnaissance. It appears likely that it was originally planned that strike would rely primarily on two different types of terminally-guided ballistic missile systems, one land-based and the other submarine-based. The SRF's capability to strike surface groupings was explicitly claimed in 1972 and, despite time-of-flight problems, there appear to be no insuperable reasons why ICBM should not be used against high-value naval targets. A submarine-based system would require the unit to remain within strike range of its target(s). A 400 n.m. tactical SLBM (SS-NX-13) was undergoing trials in 1972; it appears to have been shelved for reasons that remain unclear, and the present status of this system is uncertain.

The first two of these scenarios depend heavily on the concept of a defense perimeter and area defense. All three are applicable to high-value surface targets other than carriers.

AREA DEFENSE

Area defense is central to certain of the missions which have already been discussed. It is the way to secure the safety of home waters as SSBN bastions, and to prevent enemy carriers and missile submarines from closing Russia to launch their strikes. The success of these and several other missions will depend on the extent to which the Soviets can establish command of the sea in these areas. This is best discussed in terms of the inner and outer defense zones.

The Baltic, Black, and Barents Seas are inner defense zones and it can be assumed that the Soviets count on establishing effective command of these waters at the onset of war.[85] This will involve seizure of the Baltic and Black Sea exits and parts of arctic Norway. Command of these sea areas will automatically:

---

85 The concept of area defense is part and parcel of maritime theaters of military operation (MTVD). In a recent (unpublished) paper, James Westwood draws attention to the latter concept and its derivation from the army's land-based equivalent (TVD), and he itemized the various types of zone which can be found in such theaters. He notes that the concept of MTVDs is related to the defense of a few delimited geographical regions. He identifies five MTVDs, three of which are referred to as closed (Baltic, Black Sea and Sea of Japan), while the other two look out onto great oceans (Northern Sea and Kamchatka). There are also two ocean theaters of military operations (OTVD), one comprising the Eastern Mediterranean and the other the Arabian Sea.

—Secure the close support of army operations, including tactical landings and naval bombardment;

—Secure coastal communications, including logistic support of the land battle;

—Prevent the enemy from carrying out amphibious assaults and from providing close support to army operations ashore.

The concept of an inner defense zone also applies in the Pacific, although operational plans will be modulated by diplomatic considerations such as keeping Japan neutral. But parts of the Sea of Okhotsk are suitable for SSBN bastions, and one can also assume that effective command will be sought in the Sea of Japan and off Kamchatka.

The outer defense zone of the Baltic and Barent Seas takes in the North and Norwegian Seas, with the defense perimeter running from Greenland through Iceland and the United Kingdom. For several reasons this outer zone is of great strategic importance to the Soviet Union. Besides those which have already been discussed, the main NATO anti-submarine barrier is located in the Greenland-Iceland-United Kingdom gap, while the North Sea provides access to Denmark and northwest Germany for Western reinforcement and supply, and for Soviet amphibious hooks and flank support.

Gaining command of this area (particularly the Norwegian Sea and its approaches) has high priority, and maintaining command will be facilitated by seizing key stretches of the Norwegian coast and (more risky) Iceland and/or the Faeroes. As a way of tilting the balance in their favor, the Soviets will seek to pin down forces by mining and draw others away by diversionary attacks. It will still be a costly struggle, and there can be no certainty that shore-based air support will be available to Soviet forces.

In considering the importance of the Eastern Mediterranean as an outer defense zone, we must distinguish between peace and war. Its importance in peace-time derives from its use as a deployment area by Western strike systems targeted on Russia. This spurs the development of location and strike systems appropriate to the geographical conditions, and these would be used at the onset of war. Thereafter, there would seem to be no great urgency to gain command of the Eastern Mediterranean. In practical terms the Black Sea Fleet's outer defense zone is probably co-terminous with the Aegean, which provides a defensible perimeter and a haven for submarines operating against NATO communications.

The outer defense zone in the Pacific as importance as a shield both to the SSBN bastions and against U.S. sea-based strike systems. However, if the bastions are located in the Sea of Okhotsk, the Kuriles chain provides a natural defensive barrier. Meanwhile, there are no clearly defined geographical areas comparable to the North and Norwegian Seas and the Iceland/Faeroes gap, nor will the land battle have the same relevance. In such circumstances, effective command throughout the outer defense zone is unlikely to have a high priority.

There would seem to be a latent outer defense zone in the northwest quadrant of the Indian Ocean, covering the seaward approaches to the Persian Gulf. In the event of world war, the Soviet Union is likely to move south to control the Gulf, and will need naval forces to fend off assaults by U.S. strike carriers and amphibious groups. It may also have trouble from the regional navies.

Within the area-defense concept, anti-surface operations had overriding priority for the first 10 to 15 years after the war, but thereafter anti-submarine warfare became increasingly important, and today it has the highest priority in terms of force characteristics and research and development. The progressive improvement in the Soviet navy's ASW capability largely matched the seaward extension of the anti-submarine defense zones, and the Soviet approach to the problem continued to reflect the concepts developed when the zones were comparatively narrow. At that initial stage, shore-based systems played the major role, with the various sensors operated by the Observations and Communications Service providing detection data to shore-based helicopters carrying sonobuoys. The solution was still a combined arms one, with offshore defense units being vectored to join the hunt and to prosecute enemy contacts, while a proportion of the torpedo-armed diesel submarines (which had originally been intended to provide defense in depth against surface groups) were switched to the anti-submarine role. All these operations were controlled by the naval headquarters ashore, and procedures were developed for coordinating operations by the various types of units.

When it became necessary to extend the seaward limit of the defense zone beyond the effective radius of shore-based helicopters, it was natural to think in terms of placing them on a seagoing platform; hence the Moskva class of helicopter-carrying anti-submarine cruiser. Meanwhile, the demise of large surface ships as executors of the counter-carrier mission released destroyers to anti-submarine area defense, and methods of coordinating the various airborne, surface, sub-surface, and land-based systems continued to evolve.

While the concept of an anti-submarine defense zone is most effective in waters directly contiguous to one's coastline, similar procedures have been adopted in the outer zones, including the Eastern Mediterranean. Area defense *per se* is not possible in such areas, hence greater emphasis is placed on various types of detection barriers and on a Soviet variant of hunter/killer groups that involves surface ships and submarines working together. However, it appears that since the early seventies, priority in the employment of conventional ASW forces has been shifted from counter-SSBN operations in forward areas such as the Eastern Mediterranean, to reinforcing the security of the Soviet SSBN bastions in the North and Pacific.

There are likely to have been two main reasons for this shift in emphasis. The most significant was reports in 1967–68 that the U.S. Navy was intending to develop two new classes of submarine for service in 1973-74, one very fast and the other very silent, the latter being specifically designed to operate against Soviet SSBN. This of course had major implications in terms of the Soviet decision to embody a substantial part of the nation's strategic reserve in ballistic missile submarines, and the Deltas were due to start entering service at just that period. Meanwhile, as more anti-submarine systems became available, mounted in surface ships, submarines and aircraft, it must have become increasingly clear to the Soviets that however innovative, these traditional ASW methods had inherent limitations, and an effective solution to the Polaris/Poseidon problem would have to wait for the results of research and development still in progress. Hence the shift in emphasis to extending the inner defense zones in the Northern Fleet area and in the Pacific off Kamchatka, and to providing them with watertight ASW defenses. Because ASW forces can be brushed aside by superior force, it would be necessary to maintain command of these two sea areas. And since shored-based air was unlikely to be available after the initial exchange, *sea-based air* would be needed to deal with enemy airborne ASW systems.

INTERDICTING ENEMY SEA COMMUNICATIONS

The interdiction of enemy sea lines of communication is a traditional mission, although its importance has fluctuated with changing perceptions of the likely nature and duration of an East/West war. Recently, the possibility of protracted nuclear conflict or of conventional war has increased its relative importance, which is currently stressed by the Soviet navy. While attacks on sea lines of communication would appear to be less urgent than some other missions, they have the great advantage of tying down Western ASW forces and diverting them from posing a threat to the SSBN strategic reserve. It therefore seems certain that although the anti-SLOC mission could in principle be deferred, sufficient forces will be allocated to ensure the diversion of Western ASW forces from areas of primary Soviet concern such as the Norwegian Sea.

In the initial stages of a war, intra-theater shipping (which will be particularly vulnerable to mining) will almost certainly be more important than trans-oceanic convoys, unless the latter are already closing their destination. It should, however, be stressed that the mission of interdicting enemy sea communications is viewed in its broadest sense and includes destruction of terminal facilities and distribution networks by air or missile strike, the mining of ports and assembly areas, and attacks on all types of escort force.

PROTECTING OWN SEA COMMUNICATIONS

Traditionally, Russian sea communications have been limited to coastal shipping and this came within the general concept of area defense, although escorts were provided. Soviet forces were barely involved in the protection of the ocean convoys which brought supplies to Russia during the World War II. However, in the event of a future war with China, the Soviet Union must be prepared to supply its Far Eastern Front by sea, and to protect such shipments against attacks by Chinese naval forces, which include the world's third largest submarine force. This is a time-critical requirement and the shortest route lies across the Indian Ocean. If possible, supplies would be shipped from Black Sea ports via the Suez Canal or even round the Cape. But if the Black Sea exits were closed to them, these supplies could pass down across Iran and out through the Persian Gulf, using the route established by the Allies in the two World Wars.

BACKGROUND FACTORS

Certain persistent factors help to explain various aspects of Soviet naval doctrine. Earlier in this article we discussed the emphasis on contingency planning and the reality of world war, but the point about war fighting needs to be re-emphasized. Undoubtedly there is argument in the Soviet Union about the nature of nuclear war, whether it will be long or short, and whether or not there will be a conventional phase. But the military planner must assume the worst and has to think not only of the post-exchange phase of such a war but of subsequent phases through to its resolution. They must at the same time take account of how to handle China in such circumstances.

This emphasis on war fighting, which must allow for the disruption of supply systems and base facilities, has major implications for the employment or withholding of forces in the initial stage of a war. It also heightens the awareness that war is in large part a matter of attrition and that victory goes to the side that gives up last. This awareness leads to the principle of never allowing an enemy weapon or force a free ride, and to the continuing use of sub-optimal and obsolescent weapons in order to complicate the enemy's problems. It may also explain various aspects of the strategic infrastructure that the Soviets are seeking to establish overseas.

A second background factor is the limitations which existing capabilities impose on *the practical application of preferred operational concepts*. A substantial proportion of existing naval assets were originally designed to operate in scenarios very different from the present ones. Some were designed for tasks different from those they now discharge, and many derive from building programs that were cancelled before the first units even entered service, because the original design-concept had been outflanked by technological developments. Even when the design-concept remains appropriate, competing priorities within the national economy mean that in certain key categories, insufficient numbers are being provided. As a further source of confusion to the analyst, the contemporary evidence will reflect what the Soviets are striving to achieve at some future date, what they are attempting to do with sub-optimal capabilities to meet existing requirements, and the inertia of past concepts which have been overtaken by events. When assessing Soviet priorities it is therefore necessary to distinguish between the evidence which reflects what present capabilities *allow* them to do, and that which reflects what they are *aspiring* to do. Soviet ranking of threats and/or requirements will not necessarily reflect their ranking of corresponding missions, which will be determined partly by available capabilities.

The third factor is the emphasis on combined arms. This approach to maritime defense evolved as much from historical necessity as from proletarian revolutionary doctrine, but this fact makes it all the more persistent. In the early '30s, area defense was only practicable within a relatively narrow coastal zone and depended more on coordinated operations by submarines, torpedo boats, and aircraft than on the capabilities of the pre-World War I surface ships. As more and larger surface new-construction units joined the fleet, tactical concepts were adjusted to include them, and the defense perimeter was progressively extended. But the principle of coordinated operations remained in force. Submarine, aircraft, and shore-based systems were seen as an integral part of the main fleet, the emphasis being on defense in depth and on successive coordinated attacks. Meanwhile, the priority on providing support to ground force operations served to accentuate the "combined arms" approach, which extends beyond the different types of naval components to include other services.

The same driving principle is very much in evidence today, examples being the use of Long Range Air Force (LRAF) aircraft for maritime reconnaissance, the targeting of naval "groupings" by land-based missiles and the use of the merchant fleet for logistic support. But the most striking illustration is provided by their attempts to counter Western SSBN, which involves four out of the five branches of the armed forces, only the ground forces being exempt.

The emphasis on combined arms leads naturally to the last factor, namely that maritime operations are very much an adjunct to the battle on land, and that naval strategy is just a subset of broader military strategy.[86]

The Navy's Peacetime Role

We can say with some certainty that the navy's shift to forward deployment in 1961 was prompted by the requirements for strategic defense against sea-based nuclear delivery systems, a conclusion that is evidenced by the timing of the change of policy, the areas chosen for deployment, and the operational employment of the forces within these areas.[87] Although the original rationale remains valid and continues to underly the main pattern of Soviet naval activity, some of the specific reasons for the shift to forward deployment have been eroded or overlaid with new ones, and among the latter we find the emergence of the navy's peacetime role.

The presence of naval forces in distant sea areas provided opportunities for their political exploitation, and this coincided with a hardening of Soviet attitudes towards the United States and its overseas involvements, and a progressive shift toward a more assertive Soviet global policy. This was probably outlined at the 23rd Party Congress in 1966;[88] the 24th Party Congress in 1971 appears to have extended the trend by addressing directly the

---

86. Welander and Westwood both stress this point.
87. See my "The Soviet Navy in the Seventies", Appendix, *SNI*, pp. 653-57.
88. I am indebted to James McConnell for this point.

role of a Soviet military presence, and Grechko's recantation in 1974 would seem to confirm the general direction.

The long-term prognosis is, however, still unclear. It seems likely that the decision to commit Soviet forces to the air defense of Egypt in 1970 was finely balanced and, even if there had not already been a reversal of opinion, the eviction of these forces in 1972 must have given pause for thought. It is true that the long drawn out SALT negotiations are likely to have engendered a progressive, but fundamental shift in Soviet threat perceptions, including a downward re-evaluation of the dangers of escalation from Soviet/U.S. confrontations in the third world. But while some in the Soviet Union would have argued that this, coupled with the improving correlation of forces, required a more assertive global policy, there would have been others who argued for shifting scarce resources to the domestic economy. Meanwhile, there would remain the unconvinced. The latter might well make up the great majority of the military establishment, who would doubt whether the threat had in fact changed and would continue to press priority for the direct defense of the Soviet Union and to argue the dangers of overseas adventure, unless they were intended to strengthen the strategic infrastructure for wartime missions.

It seems possible that some compromise was reached by early 1973, whereby it was decided that direct Soviet involvement overseas would be limited to the provision of advisers, weapons, and strategic logistic support, the combat role being delegated to the Soviet-equipped forces of "revolutionary" states such as North Korea, Vietnam and Cuba. It can be argued that this policy ensures the Soviet Union the best of all worlds; namely, being able to affect the outcome of an overseas conflict with direct battlefield support, while ensuring that political commitment and liability remain strictly limited. This is achieved by facilitating the arrangements and providing the lift to bring co-belligerents to the zone of conflict; by ensuring that the client state receives adequate military supplies in the course of battle; and by remaining relatively silent about Soviet involvement until after the event.

A result of these decisions has been the increasing readiness to use a "Soviet military presence" in support of overseas objectives. Between 1967–72, as warship-days in distant waters rose year by year, so too did the trend in the political exploitation of this naval presence. And since 1973, Soviet naval forces have been used in a number of ways including crisis management, latent interposition, logistic support, political pressure, and peacetime assistance. In a major study of the employment of Soviet naval forces in peacetime,[89] Dismukes and McConnell concluded that they have been widely used for political purposes and that a significant proportion of their operations are driven by political considerations, rather than war-related requirements. However, questions still remain as to the level of political commitment behind those operations and as to the extent to which the navy's potential as a flexible instrument of overseas policy will come to determine future force requirements.

Soviet pronouncements refer to the navy's peacetime role in general terms such as "defending (or securing) state interests", a nebulous formulation, whose scope has yet to be systematically researched. They also speak of the navy's "international duty," of "increasing Soviet prestige and influence" and of "rebuffing imperialism." While not losing sight of the all-encompassing scope of "securing state interests," it is useful to distinguish between four types of objectives which underlie this peacetime employment, because each type involves a different level of risk and degree of political commitment.

At the low end of the scale of commitment, we have "Protecting Soviet lives and property." This objective is referred to but has received little priority to date. There is one clear example where a naval demonstration in February 1969 appears to have secured the release of Soviet trawlers from Ghanaian custody, but the more usual form is to have landing ships standing by to evacuate key equipment from conflict zones.

At the high end of the commitment scale we have "Establishing a strategic infrastructure to support war-related missions," which embraces the physical, political and operational aspects. This objective is not referred to directly, but can be inferred from the pattern of overseas military involvement during the last 20 years, and is implied in some of the more recent Soviet writings. I believe this task has provided the primary motive for a broad span of decisions ranging from promoting a coup in a client state, to acquiring base rights by barely concealed coercion. The pressure on Egypt (1961-67) for naval support facilities[90] provides a good example of this objective, although

---

89. B. N. Dismukes & J. M. McConnell (Eds.), *Soviet Naval Diplomacy*, Pergamon Press (forthcoming). This book has been prepared by seven members of the Center for Naval Analyses, Washington, D.C. and comprises a comprehensive survey and analysis of the peacetime employment of Soviet naval forces. This whole article reflects my debt to the analysts at CNA, although I cannot claim that we agree on all points.
90. See G. Dragnich, "The Soviet Union's quest for access to naval facilities in Egypt prior to the June War of 1967," *SNP*, 237-277.

the original impulse became obscured by wider involvements as the policy acquired its own momentum . . . and complications. Somalia is another example, with the initial interest in 1968–69 stemming from the Arabian Sea's potential as a patrol area for Poseidon submarines, but being overtaken in 1971–72 by the broader concerns of conflict with China and protracted world war. Because the task of establishing a strategic infrastructure concerns the security of the homeland, it is likely to be backed by a high level of political commitment, and the pattern suggests a willingness to incur high political costs in pursuit of this objective. However, so far the Soviets have not used military force to maintain their position when the host country has withdrawn its agreement to their presence, although on at least two occasions they have sought to engineer a coup to bring a more sympathetic regime to power.[91] Of course, once such an infrastructure has been established, it can also serve peacetime policies, as we see most clearly in the case of Soviet-built airfields in Africa.

In between these extremes we have the general objective of "Increasing Soviet Prestige and Influence." Showing-the-flag increased sharply after 1968, but since 1972 the task has assumed new dimensions, extending to port clearance and minesweeping, and to providing support for revolutionary forces or to regimes threatened by secessionist elements. The Soviets are prepared to commit substantial resources to this objective but, although the propensity for risk-taking has risen steadily, the underlying political commitment is strictly limited.

Overlapping this general influence-building objective is the more restricted one of "Countering Imperialist Aggression." Despite much bombast in declaiming this task, I believe that in terms of risking a major confrontation with the West, Soviet political commitment is low. The first clear-cut example was the establishment of the Guinea Patrol in December, 1970, since when we have the deployments to the Bay of Bengal in 1971, to the South China Sea in 1972, and to Angola in 1975, as well as the three Middle East crises in 1967, 1970 and 1973. The latter series did show a shift from a narrow concern with the strike carriers towards a more general concern for the overall capability of the Sixth Fleet.[92] But none of these examples provide evidence of Soviet readiness to actually engage Western naval forces, in

---

91. Albania and Egypt.
92. See R. G. Weinland, *Superpower Naval Diplomacy in the October 1973 Arab-Israeli War*, Center for Naval Analyses, Washington, Professional Paper No. 221, June, 1978.

order to prevent them from intervening against a Soviet client state. Indeed, it is a moot point whether the reactive deployment of a Soviet detachment during the Indo/Pakistan war in December, 1971 achieved anything at all. For instance, was the force authorized to attack the U.S. carrier group if it had launched its strike aircraft toward some unknowable target? What purpose was served by rushing a force to the South China Sea in response to the mining of Haiphong (which just hung around for a few days and then returned home), or was this the result of an inter-service argument in Moscow? And what of areas like the Mediterranean, where the urge to "counter imperialist aggression" is tempered by the dangers of confrontation and escalation to nuclear war?

What we do see is progressively greater involvement by the Soviet navy in the provision of logistic support both prior to and during third party conflicts. In 1973, Soviet landing ships carried Moroccan troops to Syria, with convoy escort. Landing ships were also used during the subsequent war to ferry military supplies from Black Sea ports to Syria. More significantly, SAM-armed naval units were stationed under the final approaches to the main resupply airfields in Syria and Egypt, as if to cover against Israeli air attack.[93] And most recently, we have the escorting of military supplies being ferried from Aden to Ethiopia.

The evidence suggests a policy of incrementalism, which explores opportunities as they occur or are created, a policy of probing Western responses and establishing precedents. The role of a "Soviet military presence" in support of overseas objectives will therefore be shaped by the scale and style of the Western response to the various Soviet initiatives. In this context the distinction which has just been drawn between the employment of Soviet naval forces to secure the safe arrival of logistic support and their employment to prevent Western intervention against a client state is important. So too is the distinction between the Soviet Union's willingness to risk hostilities with a third party state, and their continuing reluctance to engage U.S. naval forces. Meanwhile we should bear in mind that the Soviet navy's role in this more assertive overseas policy is secondary. The primary instruments are the provision of arms, military advice, and training; the transport of men, munitions, and equipment by merchant ship and long-range air; and direct participation by the combat troops of revolutionary states. The primary role

---

93. *Ibid.*

of the navy is to provide protection and support and to serve as an earnest of Soviet commitment.

This brings us to the question of whether there is some Soviet grand design driving a coordinated oceans policy in support of overseas objectives. The short answer is no, but we must distinguish here between the operational aspects and the setting of objectives. The military style organization of the merchant, fishing, and research fleets means that it is relatively simple to make use of their ships in peacetime for naval support tasks such as replenishment and forward picketing, and they all make some contribution to the generalized requirement for world-wide intelligence and information gathering. There are also the geo-strategic advantages to be gained in terms of a world-wide infrastructure, actual or potential. The latter includes the provision of improved harbor facilities in locations which would assume great strategic significance for Russia in the event of world war, as for example the fishing port at Gwardar in Baluchistan.

But when we turn to objectives, we see that the long-term interests of the three main ocean users frequently diverge. The build-up of the fishing fleet stemmed from a decision in the late forties that fishery was a more cost-effective source of protein than collective farming. The build-up of the merchant fleet reflected the post-Stalin shift in the middle fifties towards trade, aid, and arms supply, and the consequential requirement to earn hard currency and avoid dependence on foreign bottoms. The navy's shift to forward deployment reflected the qualitatively new threat to the Soviet homeland from distant sea areas. Inevitably there is some conflict between these divergent interests and, at the Law of the Sea negotiations the narrow domestic concerns of the Soviet fishing industry ran counter to the foreign policy objective of increasing Soviet influence. Similarly, national security concerns and the concept of strategic infrastructure have led the Soviet Union into political entanglements which would seem to be against its broader interests. Only the merchant fleet consistently serves these more general foreign policy goals, and I see it as the principal maritime instrument of Soviet overseas policy.

It is clear that the Soviets have progressively evolved a policy toward the employment of naval forces in peacetime, but in looking to the future the derivation of that policy becomes important. It stemmed from the availability in distant sea areas of naval forces which had been deployed forward in strategic defense. The presence of these forces was progressively exploited for political purposes and with changes in threat perception, risks, and opportunities, the peacetime political role became increasingly important. Nevertheless, only a small proportion of the Soviet fleet is deployed forward and the continuing lack of effective afloat support to sustain such operations is notable. Meanwhile, there is still no evidence that ships are being procured primarily for the projection of force in peacetime, and while the Kievs, the new landing ships and the "possible" nuclear cruiser will increase the navy's potential capability in this direction, there is a clearly defined, war-related requirement for such types in the outer defense zones or the oceanic theaters of military operation.

This presents a very different picture to the traditional British/American approach, where the navy serves as a primary instrument of foreign policy (as in Pax Britannica or the Nixon doctrine), and where this peacetime role is an important determinant of the navy's size and shape. On the evidence at hand, it does not appear that the Soviet leadership attaches a *comparable* importance to their navy's peacetime role. While they will continue to exploit its existence when possible, it does not appear that the navy is being specially developed as a primary instrument of overseas policy.

This does not, however, mean that we can ignore the possibility of naval confrontation leading to conflict. We know from Gorshkov's definition of a balanced fleet that the navy is prepared to "secure state interests in peacetime." Since an early reference to this task gave it as the particular responsibility of the submarine force,[94] one must assume that combat operations are not excluded.

Naval Arms Control

In his series of articles, Gorshkov uses the fact that there have been persistent attempts to control naval armaments as a further demonstration of the special significance of navies as instruments of state policy in peacetime. He discusses two different types of arms control: naval arms limitation at the end of a war, imposed by the victor on the vanquished; and the continual efforts to limit naval arms racing in peacetime, which can be seen as attempts by the various parties to freeze or to change existing relative strengths. Gorshkov's argument is diffuse and liable to different interpretations, but the most persuasive overall explanation is provided by Franklyn Griffiths. He points

---

94. S. Gorshkov, *Krasnaya zvezda*, 30 October 1962.

out that although at first glance Gorshkov appears to be opposed to arms control as such, he does in fact have several favorable things to say about it.[95] Griffiths concludes that Gorshkov sees arms control as a useful tactic which can: (1) inhibit the naval development of a more powerful adversary; (2) permit an internal redirection of development and building program, to reduce waste on forces and systems which are already over-provided for; (3) seal off areas of competition, such as the seabed; and (4) inhibit the growth of other branches of the armed forces. Griffiths suggests that Gorshkov's general argument runs as follows. Naval claims for additional resources for surface warship construction are being resisted. Russia cannot obtain the necessary naval capability without altering the balance between existing programs and/or achieving cutbacks in the programs of the other armed services. The United States is now basically deterred and SALT can hold offensive weapons frozen at their present level; funds can thereby be released from the SSBN force, from the SRF and the air force, and (by accepting MBFR), from the ground forces in Europe. Arms control will not stop the arms race but it will allow the navy to obtain the appropriations needed for surface construction and general purpose forces.

Running against the grain of this particular interpretation, we have the shelving of the SS-NX-13 tactical ballistic missile in 1973. There is some evidence to suggest that the Soviet navy may have intended to develop the tactical ballistic missile as a major element in its armory against carriers and missile submarines. If this were indeed the case, the Soviet leadership must have decided to sacrifice this tactical system in the wider interests of the SALT accord.

Moving on from naval arms control as an instrument of bureaucratic politics to wider international considerations, we are faced by the reality of Soviet contingency plans. The inventory of inter-continental systems will depend primarily on Soviet requirements to fight and win a war, and only incidentally on what might be considered sufficient to deter an American attack. The number of SLBM will depend on the global concept of operations, on the need for strategic reserves and on the demands of theater warfare. The requirement to defend SSBN bastions and secure their deployment will partly determine the number of general purpose forces. Although defense against Western sea-based systems is moving steadily towards satellite and

---

95. F. Griffiths, "The Tactical use of Naval Arms Control," *SNP*, pp. 637-660.

land-based systems, any surplus naval forces will be needed to bolster area defense. The asymmetry of naval warfare, where like does not fight like, means that as a general rule mutual force reductions make no sense if restricted to single systems, and must be across the board.[96] And this will have to wait on changes in risk and threat perceptions.

The same type of strategic considerations are likely to apply to the limitation of naval operations in distant waters. If this would achieve an effective reduction of the threat to the homeland, that factor would weigh heavily against the possible political gains from a naval presence in the area. But the world wide strategic infrastructure is a different matter and brings us back to perceptions of the risk of world war. While it may be possible to demonstrate a net gain to Russia from a mutual renunciation of certain overseas facilities, the Soviet Union can expect to secure unilateral U.S. withdrawals by generating local political pressures. Meanwhile, there remains the problem of China, which reinforces the need for a strategic infrastructure outside Russia's borders, particularly in the regions of West, South, and South East Asia.

Overview

It is useful to distinguish between the navy as a separate force, the navy as part of the Soviet Union's broader capability, and the navy as an indicator of Soviet intentions and aspirations.

Looking at the navy in isolation, we see that Admiral Gorshkov and his supporters have been dissatisfied with the strength and structure of the fleet. The failure of the military leadership to appreciate fully the navy's role,

---

96. The exception involves submarines, where three opportunities for controlling armaments make up a package that could engender a shift by both sides from war-fighting doctrines (which fuel the arms race), to policies of mutual assured destruction based on MAD. The leverage in this direction is provided by the role of sea-based systems as strategic reserves in world war. Two of these opportunities, the mutual limitation of SSN forces and foregoing the strategic version of Tomahawk, form a conventional arms control arrangement, built on reciprocal fears and interests. The third and more important opportunity involves transferring the Trident SLBM inventory from large nuclear submarines to a spartan diesel submarine force, operating within the protection of American home waters. Besides improving U.S. security, such an initiative would have a catalytic effect, and the synergism of the complete package could induce a major change in attitudes, break the action/reaction cycle, and open up new opportunities in other fields, including strategic weapons. The domestic obstacles to such an initiative are probably insuperable, but if they could be overcome, it would start the essential process of bringing the military postures of the two sides into line with middle-of-the-road political perceptions of threat.

coupled with prejudices concerning particular weapons and platforms, has meant that on the one hand, the fleet is configured for a relatively narrow span of specific missions, with a consequential loss of flexibility, and on the other hand, it lacks the full range of forces to discharge these missions effectively. Gorshkov's judgment parallels official U.S. assessments, which note that the Soviet navy's mission is limited to sea denial, whereas Western navies have the broader strategic mission of securing the use of the sea for a wide range of purposes.

Gorshkov's most persistent criticism has been of the military leadership's inability to grasp the fact that task-specific naval forces, however deadly in themselves, require the support of general purpose forces (particularly surface ships), if they are to be able to accomplish their task.[97] He argues that army dominance has meant that the importance of seapower in a struggle with maritime states has been de-emphasized and a relatively low priority has been given to traditional naval roles in war and in peace. Again, this would seem fair comment, and is supported by official U.S. assessments which continue to credit an overall margin of superiority to the U.S. Navy, while making no allowance for the other forces that the Soviet navy would have to face in the event of world war.

It is difficult, therefore, to agree with Western assertions that the Soviet Union, as a land power, has more naval forces than it "needs." Any attempt to allocate Soviet forces among the various missions that have to be discharged at the onset are seen as essential requirements.[98] What constitutes "enough" is more difficult to determine, but it is relevant that when the procurement of present-day Soviet forces was originally decided, the U.S. Navy was substantially larger than it is today, hence it is largely fortuitous that the Soviet deficit is not greater. What is more, over the last twenty years, the West (i.e., NATO, including France, plus Australia, New Zealand, and Japan) has taken delivery of two to three times as many distant-water surface ships as the Soviet Bloc, and the imbalance is even greater when account is taken of ship size and capability.[99] Until 1967, the West was even building

---

97. Gorshkov alluded to this problem when talking with Western naval officers in late 1971/early 1972. In the course of a discussion on "balanced forces," Gorshkov commented that while it was easy to defend the requirement for submarines, it was much harder to justify the need for surface ships.

98. I make such an attempt in Chapter 4 of *Securing the Seas,* by Paul Nitze and Leonard Sullivan (Boulder: Westview Press, 1979).

99. It must be stressed that this is *not* the same as comparing naval strengths, but it does help one to appreciate the Soviet perception of Western naval programs. The comparative figures are

more nuclear submarines than Russia; after that time the Soviet delivery rate doubled. But the subsequent disparity was concentrated in SSBN, and the production rate of attack boats remained roughly in balance. Meanwhile, the Soviet Union must take account of other countries, including Sweden, Spain, South Korea, and Taiwan. Although both the United States and the Soviet Union must be wary of the large Chinese submarine force, the threat to Soviet interests is more direct. Indeed, a Soviet contingency planner has to allow that, with few exceptions, all the significant navies in the world are at best neutral and most of them must be included in the list of potential adversaries.

This is not to underestimate the very real strength of the Soviet navy, or the threat that it could pose to Western interests. The most disturbing aspect is the size of the submarine force which, by the end of 1985, could include about 230 nuclear units, of which some sixty to seventy might be SSBN, the remainder being configured for ASW and surface attack. There will also be a diesel submarine force of between 135 and 185 boats, of which about seventy will be more than twenty years old. If current building rates persist, the distant-water surface force at that date will comprise six to seven air capable ships, about thirty-five cruiser-sized large anti-submarine ships (all less than twenty years old), about sixty-five of the destroyer-sized units (including twelve conversions, whose hulls will be over twenty-five years old), four rocket cruisers (twenty-one to twenty-three years) and perhaps five command ships. By 1985, satellite surveillance systems will provide the Soviet Command with a real-time global surface plot and may have some capability against submarines. The backfire bomber will have replaced Badger, and this capacity for long-range maritime strike will be supplemented by land-based ballistic missiles.[100] Bulked together in this way, it adds up to a formidable capability, of which the global systems are the most disturbing. But when we divide these forces between the four widely-separated fleet areas, Soviet naval strength falls into perspective, particularly when we take account of the demanding requirements for area defense, and consider the scale of forces that can be assembled against them.[101] Certainly, the West has cause to be concerned. Nevertheless, Gorshkov was justified in his

---

drawn from my chapters on comparative naval building programmes in *SND* (Ch. 12, pp. 144- 50) and *SNI* (Ch. 17, pp. 327- 336).

100. See my "Soviet Naval Programmes" in *SNI* pp. 355- 6.

101. See Nitze and Sullivan, *op cit.*, Chapter 9, "Comparative Force Levels . . ." by Leonard Sullivan for an extended discussion of these points.

complaints about the serious limitations on the navy's capabilities—which flow from its narrow mission structure—its task-specific characteristics, and the paucity of general purpose forces.

However, the fact that Gorshkov's complaints were justified does not mean that the Soviet navy is not reasonably well equipped for the particular role in which it has been cast by the army-dominated leadership. There is a trade-off between effectiveness in discharging a specific mission and general purpose flexibility. The Soviets' concentration on the former at the expense of the latter reflects the ordering of their priorities.

In the event of war with the West, the primary purpose of maritime operations is to contribute to the success of the battle on land. The Soviet Union has a unified strategy for the conduct of such a war, whereby each branch of the armed forces makes a specific contribution, so that the effect of the whole is greater than the sum of its parts. Distant water operations are only "independent" in the sense of being separate from the coastal zone command structure, and the emphasis on a combined arms approach to military problems means that, as a general rule, naval forces are partly dependant on other branches of service in discharging their missions.[102] This of course diminishes their general purpose capability. However, an exception to this rule is provided by the forces assigned to protect Soviet SSBN from enemy interdiction. Since the importance of the SSBN force stems from its invulnerability to attack on Soviet territory, its security cannot be allowed to depend on shore-based air support and other weapon and sensor systems. This means that naval forces must possess an equivalent capability, and justifies the procurement of a ship like the Kiev anti-submarine carrier. The mission of strategic strike also breaks the general rule that naval operations are subordinated to the requirements of the land battle. The importance of the SSBN force as a strategic reserve and the increased security that command of the Norwegian Sea would provide would appear to have generated a new requirement to seize key islands and stretches of coast at the onset of war. The effect of these two exceptions is to improve the Soviet navy's general purpose and projection capabilities.

Turning to consider the navy as a part of the Soviet Union's broader capability in peacetime, we start from the fact that Russia is Mackinder's heartland. It stretches from Western Europe to Japan and borders twelve

---

102. The first part of this paragraph draws on Welander, *op. cit.*, note 83.

states, with another seven directly accessible across short stretches of water. The country spans 170° of longitude (a full 180° if we include the Warsaw Pact), and thus looks south at half the globe, where 85 percent of the world's population lives within 300 miles of Soviet territory. Western Europe, North Africa, the Middle East, and the Indian subcontinent are all within 2000 miles of the Soviet Union, while the territories of its national security zone are continuous. With this type of strategic access, it is to be expected that Soviet perceptions of the navy as an instrument of overseas policy would have differed from America's.

At the time of the post-Stalin shift toward a more outward-looking foreign policy, the Soviet Union chose to cut back its naval programs and shift resources to commercial shipbuilding, relying on trade, aid, and arms supply to circumvent the West's world-wide maritime preponderance, rather than trying to compete with it. Despite the decision in 1961 that the navy should move forward in strategic defense, the political exploitation of this Soviet presence in distant sea areas developed very slowly, and it was not until after the Arab-Israeli war in June, 1967 that the process took hold. This delay can probably be ascribed to a preoccupation with the war-related role, coupled with uncertainty as to U.S. reactions to this intrusion into Western-dominated water. Furthermore, in the 1960s the Soviet navy still lacked a true distant-water capability, and it was only after access to Egyptian ports had been granted in the wake of the June war that the Soviets were able to maintain an effective naval presence in the Mediterranean. Thereafter, the political employment of Soviet naval forces steadily increased, but it seems likely that forward deployment was still seen as an interim phase, pending the development of global systems designed to counter Western strike carriers and SSBN. However, by 1971-72, Gorshkov and his supporters were advocating a more assertive use of naval forces for political purposes, in opposition to those whose primary concern continued to be the danger of war with the West. This was part of a broader argument about the importance of navies in peace and war, and while the final outcome is still unclear, it appears that Gorshkov's call for an improved surface capability to support the SSBN force was heeded. This has the co-lateral effect of increasing the navy's peacetime capability to project power in distant sea areas, although the extent to which these forces can be diverted from their primary areas of concern remains to be seen.

As we come to the end of the 1970s, the naval prognosis must remain uncertain until we know what the new building programs will bring. On past practice, a new generation of nuclear submarines was to be expected in 1978, and while one such class has duly appeared,[103] the full picture is far from clear.[104] Similarly, the two main surface warship programs are nearing the end of their production runs and new classes are awaited.[105] When we have been able to evaluate this new construction, we will be in a position to assess the extent to which Gorshkov's advocacy was successful and to plot the future development of the Soviet navy.[106] Meanwhile, we can usefully consider what the past record of Soviet naval developments has to suggest about the wider questions concerning the Soviet Union's military posture and its willingness to wage nuclear war.

Professor MccGwire will address these questions of the Soviet approach to nuclear war in the next issue of *International Security (Vol 4, No 1)*.

---

103. *The New York Times*, 20 March 1979, p. 7 This is probably the Alpha that has been under development since the early seventies.

104. There has been a single mention of a 60,000 ton nuclear powered surface unit being built at Severodvinsk (*The New York Times*, 17 March 1979, p. 3) but this has yet to be confirmed.

105. *Krivak* began delivery in 1970. *Kara* began in 1972, but the *Kresta II* program completed at Zhdanov by 1978, and a new cruiser-sized class could be expected. A new 25,000 ton 650 foot well-armed surface unit is building at Leningrad (*The New York Times*, 20 March 1979, p. 7) that could be intended as a Command ship.

106. It is relevant that at this same stage ten years ago, official U.S. estimates put the new nuclear submarine building programs at twenty p.a., whereas they turned out to be only ten p.a. See *Status of Naval Ships*, Seapower Sub-Committee Report dtd 19 March 1969, Armed Services Committee, House of Representatives.

Part II:
Naval Technology

Technology and the Evolution of Naval Warfare

Karl Lautenschläger

$\mathbf{T}$he perennial concern of military planners is that technological surprise will give an opponent a decisive advantage in event of war. Technological developments combined with tactical innovation can bring about fundamental change in fighting capabilities. The concern is over how to anticipate such change, particularly if it comes suddenly.

This paper suggests revising some current assessments of naval developments on the basis of recent historical trends. It reconsiders the evolution of warfare at sea since 1851, when technology produced fundamental changes in capabilities and tactics every ten to fifteen years. In an age of systems analysis it may seem a florid diversion to review a century of history before assessing the present and speculating about the future. Yet, debate over naval policy is encumbered by fanciful history that is more popular than useful. Therefore, reconsideration of the long term could bring needed perspective to the problem. The results are two: the historical review provides case studies in how technology can affect warfare, and the analysis highlights basic trends that could be useful in predicting future developments. Together, these form a conceptual approach that breaks with the prevailing method of projecting current naval trends into the future.

That prevailing method stresses overt physical features and the prospect of a revolutionary breakthrough in naval technology. Changes in the external appearance of ships, aircraft, and hardware dominate our perception of past and present technical developments in the naval sphere. If new systems look exotic, it is assumed that they must have important new capabilities. If the Soviets are building larger warships, then their naval capability is said to be expanding dramatically. The cruiser *Kirov* and the Typhoon-type submarine are cases in point. They are very large compared to their predecessors, but they have been a long time coming and they represent evolutionary rather

Karl Lautenschläger is a Staff Defense Analyst at the Los Alamos National Laboratory. He has been an Advanced Research Scholar at the Naval War College and a Visiting Faculty Member at the Fletcher School of Law and Diplomacy, and was a Naval Officer for five years with two combat deployments to the Tonkin Gulf.
Graphic artwork is by Dennis Olive.

International Security, Fall 1983 (Vol. 8, No 2) 0162-2889/83/020003-49 $02 50/1

than revolutionary change in capability.[1] We have also misinterpreted the nature and significance of technological change in contriving a single spectacular breakthrough at each stage. This bias is also persistent. Today, respected professionals worry openly about a single breakthrough in antisubmarine warfare technology that will seemingly make the oceans transparent.[2]

The thesis of this paper, by contrast, is that the idea of a single technological breakthrough in the military sphere is popular mythology. Important advances in naval weaponry have not come with the introduction of spectacular new technology, but with the integration of several known, often rather mundane, inventions. Developments in warship and aircraft design have tended to be evolutionary rather than revolutionary. But there have been several instances when combinations of technology were brought together to produce rapid change so significant that all existing combat fleets had to meet the new standard in fighting capabilities or remain hopelessly ineffective. The extent of these new capabilities was seldom reflected in obvious physical changes. In sum, the key to identifying important developments for the future is to concentrate on the synthesis of different technologies and how that synthesis can produce fundamental change in mission capabilities.

Another area of confusion in today's assessment of technological developments is in the use of benchmarks of change. The many familiar measures range from conception of a scientific principle and validating experiments to practical civilian applications or operational military capability. These measures are poorly defined and often mixed indiscriminately in making comparisons. The case studies of this historical survey use only three bench-

---

1. The *Kirov* is a very large guided missile cruiser with a hybrid nuclear/oil-burning propulsion plant. She carries twenty anti-ship missiles with an estimated range of 300 nautical miles (nm) and an impressive array of defensive systems. The *Kirov* first went to sea in 1980. Only one sister ship is known to be under construction. Nearly two decades earlier, in 1962, the first of four Kynda-class missile cruisers went into service, armed with sixteen 220 nm anti-ship missiles and a large array of defensive systems. See Jean Labayle Couhat and A.D. Baker III, *Combat Fleets of the World 1982–83* (Annapolis, Md.: Naval Institute Press, 1982), pp. 584, 616–619, 625–626. James W. Kehoe and Kenneth S. Brower, "Their New Cruiser," *U.S. Naval Institute Proceedings*, No. 106 (December 1980), pp. 121–126. The first Soviet Typhoon-class ballistic missile submarine went to sea in 1981. She will carry twenty 4,000 nm missiles when operational in 1983 or 1984; converted Poseidon boats carry 16. Couhat and Baker, *Combat Fleets*, pp 584, 602, 696, 720–721.
2. William J. Perry, "Can't Miss Weapons—Revolution in Warfare," *U.S. News and World Report*, September 8, 1980, p. 61; Norman Polmar, "Soviet ASW: Highly Capable or Irrelevant?" *International Defense Review*, No. 12 (1979), p. 729.

marks: 1) the first practical demonstration of the principle, 2) the first complete set of basic components put into service, and 3) the first complete combat unit such as a squadron of ships or aircraft. The comparative time span used here for service adoption is measured from the first set of basic weapon components to the first combat unit. One might be called initial operational capability, familiar as IOC; the other represents a deployed fighting unit or actual operational capability. Each benchmark is a world first, without regard to nation.

In order to develop these concepts in the space of this paper, the focus here will be on the evolution of battle fleets, the principal combat arm of big navies. This does not presume that naval strategy must be based on the ideas of Alfred T. Mahan, Philip Colomb, and Julian Corbett. Nor does it indicate that a "balanced" fleet necessarily includes capital ships. The proper balance of components in a force depends on its intended function.

The term "combat" is used here in its more restricted sense to mean direct engagement of at least moderate duration. Both opposing forces concentrate their offensive power as well as exercise the capacity for sustained defense. This is in contrast to what the Soviets call "strike warfare," which assumes a short one-way assault on land or sea objectives and depends upon surprise, because the strike force has little or no means of defense. Ballistic or cruise missile submarines are ideal components of strike as opposed to battle forces. Commerce destruction is really another form of strike warfare, since the attacker has little means of defense and his objective is unarmed merchantmen. Commerce protection emphasizes defense of numerous dispersed convoys, as opposed to concentrated offense, and the objective is just as readily gained by avoiding contact with the enemy altogether.

The focus here on combat at sea necessarily ignores these other essential forms of naval warfare, as well as amphibious operations, mine warfare, and coastal defense. However, the evolution of the battle fleet provides a long-range view of how technology produces change in combat capability. Naval warfare in general is sensitive to changes in technology, because it is platforms as well as weapons that are necessary for combat at sea. Whereas armies have historically armed and supported the man, navies have essentially manned and supported the arm. The battle fleet is of particular interest because it is the one important element of navies that existed before the industrial revolution but continues to have important functions in international power politics today. Even in the age of thermonuclear weapons and

	IOC – AOC	
Ocean Area C³I	1974–c.1987	├──┤·
Guided Ordnance	1956–1961	├──┤··········
Tactical Nuclear Weapons	1953–1961	├──┤·········
STRATEGIC NUCLEAR WARFARE	1945–1956	├──┤ ·········
Aircraft & Electronics	1940–1943	├┤········
Three–Medium Combat	1918–1932	├───┤······· ···· ··
Integrated Systems	1907–1914	├─┤·········
High Seas Combat	1896–1900	├┤· ········· ··
Fuel Dependence	1873–1889	├─────┤········ ·····
Long-Range Weapons	1860–1863	├┤·················
Auto Maneuver	1851–1854	├┤················

Los Alamos 1850 1900 1950 1990

intercontinental delivery systems, carrier battle groups have played prominent roles in superpower interaction, in regional conflict, and in global relations.

This is not a treatise on the workings of naval technology. It could not even presume to catalog all of the relevant technologies. It is an attempt to discover how technology has changed combat capabilities in elemental yet significant ways. Rather than looking first at technology, one must determine which changes in operating capability had the most far-reaching effects. These discontinuities in the otherwise gradual evolution of naval warfare provide points of reference for identifying sets of technology that were essential for the change and for describing the dynamics of the change.

Auto Maneuver

The steam engine and screw propeller ushered in what Bernard Brodie called the machine age in naval warfare.[3] For centuries, technology had made naval warfare possible by taking arms and men to sea. But the technology and tactics adopted during the early 1600s did not change for two hundred years. The introduction of steam-powered ships-of-the-line brought the first of several fundamental changes in fighting capabilities that were to occur every ten to fifteen years.

The transition from sailing warships to steam ships-of-the-line exemplifies two conditions present at the start of the few rapid transitions in the otherwise gradual evolution of modern fighting fleets. First, it came rather suddenly, after technical inventions that had existed for some time were refined and combined. And second, it introduced new dimensions to the conduct of naval warfare.

The essential technologies for a self-propelled battle fleet were a reliable steam engine of several hundred horsepower and the screw propeller. The principles of the steam engine had been known since ancient times. Engines had wide use in the mining industry by 1725, but the first ship did not steam across the Atlantic until 1838. Paddle wheels used on the first steamships were large and easily smashed by gunfire, and the necessary drive mechanism was delicate and exposed. Thus, the first naval steamships were ancil-

---

3. Bernard Brodie, *Sea Power in the Machine Age* (Princeton: Princeton University Press, 1941), pp. 17–91. This work is impressive for its penetrating insights and useful concepts, but more recent research revises many of its historical details

lary craft such as gunboats and dispatch vessels. Perfection of the screw propeller allowed both power plant and drive train to be placed below the waterline where they were relatively safe from gunfire. The principle of the screw propeller was first advanced by Daniel Bernoulli in 1752. Yet, it was not refined sufficiently for naval use until the 1840s.[4]

Once the technology was refined and adopted, however, the transformation of fighting fleets was rapid. The first steam battleship, HMS *Sans Pareil*, entered service in 1851. In 1854, the British battle fleet sent to fight Tsarist Russia in the Baltic had 10 steamers out of 14 ships-of-the-line. The next year it became the world's first all-steam fighting force.[5] The change would not have been more dramatic if, after the *Nautilus* (SSN-571) first went to sea in 1955, the active U.S. submarine fleet had been converted to nuclear propulsion before the end of President Eisenhower's second term. As it was, the United States, although the leader in the field, did not have nuclear propulsion in even half of its submarines until 1969 when 82 out of 161 in commission were nuclear powered.[6]

The new dimension to naval warfare was maneuver independent of the wind for extended periods. Steam completely changed fundamentals of battle tactics that had prevailed for two centuries. It made existing fleets of sailing battleships obsolete, and it introduced a basic characteristic to naval weapon platforms that persists to this day. Whether surface, subsurface, or airborne, their fuel-burning engines make tactical and strategic mobility two different problems. Since tactical mobility influences combat effectiveness, its critical elements are speed and maneuverability. Strategic mobility, on the other hand, determines the distance and duration that a force can be deployed from its base. Endurance then becomes more important. Self-propulsion gave independence from the wind only for tactical mobility at first. Fuel consumption was too high and coal fuel too bulky to permit long-range steaming. In steam ships-of-the-line and seagoing ironclads that followed, the solution

4. Georges G.-Toudouze et al., *Histoire de la Marine* (Paris: Baschet, 1966), pp 323–369; John Bourne, *Treatise on the Screw Propeller, Screw Vessels and Screw Engines, as Adapted for Purposes of Peace and War*, 3rd ed. (London: Longmans, Green, 1867); David Brown, "The Introduction of the Screw Propeller into the Royal Navy," *Warship*, No. 1 (January 1977), pp. 59–63.
5. Great Britain, Royal Navy, *The Navy List*, for the years 1846 through 1855. Steam ships-of-the-line do not include eight sailing battleships converted to "steam block-ships" of low power and reduced rig. See Hans Busk, *The Navies of the World* (London: Routledge, Warnes, and Routledge, 1859), p. 58.
6. Raymond V.B. Blackman, ed., *Jane's Fighting Ships, 1969–70* (New York: McGraw-Hill, 1969), pp. 388–391, 439–457.

to the problem was having two propulsion systems. Steam was used for tactical maneuver, but sailing rig was retained for movement over long distances.

The separation of tactical and strategic mobility factors has persisted. It is significant that today over two-thirds of the surface combatants in the Soviet navy lack the endurance, not to mention the seakeeping characteristics, necessary for ocean operations. The entire navies of many smaller countries are constrained by the endurance of their ships. In terms of strategic mobility, they are limited to a coast defense role.

Long-Range Weapons

The next transformation was essentially a revolution in weaponry. Ironclads replaced wooden ships-of-the-line as the mainstay of fleets, but ordnance, not armor, brought the important change in naval warfare at this stage. Americans think of the *Monitor* and the *Merrimac* as the ships that ushered in the age of ironclads.[7] Europeans consider that explosive shells made wooden battleships vulnerable and obsolete. They see the adoption of iron armor in *la Gloire* and HMS *Warrior* as an antidote to the shell gun and thus the critical advance of the period.[8]

With today's perspective, we can see that neither viewpoint captures the essence of the technical revolution of the 1860s. Important as ironclads were in the American Civil War, not one used on either side was suitable for navigating, let alone fighting, on the high seas. They were river gunboats and floating batteries, not the components of a first-class battle fleet, nor the engines of transition in the world's big navies. The traditional European interpretation is equally skewed. Iron armor was not adopted in Europe to defeat the shell gun, nor was it the first means of protecting a warship from the effects of gunfire. Shell guns had been adopted and used in action decades before anyone thought seriously of retiring the wooden three-decker. Spherical explosive shells gave wooden warships a better capability to start lethal fires in one another, but they were unreliable and represented a fire

---

7 Brodie, *Sea Power*, p. 171; Harold and Margaret Sprout, *The Rise of American Naval Power, 1776–1918* (Princeton: Princeton University Press, 1939), pp. 158–161.
8. William Hovgaard, *Modern History of Warships* (London: E. and F.N. Spon, 1920), pp. 4–8; James Phinney Baxter III, *The Introduction of the Ironclad Warship* (Cambridge: Harvard University Press, 1933), pp. 17–32 Baxter provides an excellent review of European writing on this matter, but he focuses on the shell gun at the expense of developments in rifled ordnance.

danger themselves when stowed in a magazine. The bulk of ammunition carried by capital ships continued, therefore, to be solid roundshot.[9] Armor was not really new either. Thick oaken sides afforded considerable protection against solid projectiles when fired at normal battle ranges of 200 to 600 yards. Wooden ships-of-the-line were in effect "armored" and had been for more than a century.[10]

It was rifled ordnance that changed fighting capabilities dramatically and, at the same time, forced the adoption of ferrous armor. The new type of naval artillery had significantly greater range, accuracy, and penetrating power than the old smoothbores. The concept of rifling had been applied to small arms since the time of Columbus, but the technology necessary to produce rifled artillery was not available until the mid-nineteenth century. Improvements in metallurgy made it possible to build guns that were both much larger and able for the first time to withstand higher internal pressures inherent in efficient rifles. Advances in machining techniques made it possible to reduce irregularities and thus "windage" in the bore.[11] The rifle fired a pointed, cylindrical projectile with three times the mass of a cannon ball of similar diameter. The spin imparted to the projectile gave it stability in flight. Greater mass and streamlining gave the rifle projectile more momentum by an order of magnitude and thus far greater hitting power. Stability in flight and momentum meant much greater accuracy.[12] The effect was to extend the maximum fighting range at sea from 600 yards to between 1,500 and 2,000 yards.

Iron armor became essential because wood could not stand up to the smashing power of rifled ordnance. Iron plates were first used as hull protection on floating batteries bombarding stone forts in 1855 during the Crimean War.[13] With the success of these batteries, it was natural that the most important units of battle fleets would be given the advantage of ferrous

---

9. Howard Douglas, *A Treatise on Naval Gunnery*, 5th ed. (London: John Murray, 1960), pp. 184–196, 240–340, 633–636.
10. Brodie, *Sea Power*, pp. 172–174.
11. R.A. Stoney, "A Brief Historical Sketch of Our Rifled Ordnance from 1858–1868," *Minutes of Proceedings of the Royal Artillery Institution*, No. 6 (1870), pp. 89–119; Charles Leopold Gadaud, *L'Artillerie de la Marine Francaise en 1872*, 2nd ed. (Paris: Arthus Bertrand, 1872), pp. 25–80.
12. Douglas, *Treatise on Naval Gunnery*, pp. 234, 236; Andrew Noble, *Artillery and Explosives* (New York: E.P. Dutton, 1906), pp. 499–501.
13. G. Butler Earp, ed., *The History of the Baltic Campaign of 1854* (London: Richard Bently, 1857), pp. 166, 184–185, 188–197; Edgar Anderson, "The Role of the Crimean War in Northern Europe," *Jahrbucher fur Geschichte Osteuropas*, Vol. 20 (March 1972), pp. 43–45, 61; Baxter, *Introduction of the Ironclad Warship*, pp. 69–91.

plating eventually. Tests of the new rifled ordnance speeded the process and convinced the French in 1857 that iron armor was the only way to keep their ships afloat in combat. The next year they began building seagoing ironclads, and the British soon followed.[14] Since the largest guns could perforate the armor of most battleships, iron armor was about as effective against rifles as oak had been against smoothbores. The most significant aspect of this transition, then, was a dramatic improvement in the range of weapons.

Fuel Dependence

For the next three decades, technical change in navies was evolutionary. Tactical mobility was provided by steam propulsion, but strategic mobility continued to be under sail. The first of a few seagoing monitors went into service in 1873.[15] But another 16 years passed before they were joined by entire squadrons of their all-steam cousins.[16] Guns gradually grew in size and were mounted in fewer numbers. Turntables allowed these monster pieces to be trained inside armored turrets or fixed barbettes for protection. Armor became thicker and was made of simple steel or a sandwich of steel and iron called compound armor.[17] With these features the "armorclad" gradually supplanted the broadside ironclad with its sailing rig.

Self-propulsion now gave independence from the wind for both tactical and strategic mobility, but this meant complete dependence on fuel. In essence, fuel dependence made logistics an important aspect of naval warfare. A fleet's endurance now depended on its fuel supply. Its area of operations depended on the proximity of bases. Complex munitions, diverse provisions, and spare parts have since added to the logistics problem, but

---

14. Paul M. Dislere, *La Marine Cuirassee* (Paris: Gauthier–Villars, 1873), pp. 6–20; Oliver Guiheneuc, "Les Origines du Premier Cuirasse de Haute mer a Vapeur," *La Revue Maritime*, No. 100 (Avril 1928), pp. 459–82; No. 104 (Aout 1928), pp. 183–202; Oscar Parkes, *British Battleships*, 2nd ed. (London: Seeley Service, 1966), pp. 2–6, 11–24.

15. Thomas Brassey, *The British Navy: Its Strength, Resources, and Administration* (London: Longmans, Green, 1882), Vol. 1, pp. 343–359; J.W. King, *Report of Chief Engineer J.W. King, United States Navy, on European Ships of War and Their Armament, Naval Administration and Economy, Marine Constructions, Torpedo-Warfare, Dock-Yards, Etc*, 2nd ed. (Washington, D.C.: U.S. Government Printing Office, 1878), pp. 53–69; Parkes, *British Battleships*, pp. 191–202.

16. Commissioning dates and squadron assignments compiled from *The Navy List* and Parkes, *British Battleships*.

17. Hovgaard, *Modern History of Warships*, pp. 43–51, 54–69, 73–80, 456–464; F. Singer, "A Graphic History of Armor Protection and Distribution on War Vessels," U.S. Office of Naval Intelligence, *General Information Series*, No 8 (June 1889), pp. 82–86.

fuel first made it significant. Consumption of fuel for both tactical and strategic mobility also brought about the trade-off between speed and endurance. Most warships built since the 1880s and all naval aircraft compromise one for the other. Nuclear power and underway replenishment have reduced design sacrifices for long endurance in some warships. However, the trade-off is even more critical in aircraft, and they have become essential elements of naval forces.

High Seas Combat

Between 1895 and 1900, a series of technical innovations improved the fighting capabilities of fleets dramatically. Technology transformed the battleship into a blue water gun platform. Chemical propellants and scientific gunnery made it lethal at four times the fighting range of its armorclad predecessor. Efficient steam engines and lighter armor made of steel gave cruisers the speed advantage they needed to serve as scouts for the battle fleet. The wireless telegraph obviated the need for visual communication. A commander could concentrate widely dispersed units of his fleet, and cruisers could report the findings of their reconnaissance from over the horizon. Finally, the rise of a novel threat to the battle fleet, in the form of surface torpedo craft, was neutralized with adoption of quick-firing torpedo defense batteries, the first specialized defensive weapons in warships.

Four sets of technology transformed the battleship by producing 1) high velocity heavy ordnance, 2) telescopic gunsights, 3) face-hardened, alloy steel armor, and 4) quick-firing medium and light caliber guns. High velocity guns were a significant advance over earlier naval artillery, because they gave heavy projectiles the momentum required for high striking velocity and small dispersion at much greater range. This meant substantial gains in both destructive power and accuracy. The essential technology was propellant made from a chemical compound of nitrogen and cellulose. The new chemical propellants were much more efficient than black powder, which is a mechanical mixture. Nitrocellulose propellants produced muzzle velocities 50 to 100 percent higher than had been possible with black powder.[18] Although

---

18. J. Corner, *Theory of the Interior Ballistics of Guns* (New York. John Wiley, 1950), pp. 24–25, Noble, *Artillery and Explosives*, pp. 405–438, 462–481; Thomas J. Hayes, *Elements of Ordnance* (New York: John Wiley, 1938), pp. 1–28, Charles Singer et al., *A History of Technology* (Oxford: Clarendon Press, 1958), Vol. 5, pp. 284–298.

the development of chemical propellants is usually lost in the myriad of inventions made in the 1880s, it probably ranks with steam and iron in revolutionizing naval warfare. Naturally, the potential of chemical propellants could only be realized when they were introduced in combination with other innovations such as long gun tubes and more efficient mountings.

The second set of technologies to transform the battleship brought telescopic fire control to naval gunnery. The limitations of the human eye in aiming a gun using an open sight precluded accurate shooting beyond about 2,000 yards. Normal ironclad and armorclad fighting range was considered to be 1,500 yards in spite of vast improvements in ordnance. The telescopic sight enabled gunners to shoot with consistent accuracy out to 6,000 or 7,000 yards.[19] It was the first dramatic improvement in naval fire control and a portent of how future advances in weaponry would have to be accompanied by commensurate improvements in target acquisition and weapon control.

Technological advances in the area of metallurgy finally allowed a well-protected warship to fight effectively on the high seas. Before 1900, all major naval battles but one took place within sight of land, usually in sheltered waters.[20] Armorclads, like sailing ships, steamships, and ironclads before them, could fight only on moderate seas. Earlier types had to close their broadside gunports in rough water. Armorclads were hampered by similar limitations. Many were built with low freeboard because of the great weight of their armor. This meant that waves and spray interfered with the loading and aiming of their guns. The French built high freeboard armorclads, but their upper sides were unprotected, making them liable to flooding and capsizing when waves poured into shell holes.[21]

The advent of true ocean-fighting capability, hardly noted in traditional histories, came with the revolution in armor technology. The introduction of

---

19. William F. Fullam and Thomas C. Hart, *Textbook of Ordnance and Gunnery*, 2nd ed. (Annapolis, Md.: United States Naval Institute, 1905), pp. 248–256; Bradley A. Fiske, "Progress in the Naval Use of Electricity," U.S. Office of Naval Intelligence, *General Information Series*, No. 14 (July 1895), pp. 119–122; Bradley A. Fiske, *From Midshipman to Rear Admiral* (New York: Century, 1919), pp. 123–128, 177–180, 213; Percy Scott, *Fifty Years in the Royal Navy* (London: John Murray, 1919), pp. 30–32, 81–82, 92–93.

20. The "Glorious First of June" (Battle of Ushant), May 28 to June 1, 1794, was fought between British and French fleets in the Atlantic about 500 miles west of Brest, France. For a survey of naval battles, see Helmut Pemsel, *Atlas of Naval Warfare* (London: Arms and Armour Press, 1977).

21. Hovgaard, *Modern History of Warships*, pp. 54–56; Parkes, *British Battleships*, pp. 189–194, 199, 354, 357, 373–374; Frederic Manning, *The Life of Sir William White* (London: Murray, 1923), pp. 290–291.

alloy steel armor provided a tougher armor material, and cementing (carburizing) made the face harder. It was, in effect, lighter, because it was twice as effective as an equal thickness of compound armor. Now the hull and ordnance of a high-freeboard battleship could be protected adequately without a dangerous excess in topweight. High sides and high-mounted guns allowed aiming without interference from ocean spray. Harvey process armor was introduced in 1890, and Krupp cemented alloy steel followed in 1895.[22] Two years later, Great Britain had in service a squadron of battleships employing cemented steel armor in the new scheme of protection. No nation could challenge British naval supremacy without adopting the new technology.

Quick-firing guns combined a fourth set of technologies that transformed the battleship. They were made possible with the introduction of fast-operating breech mechanisms (1866), cartridge cases of cannon caliber (1877), hydraulic recoil mechanisms on pivot mountings (1881), and smokeless (chemical) propellants (1886).[23] From the 1890s until the advent of dreadnought battleships, the main battery of a capital ship consisted of both heavy guns and quick-firers. Medium caliber pieces (5- to 8-inch) augmented the offensive capability of the new type of battleship. They could wreck another battleship's upperworks with a torrent of high explosive shells, while the heavy guns punched holes in its heavy armor. Light quick-firing guns represented something fundamentally new in warships. They formed a specialized defensive battery intended specifically to counter assaults by craft which depended on surprise and stealth. The impetus for developing quick-firing guns for naval use had in fact come from the torpedo boat. The lighter weapons (4-inch and below) provided big ships with an effective defense against this threat.

The rest of the fleet underwent a transformation at about the same time. The essence of the change was in reconnaissance capabilities, and it came about in two ways. First, improvements in propulsion and armor technology

---

22. Roland I. Curtin and Thomas L. Johnson, *Naval Ordnance* (Annapolis, Md.: United States Naval Institute, 1915), pp. 323–338; "Armor," U.S. Office of Naval Intelligence, *General Information Series*, Vol. 5 (June 1886), pp. 239–245; Vol. 6 (June 1887), pp. 322–331; Vol. 10 (July 1891), pp. 279–337; Vol. 11 (July 1892), pp. 271–312; Vol. 13 (July 1894), pp. 134–154, Vol. 14 (July 1895), pp. 83–92; Vol. 19 (July 1900), pp. 175–194; Vol. 20 (July 1901), pp. 247–267; Vol. 21 (July 1902), pp. 121–134.

23. H. Garbett, *Naval Gunnery* (London: George Bell 1897), pp. 136–147, 159–187; Great Britain, War Office, *Treatise on Ammunition* (London: His Majesty's Stationery Office, 1905), pp. 111–138.

finally gave cruisers the speed they needed to maintain contact with the enemy at a distance and to escape if pursued. Second, the invention of wireless telegraphy enabled scouts to report to their fleet command without steaming all the way back to within visual signaling distance of the flagship.

Cruisers were the workhorses of late nineteenth-century navies. They served on the foreign stations of Europe's worldwide empires, and showed the flag on extended cruises. Had there been a war between the major naval powers, the French and Russian navies intended their cruisers for destroying commerce, while the British navy would have used its cruisers to protect shipping and to hunt commerce raiders.[24] The mission absent in cruisers since the age of sail was reconnaissance. In early steam navies, technology prevented cruisers from serving usefully with the battle fleet, because they lacked the speed advantage necessary to serve as the eyes of the fleet.[25]

By 1882, ship propulsion plants were reliable and powerful enough for sustained high-speed steaming. The advent of steel armor and the curved protective deck made it possible to protect cruisers against the gunfire of other cruisers without excessive weight. In 1889, the Armstrong firm of Britain delivered the first of its many fast "protected" cruisers.[26] From that time until the rise of fast carrier task forces during World War II, cruisers could be built with a 20 percent speed advantage over existing battle fleets. Although cruiser designs continued to stress commerce destruction or protection roles for a time, navies could employ their faster cruisers as scouts for the battle fleet. In 1901, the Royal Navy introduced armored cruisers with heavy ordnance and extensive armor belts. These cruisers were specifically intended for both fleet reconnaissance and as a fast reinforcing wing of the battle line.[27]

The wireless telegraph, however, was ultimately far more significant than improved cruiser design for fleet operations. In 1899, eleven years after the results of Heinrich Hertz's experiments with electromagnetic waves were published, the Royal Navy tried out its first wireless sets on fleet maneuvers.

---

24. Theodore Ropp, "Development of a Modern Navy: French Naval Policy, 1871–1904" (Ph.D. dissertation, Harvard University, 1937), pp. 33–35, 68–72; Brassey, *The British Navy*, Vol. 1, pp. 477–509, 513–522.

25. Robert Gardiner et al., eds., *Conway's All the World's Fighting Ships 1860–1905* (Greenwich: Conway Maritime Press, 1979), pp. 41, 61.

26. Peter Brook, "The Elswick Cruisers," *Warship International*, Vol. 7 (1970), pp. 154–176; Vol. 8 (1971), pp. 246–273; Hovgaard, *Modern History of Warships*, pp. 174–178.

27. Manning, *William White*, pp. 356–368; Parkes, *British Battleships*, pp. 441–450.

The next year, the British fleet was the first to be equipped with wireless telegraph, beginning with an order for 32 Marconi sets. The other major navies soon followed. At the turn of the century, all ships fitted with wireless equipment were capable of communication with other ships or shore stations at ranges of 50 to 70 miles.[28]

With the wireless link, cruisers became forward-based sensor platforms that could provide the fleet commander with information about the movements of enemy forces. The system was crude by today's standards—its sensor was merely human eyesight aided by binoculars, and, at its best, the communication link took many minutes to relay information. Yet this was a fundamentally new capability. For the first time, a tactical commander was served by immediate, continuous reconnaissance, extending far beyond the limit of his vision. Such a capability is considered to be essential in today's naval operations.

Surface torpedo craft also came into use during the late nineteenth century. The torpedo boat was one of those novel weapons that comes along every decade or so; at such times, it becomes popular to declare that the novel weapon can make battle fleets obsolete. The idea that new technology can make battle fleets suddenly obsolete has been fashionable many times. In most of these cases, when technology has been employed to produce a small, inexpensive device that can sink a capital ship, the mere possibility of sinking those big ships is often assumed to make them immediately obsolete. However, probability, as opposed to theoretical possibility, brings the operational utility of the novel weapon into question because probability depends on numerous factors related to operating conditions, fleet defenses, and tactics.

In the case of the torpedo boat, there is no doubt that a new dimension was added to naval warfare—but the demise of the battle fleet was never a real possibility. Although a torpedo certainly could be potent, it had to be launched at close range, well within the lethal reach of a warship's gun batteries. This meant that the torpedo carrier had to be small to avoid early detection and fast to avoid being hit by gunfire. These requirements precluded the design of robust, seaworthy craft. The result was that torpedo boats could threaten fleet operations in confined waters along a coast. But

---

28. Arthur Hezlet, *Electronics and Sea Power* (New York: Stein and Day, 1975), pp. 26–52; W.L. Howard, "Wireless Telegraphy," U.S. Office of Naval Intelligence, *General Information Series*, No. 18 (November 1899), pp. 277–287; Linwood W. Howeth, *History of Communications-Electronics in the United States Navy* (Washington, D.C.: U.S. Government Printing Office, 1963), pp. 11–112.

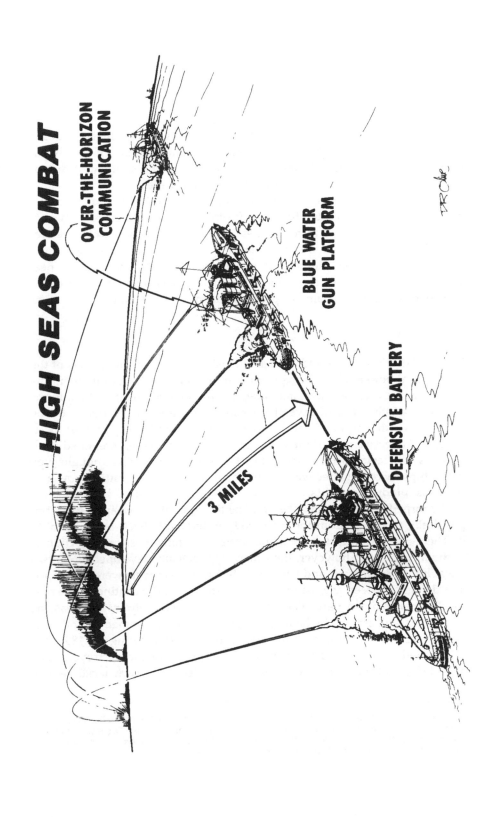

HIGH SEAS COMBAT

OVER-THE-HORIZON
COMMUNICATION

BLUE WATER
GUN PLATFORM

DEFENSIVE BATTERY

3 MILES

the fleet adopted quick-firing guns as a defense, and with new high-freeboard battleships, it moved farther out to sea where it could operate effectively but torpedo boats could not.[29] This is the period when the strategy of distant blockade began to replace the traditional close blockade.

Thus, in the late 1890s, the fleet became a blue water combat force. Its primary gun platforms could function effectively on the oceans in rough seas, and they were effective at four times the range thought possible before. Extended reconnaissance was possible using fast cruisers and wireless communication. This over-the-horizon communication link also enabled fleet commanders to disperse elements of their forces over many miles of ocean and concentrate them again at opportune moments for action.

Integrated Systems

In the relatively short period from about 1907 to 1914, the battle fleets of the major naval powers went through another fundamental transformation. The essence of the change was system integration—integration both of platforms with one another and of several systems aboard a single platform. In the former case, the torpedo boat destroyer matured and joined the fleet as an offensive/defensive arm of the battle line. In the latter case, integrated components of a centralized fire control system gave the new dreadnought-type battleships twice the effective fighting range and twice the hitting power of the latest predreadnoughts.

The torpedo boat destroyer became the first specialized weapon platform, physically separated from the main body of the battle fleet, but integral to it. The destroyer, in fact, took its name from its initial mission as a counter to the torpedo boat; however, the generic type actually developed along two lines. In the Royal Navy and later the U.S. Navy, destroyers were developed as a defensive screen against torpedo boat attacks. In the continental navies, the German navy in particular, destroyers were primarily intended as the torpedo assault arm of the battle fleet. Each approach assumed the other as a secondary role, but the differing emphasis was clear. During the 1890s the British and German navies led in the development of the destroyer. Each intended to integrate the type into its battle fleet, and each faced serious

---

29. S.A. Staunton, "The Naval Maneuvers of 1888," U.S. Office of Naval Intelligence, *General Information Series*, No. 8 (June 1889), pp. 57–58; Edgar J. March, *British Destroyers* (London: Seeley Service, 1966), pp. 21–26, 38–39; Harold Fock, *Schwarze Gesellen* (Herford: Koehlers, 1979, 1981), Vol. 1: *Torpedoboote bis 1914*, Vol. 2: *Zerstörer bis 1914*.

problems of seakeeping and endurance that plagued torpedo craft at that time. These problems were eventually solved by strengthening the hull and increasing the displacement to over 700 tons, by raising and flairing the forecastle, and by replacing reciprocating engines with steam turbines. Together, these measures permitted destroyers to operate efficiently at sea with the battle fleet.[30]

By 1907, the larger units of improved design were entering service. Armed with quick-firing guns, as well as torpedo tubes, the destroyer in effect took the battleship's torpedo defense battery out thousands of yards from the battle line, extending its perimeter of defense. At the same time, the torpedo was placed in a forward position where it could be used as a preliminary assault weapon, supplementing the battleship's big guns. The high-powered electric arc light with movable shutter allowed day and night communication between ships and thus tactical integration of the now disparate units of the battle fleet. No longer were all the essential weapons of the force mounted on one type of unit operating in close proximity with its companions.

Integration of shipboard devices brought something new to the fleet and capped the development of a new type of battleship that has come to be called the dreadnought. Although generally considered to be any battleship with all heavy guns of uniform caliber (that is, no medium-caliber, quick-firing guns in the main battery), the dreadnought as a technological advancement represented much more. It was the first linking of several dissimilar components on a warship to give its weapons significantly improved capabilities. At the same time, each component became essential to the whole. Whereas the many guns of the predreadnought operated independently, the big guns of the dreadnought functioned as a single system.

In the dreadnought, the system was called central fire control. With it, effective shooting with several heavy guns was possible far beyond the range of medium caliber ordnance on predreadnoughts. This type of gunnery control consisted of a telescopic central director, optical range finders, an observer aloft to spot the fall of shot, and electrical communication to link these components with the guns and with each other. Furthermore, guns and instruments had to be calibrated according to common references in train and elevation.[31]

---

30. March, *British Destroyers*, pp. 72–75; David Lyon, "Torpedo Craft," in *Conway's All the World's Fighting Ships*, pp. 87, 99; Norman Friedman, *U.S. Destroyers* (Annapolis, Md.: Naval Institute Press, 1982), pp. 11–29; Fock, *Schwarze Gesellen*, Vol. 2, pp. 119–158.
31. Peter Padfield, *Guns at Sea* (New York: St. Martin's, 1974), pp. 244–250; Scott, *Fifty Years in the Royal Navy*, pp. 179–186, 204, 242–247, 251–255, 258–267.

Three things are notable about this development. First, the individual technologies were simple and had existed for decades before they were first combined into an effective weapon system in 1911. Second, the dreadnought could shoot with consistent accuracy at twice the effective range of predreadnoughts, extending normal fighting distance from between 5,000 and 7,000 yards to between 10,000 and 14,000 yards. Third, an all-big-gun battleship with central fire control was capable of several times the hits possible for a sistership without the system. Yet, the differences between the two were invisible to all but a well-informed observer.

While the fleet was undergoing this transformation, the submarine entered the scene as the next of the novel weapons. During World War I, it supplanted the cruiser as the primary commerce raider. Once employed in an unrestricted campaign, German U-boats threatened to sever the lines of military supplies and sustenance to the British Isles. The situation remained grave until the reintroduction of the old convoy system of sailing ship days.[32]

In spite of its performance as a commerce destroyer, the submarine neither supplanted the battle fleet nor drove it from the seas. Its success against warships was in more confined waters where it could lie in ambush. During the war, most big ships were torpedoed at the approaches to naval bases, such as Heligoland Bight, or in the area of shore bombardment, such as the Dardanelles. No first line capital ship was sunk by a submarine. Of the six dreadnoughts torpedoed, five were repaired in one to three months. Nine old predreadnoughts were sunk by submarines, demonstrating the necessity for underwater protection included in dreadnought designs.[33] However, the main difficulties for submarines operating against first-line naval forces were not the protection of battleships and defensive armament of their accompanying destroyers. The problems were finding the fleet and then overcoming a gross disparity in speed. Once in contact, the submarine usually could not get into position to launch its relatively short range weapons.[34] Subma-

---

32. John Rushworth Jellicoe, *The Crisis of the Naval War* (London: Cassell, 1920); Arthur J. Marder, *From Dreadnought to Scapa Flow* (London: Oxford University Press, 1969), Vol. 4: *1917. The Year of Crisis*; Bodo Herzog, *60 Jahre Deutsche Uboote 1906–1966* (Munchen: J.F. Lehmanns, 1968), pp. 101–126.
33. *Jean Bart*, torpedoed December 21, 1914 (103 days to repair); *Moltke*, August 19, 1915 (32 days); *Grosser Kurfürst*, November 5, 1916 (97 days); *Kronprinz*, November 5, 1916 (31 days); *Westfalen*, August 18, 1916 (46 days); *Moltke*, torpedoed April 25, 1918 while under tow after major machinery casualty and flooding (137 days). Note that five were German dreadnoughts torpedoed by British submarines, while German submarines did not torpedo any dreadnoughts.
34. Arthur Hezlet, *The Submarine and Sea Power* (London: Peter Davies, 1967), pp. 25–26, 29–42, 67–84.

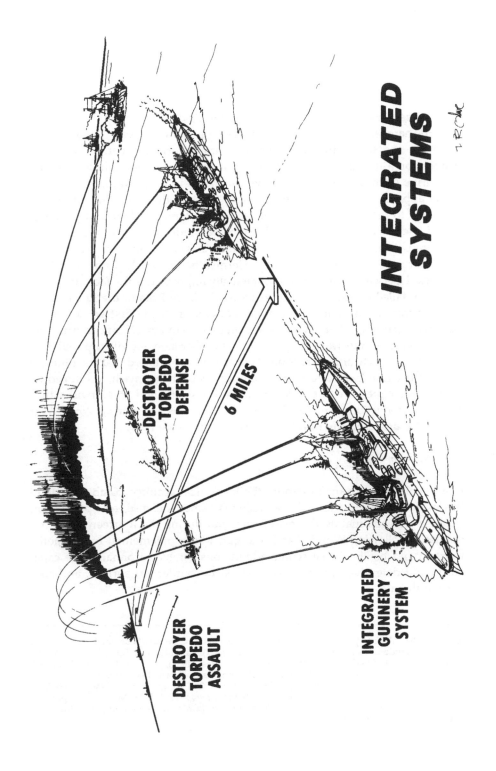

DESTROYER
TORPEDO
DEFENSE

DESTROYER
TORPEDO
ASSAULT

6 MILES

INTEGRATED
GUNNERY
SYSTEM

INTEGRATED
SYSTEMS

rines caused great concern among fleet commanders, but they did not particularly hamper operations of battle fleets on the high seas. Used against warships, the submarine was, like the surface torpedo boat, a defensive weapon or a means of harassment. As a commerce raider, however, the submarine opened an important new dimension in naval warfare.

The apparent inactivity of dreadnought fleets in the First World War has led many to the conclusion that the leviathans served little useful purpose. The single major fleet action of Jutland was indecisive, and the struggle between U-boats and convoys was awesome in its intensity. Yet the convoys could not have been successful if the German battle fleet had not been kept in check. Less widely recognized, but equally important, is the fact that while Germany failed to impose a submarine blockade on Britain, the Royal Navy put Germany in the grip of a surface blockade from the first weeks of the war, with crushing effect. Vital raw materials were cut off, and during the last two years of the war, the greater part of Germany's civilian population was in a state of chronic starvation.[35] Liddell Hart concluded that the ultimate German collapse "was due more to emptiness of the stomach, produced by the economic pressure of sea power, than to loss of blood."[36] Notwithstanding the introduction of novel weapons in the 1890s and during World War I, the final arbiter in the naval war was still the battle fleet.

Three-Medium Combat

In an evolutionary transformation, the battle fleet acquired combat capabilities in three mediums for offense, reconnaissance, defense, and protection. Offense was still mainly the heavy gun, but during the fourteen years immediately following World War I, the gun was supplemented by the aerial torpedo and bomb. Aerial scouts added considerable reconnaissance capability, and defensive systems were refined to counter air, surface, and submarine threats.

---

35. A.C. Bell, *A History of the Blockade of Germany and of the Countries Associated With Her in the Great War* (London: His Majesty's Stationery Office, 1937); Ralph Haswell Lutz, *The Causes of the German Collapse in 1918* (Stanford: Stanford University Press, 1934), pp. 180–187; Ernest L. Bogart, *Economic History of Europe 1760–1939* (New York: Longmans, Green, 1942), pp. 501–505.
36. Basil H. Liddell Hart, *Strategy: The Indirect Approach*, 2nd ed. (New York: Praeger, 1967), p. 358. See also F. Lee Benns and Mary Elisabeth Seldon, *Europe 1914–1939* (New York: Appleton-Century-Crofts, 1965), p. 100; Theodore Ropp, *War in the Modern World* (Durham: Duke University Press, 1959), p. 251.

The obvious change was the advent of fleet air power. From 1924, aircraft contributed directly and importantly to the offensive, defensive, and reconnaissance capabilities of the main battle force. During World War I, flimsy aircraft had been used to drop small bombs and torpedoes against a few ships. Aircraft had also been used for fleet reconnaissance and employed defensively against submarines and zeppelins. However, the initial obstacle to conducting air operations in conjunction with the fleet was launching and recovering aircraft at sea. Other than zeppelins, the first naval aircraft were floatplanes. They could be launched from inclined shipboard platforms, but they could not be recovered until the carrier stopped to hoist them back aboard. HMS *Argus*, the first aircraft carrier with a flight deck that permitted both launch and recovery of wheeled aircraft, entered service at the very end of the war. However, like all early carriers, she did not have enough speed to rejoin the main force after she had turned into the wind to launch or recover aircraft.[37] The first carriers with ample speed for fleet operations went into service in 1924, and by 1929, the British, American, and Japanese navies had at least two each.

As an offensive system, the carrier operated torpedo planes from the first. A single torpedo hit could disable all but the largest ships, but like surface and subsurface torpedo craft, the airplane had to launch its weapon from close range to have a reasonable chance of success. During the long, straight run in, the relatively slow biplane would be exposed to concentrated fire from anti-aircraft guns that were being installed in ever-increasing numbers. Development of the dive bomber gave the carrier an alternative means of attack. Bombing from level flight was unlikely to hit a moving ship, but a high speed dive could be consistently effective. Aircraft of the 1920s could not stand the stress of pulling out of a dive. Gradually, airframe structures were refined, however, and in 1932 the U.S. Navy introduced the first attack aircraft that could deliver a 1,000-pound bomb in a dive.[38]

An aerial platform was clearly a boon to reconnaissance. It could provide a superb overall view of the situation, and inherently greater speed permitted more area to be scouted in a given time. Floatplanes, carried aboard cruisers

---

37. Stephen W. Roskill, ed., *Documents Relating to the Naval Air Service* (London: Navy Records Society, 1969), Vol. 1: 1908–1918; Norman Polmar, *Aircraft Carriers* (Garden City, N.Y.: Doubleday, 1969), pp. 27, 31–37, 41–43, 60–61; H.H. Smith, *A Yellow Admiral Remembers* (London: 1932).
38. Ray Wagner, *American Combat Planes*, 2nd ed. (Garden City, N.Y.: Doubleday, 1968), pp. 329–330; Gordon Swanborough and Peter M. Bowers, *United States Navy Aircraft Since 1911* (London: Putnam, 1976), pp. 314–315.

and launched from catapults after 1922, extended the coverage of the fleet reconnaissance force. Carrier-based scouts offered even more flexibility, because they could be launched and recovered under a broader range of conditions. As an element of air defense, the carrier brought another improvement in existing capabilities. Carriers allowed the fleet to take its own fighter-interceptors to sea for continuous operation.[39] The fighter force operated as an extended line of air defense, analogous to the destroyer screen against torpedo craft. Of course, its time on station was severely limited by fuel endurance.

As a second line of air defense, capital ships were fitted with their own anti-aircraft batteries. During the Great War a few high-angle, quick-firing guns were fitted as a defense against zeppelins. By the end of the twenties entire batteries of director-controlled AA guns were standard. These were strengthened with automatic, multi-barreled weapons for close-in defense.

The big gun remained the principal offensive weapon of the fleet, because it could deliver the most destructive power. By the 1920s, battleship guns in the major fleets fired armor-piercing and high explosive projectiles weighing from 1,400 to 2,100 pounds. Only this kind of firepower had a good chance of sinking the latest capital ships. In destructive power and accuracy (although not in striking range) the battleship was superior to the torpedo plane and dive bomber for some time to come. As late as 1941, a single dreadnought could deliver more ordnance in an hour than all the aircraft in the largest fleet could carry.[40]

Improvements in fire control permitted consistently accurate shooting almost to the horizon, more than double the fighting range of the first generation of dreadnoughts. Advanced surface gunnery was made possible by long-based range finders, gyro-stabilized optics, mechanical analog computers, and synchronous data links. The analog computer is of particular interest because it allowed reasonably accurate prediction of where the target would

---

39. Arthur Richard Hezlet, *Aircraft and Sea Power* (New York: Stein and Day, 1970), pp. 41–84.
40. In December 1941, 162 SBD, 63 SB2U, and 75 TBD dive and torpedo bombers on six U.S. carriers could carry 251 tons of ordnance. William T. Larkins, *U.S. Navy Aircraft 1921–1941* (Concord. Aviation History, 1961), pp. 300–314; Hal Andrews and John C. Reilly, "U.S. Navy Airplanes, 1911–1969," *Dictionary of American Naval Fighting Ships* (Washington, D.C.: U.S. Government Printing Office, 1970), Vol. 5, pp. 541–543. In December 1941, 135 D3A, 174 B5N, and 8 B4Y dive and torpedo bombers aboard 9 Japanese carriers could carry 215.5 tons. Rene Francillon, *Japanese Aircraft of the Pacific War* (London: Putnam, 1970), pp. 41–42, 276, 415, 451. One U.S. *Colorado* class battleship, firing eight-gun broadsides every two minutes would deliver 268.8 tons of Mk. V, 16-inch projectiles in an hour.

be when the projectiles arrived, even when both target and firing platform maneuvered. One contemporary gunnery manual noted that fire control now followed tactics rather than determining them.[41] Floatplanes were carried aboard battleships as aerial spotters to extend gunnery range over the horizon. Although much was claimed for this capability, it did not work well against a maneuvering force.

Protection of the heavily armored capital ship took a significant turn. In the twenties, horizontal armor became more important than vertical. Naturally, armored decks were needed as protection against aerial bombs. Somewhat earlier, advanced surface gunnery made them essential. Since shooting was now accurate at over 25,000 yards, engagements would be fought and decided at long range. Even a large, high-velocity projectile tends to plunge at the end of a long trajectory. At these new ranges, there was thus a greater chance of hitting the decks of a ship than its sides.[42] The newest and most heavily armed battleships were therefore "modernized" by having thick armor placed over their decks and all subsequent designs paid particular attention to this feature. Underwater protection was also improved considerably. One approach was to install layers of shock-absorbing cells between the outer hull and the ship's vital compartments. The other was to extend the underwater hull outward in a blister or bulge that ran most of the ship's length. In several designs the two techniques were used in combination, and variations are still used today.[43]

Not only were later dreadnoughts designed with extensive protection against underwater explosions, the battle fleet as a whole was equipped to deal with submarines. During World War I, the destroyers of the torpedo boat screen became an extended defense against submarines as well. They could easily outrun a submarine on the surface and smother it with gunfire. Underwater, a submarine had little chance of getting into position to launch torpedoes against a fast-moving battle fleet. Even so, lookouts combed the seas for periscope wakes, and destroyers carried depth charges to attack their prey underwater. After the war, fleet destroyers were equipped with active

41. United States Naval Academy, *Notes on Fire Control 1941* (Washington, D.C.: U.S. Government Printing Office, 1941), pp. 106–108

42. One detailed series of tests is summarized in U.S. Navy, Bureau of Ordnance, "Battleship and Scout Armour," Report 34631 (A2)-0 of 3 June 1918, Box 861, Series Entry 88, Record Group 19, Old Military Branch, U.S. National Archives.

43. Great Britain, Admiralty, *Experiments on Underwater Protection and Its Application to Warships, 1913–1920* (London: His Majesty's Stationery Office, 1920); Norman Friedman, *Battleship Design and Development 1905–1945* (New York: Mayflower Books, 1978), pp. 75–83.

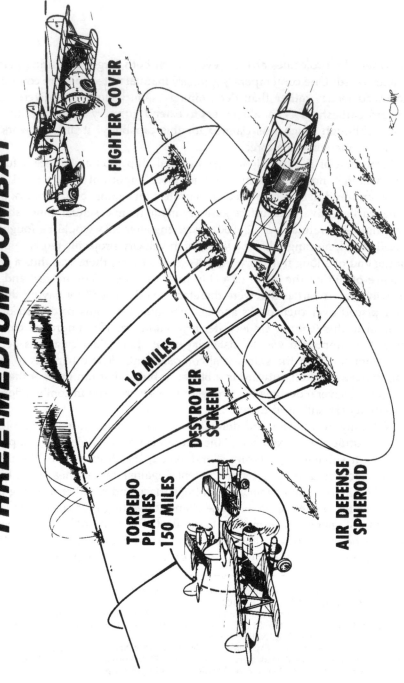

THREE-MEDIUM COMBAT

FIGHTER COVER

16 MILES

DESTROYER SCREEN

TORPEDO PLANES 150 MILES

AIR DEFENSE SPHEROID

acoustic detection devices developed first by the Royal Navy as ASDIC and later by the U.S. Navy as Sonar.[44]

Thus, the major fleets entered the second decade of peace able to attack and defend themselves in the air, surface, and subsurface mediums. The introduction of naval air power brought new capabilities, but the importance of this development at this stage should not be exaggerated. We now know that the aircraft carrier would eventually supplant the armored gun platform as the capital ship of battle fleets; however, the necessary technology came slowly.

The battleship-aviation controversy is a good example of the gulf that often occurs between the extremes of wild prediction and conservative skepticism when dealing with the future. Aviation pioneers like Billy Mitchell claimed in 1920 that airplanes had made traditional fleets obsolete, but this was a case of general prediction. Although ultimately correct, it was hardly a useful insight at the time. At the other extreme, the admirals saw carriers and their aircraft as nothing more than another component in a combat fleet organized around the battleship. Their conservatism can be blamed for a late appreciation of carrier air power during World War II, but in 1920, they were right. The seaborne strike aircraft would not be ready to assume primacy in naval warfare for two or more decades.

Aircraft and Electronics

Only the first of the transitions described so far occurred during a war. The next transformation of combat fleets also happened in wartime, although by the standards of measurement used here, it was no faster than the most dramatic to date. Two separate and profound changes took place. Airborne platforms became the main combat systems for both offense and defense in big navies, and electronics came to dominate weapons and sensors.

Manned flight was given a practical demonstration in 1903. It was two decades before aircraft served in regular functions with the fleet, and it would be another two decades before they became the essential offensive and defensive elements of naval warfare. The basic reasons were technological. In an offensive role airplanes needed the range, payload, and speed to be

---

44. Stephen Roskill. *Naval Policy Between the Wars* (New York: Walker, 1968). Vol. 1, pp. 345–347; Peter Hodges and Norman Friedman, *Destroyer Weapons of World War 2* (Greenwich: Conway Maritime Press, 1979), p. 136.

able to search for and find the enemy fleet, sink its ships, and return to their carriers. Before this was possible, the airplane had to progress considerably in its evolution. Engines needed much higher power-to-weight ratios than were possible in the twenties, and variable pitch propellers would make an important contribution to their efficiency. The high-drag biplane had to be supplanted by aerodynamically efficient monoplanes. And lighter, stressed-metal airframe construction was needed to reduce structural weight and increase fuel and payload capacities.

To be a main offensive system for combat and not merely a harassing agent during the preliminaries, carrier strike aircraft would have to carry a reliable torpedo or 1,000-pound bomb. Less of a weapon load was little threat to a large armored warship, including a carrier. The strike aircraft would have to cruise at over 120 knots, combat-loaded. The normal cruising speed of most combat-loaded biplanes was less than 100 knots, giving little margin over headwinds and fast-streaming battle fleets. A combat radius of at least 250 miles was essential. Peacetime maneuvers showed that range was needed to search for the enemy and to maintain separation necessary to prevent enemy battleships from pouncing on one's carriers. Finally, the strike aircraft needed a top speed near that of contemporary fighters. Slow speed over the enemy fleet made biplane bombers easy targets for defending fighters and director-controlled anti-aircraft batteries. All carrier aircraft in service before 1938 and several later models were deficient in at least two of these four requirements.

The first carrier aircraft that would seriously threaten capital ships at sea was Japan's B5N torpedo bomber. Codenamed "Kate" by Allied intelligence the Nakajima attack plane entered service in 1938. It had the combination of payload, range, and speed to make it a first-line offensive weapon. It could carry an 1,800-pound torpedo at 140 knots to a target 250 miles away and return to its carrier. Its top speed was 200 knots at sea level, only 38 knots slower than the best enemy land-based fighter until 1943. It was superior in speed to most carrier fighters for the next four years. An element critical to this formidable system was the robust Type 91 aerial torpedo, which could be dropped from hundreds of feet by a B5N flying at full speed.[45]

American and British torpedoes were fragile and mechanically temperamental by comparison. They could only be dropped at low speeds from less

---

45. Francillon, *Japanese Aircraft*, pp. 411–416; M.F. Hawkins, *The Nakajima B5N "Kate"*, Aircraft Profile No. 141 (Windsor: Profile, 1972); James H. and William M. Belote, *Titans of the Seas* (New York: Harper and Row, 1975), pp. 31–32.

than 200 feet. This influenced the design of aircraft intended to carry them and helped perpetuate poor overall performance in these aircraft. The U.S. Navy inadvertently sidestepped its torpedo problem by perfecting its dive bomber. By the outbreak of the Pacific War, U.S. carriers were equipped with the SBD Dauntless. Capable of carrying a 1,000-pound bomb, this dive bomber was roughly comparable in performance to the Japanese B5N.[46]

In 1940, the Imperial Japanese Navy put the new Mitsubishi A6M, Type 0 carrier-fighter into service. The "Zeke" (later "Zero") put naval aviation on a new footing in two respects. It had the range to escort the torpedo bombers all the way to their targets and back. The "Zero" fighters could thus engage defending interceptors and give the torpedo planes a much better chance to get through. Second, the "Zero" was the first carrier-based fighter whose performance was on a par with any contemporary land-based fighter.[47] The threat from land-based aviation was thereby greatly reduced.

The "Kate" and "Zero" marked the beginning of a significant and rapid change in naval aviation. In September 1939, when World War II began in Europe, three-quarters of the combat aircraft in the U.S. fleet and all in the Royal Navy were biplanes or underpowered monoplanes with biplane performance.[48] Japan led in carrier aviation, but one-third of its strength was still made up of obsolete biplane dive bombers until late 1940. By the time of the Pearl Harbor attack, the SBD Dauntless dive bomber and the F4F Wildcat fighter put American carrier air power in a league with Japan's First Air Fleet. The Royal Navy, however, did not improve its carrier air arm substantially until the adoption of American aircraft in late 1942.

Although the Japanese and American navies had modernized their carrier air forces by 1941, the supremacy of the battleship still dominated their tactical doctrine. American war plans and construction programs reflected the notion that the best approach was for carriers to provide striking range and flexibility while the battleships would deliver the heaviest punch. American carriers had practiced in peacetime as separate task forces, but doctrine called for their operation in conjunction with the battle line.[49] Japanese doc-

---

46. Barrett Tillman, *The Dauntless Dive Bomber of World War Two* (Annapolis, Md.: Naval Institute Press, 1976), pp. 4–15.

47. Jiro Horikoshi, *Eagles of Mitsubishi* (Seattle: University of Washington Press, 1981); Francillon, *Japanese Aircraft*, pp. 362–377.

48. Stephen W. Roskill, *The War at Sea* (London: Her Majesty's Stationery Office, 1954), Vol. 1, p. 31; Larkins, *U.S. Navy Aircraft*, pp. 243–244, 256–288.

49. Patrick Abbazia, *Mr. Roosevelt's Navy* (Annapolis, Md.: Naval Institute Press, 1975), pp. 33–

trine was for carriers to deploy forward for a preliminary engagement before the decisive gunnery duel.[50] Admirals on both sides recognized the essential role of carriers, but they were not ready to do away with the battleship because it was still both the most powerful and best protected weapon platform in the fleet. It should also be kept in mind that during the first year of the war in Europe, carriers had produced disappointing results. Of 80 sorties flown against capital ships, only four hits were made on two ships in port. No battleship was sunk, and none was hit at sea. By contrast, a carrier was sunk by the guns of two battleships.[51]

The first carrier battles in the Coral Sea and west of Midway Island accelerated what had been a gradual reorientation in thinking. The Battle of Midway was not only a strategic turning point in the Pacific War, it precipitated a sudden change in the makeup and tactics of combat fleets. It is often assumed that the Japanese went into that action with fully developed tactics emphasizing the carrier. In fact, Admiral Nagumo's carrier force was merely an advanced striking arm of Admiral Yamamoto's main battleship force. According to the Japanese plan, the carriers would operate against Midway, a fixed land objective, and then soften up U.S. naval forces operating nearby. Finally, the battleships would move in for the coup de grâce. In the event, the battle was decided and Japanese carrier forces were smashed without a heavy gun being fired.[52] The shift in force structures took place immediately after Midway. Within three weeks, the Japanese cancelled all battleship construction and implemented a new construction and conversion program that emphasized carriers.[53] The Americans already had 23 carriers under

50; Clark G. Reynolds, *The Fast Carriers* (New York: McGraw Hill, 1968), pp. 5–6, 10, 17–18, 39–40; Polmar, *Aircraft Carriers*, pp. 54–60.
50. Minoru Genda, "Evolution of Aircraft Carrier Tactics of the Imperial Japanese Navy," in Paul Stillwell, ed., *Air Raid: Pearl Harbor!* (Annapolis, Md.: Naval Institute Press, 1981), pp. 23–27; Mitsuo Fuchida and Masatake Okumiya, *Midway: The Battle That Doomed Japan* (Annapolis, Md.: Naval Institute Press, 1955), pp. 77–90, 97–98, 240–243.
51. Roskill, *War at Sea*, Vol. 1, pp. 194–197, 245, 298–299, 317; Ian Stanley Ord Playfair, *The Mediterranean and Middle East*, History of the Second World War (London: Her Majesty's Stationery Office, 1954), Vol. 1, pp. 136–137; Paul Auphan and Jacques Mordal, *The French Navy in World War II* (Annapolis, Md.: United States Naval Institute, 1959), p. 133.
52. Fuchida and Okumiya, *Midway*, pp. 145–204; Samuel Eliot Morison, *History of United States Naval Operations in World War II* (Boston: Little, Brown, 1947–1958), Vol. 4, pp. 102–156; U.S. Office of Naval Intelligence, *The Japanese Story of the Battle of Midway* (Washington, D.C.: U.S. Government Printing Office, 1947).
53. Anthony J. Watts and Brian G. Gordon, *The Imperial Japanese Navy* (Garden City, N.Y.: Doubleday, 1971), pp. 70, 73, 194–196, 200–203, 517–518; Polmar, *Aircraft Carriers*, pp. 70, 132–134, 235–239.

construction, but within a month of Midway, Congress authorized 13 more and 10 of those were ordered from shipyards within thirty days. The 18 U.S. capital ships under construction were eventually cut to 10.[54]

Less than two years after the Battle of Midway, U.S. carriers returned to the Western Pacific in force. When they did, the battle fleet had a completely different character. The new unit central to American combat forces was the *carrier task group. Primary offensive capability lay with the torpedo and dive* bombers aboard two to four carriers in each group. Primary defensive capability continued to be carrier-based fighters. Integral to the unit were cruisers and destroyers. If battleships accompanied the fleet at all, they were the new, fast type. Cruisers (and the few fast battleships) provided a major portion of the carrier group's close-in air defense. Specially designed antiaircraft cruisers were also part of the carrier forces. They were predecessors of today's guided missile cruiser. Outside the concentric "boxes" of carriers and cruisers was a ring of destroyers for anti-submarine and modest antiaircraft defense. Carrier task groups were joined to form carrier task forces, but the formation and command integrity of each group was maintained for flexibility. The carrier group, not individual ships, formed the basic unit of the new American battle fleet.[55]

In parallel with the refinement of aircraft technology, electronics assumed an essential role in combat at sea. By the end of 1943, the battle fleet had become an integrated set of air and surface platforms which depended on a complete range of electronic equipment. The principle of radar had been demonstrated in 1904. However, the German navy did not put the world's first crude search radar on a warship until 1936, and the introduction of dependable electronic systems did not come until 1940, when they were refined by the British and American navies. After that, it took less than three years for radar and associated devices to become important parts of every aspect of naval combat.

Shipboard search radar (1940) extended early warning of enemy air attacks out to 100 miles, day or night. Airborne search radar extended warning time further and could be used to detect surfaced submarines at night (1941) or low-flying aircraft (1942). Coded transponders called IFF (Identification

54. "Aircraft Carriers 1908–1962," *Dictionary of American Naval Fighting Ships*, Vol. 2, pp. 464, 472–473; James C. Fahey, *Ships and Aircraft of the U.S. Fleet*, 5th ed. (New York: Ships and Aircraft, 1945), pp. 4, 9–10, 14.
55. Reynolds, *The Fast Carriers*; Belote and Belote, *Titans of the Seas*.

Friend or Foe, 1940) helped to sort out the electronic blips and coordinate air defenses. Surface-to-air voice radio and height finding radar allowed fighters to be directed to the best position in time to defend the fleet against mass air attacks. This system of air intercept control (1943) was soon supplemented by night fighters equipped with their own small radar. Radar and communications jamming known as ECM (Electronic Countermeasures) introduced a means of disabling an enemy's sensors or weapons without having to destroy them. During World War II, ECM blossomed into one of the important fields of weapon technology.[56]

Radically different technology does not always mean significant change in military capability. An interesting example is the advent of turbojet engines in combat aircraft. The change from reciprocating engines to turbojets was revolutionary in a technical sense. The principles, design problems, and operation of each are completely different. Yet the effects on payload, range, speed, and maneuverability, the critical elements of military aircraft performance, were small. True, the piston engine had just about reached the limit of its potential for improvement, and the jet opened significant possibilities to improve aircraft performance, particularly, at first, in the realm of speed. However, the early jets were only incrementally faster than the fastest propeller planes. Moreover, fuel consumption and low power made jets poor competitors in range and payload.[57] In another decade, all of this would change, but until then, there was only gradual improvement in the combat capabilities of carrier aircraft.

Thus, between 1940 and 1943, the battle fleet exchanged the armored gun platform for the carrier strike aircraft. The essential weapons and sensors for both offense and defense were now carried by aircraft. Offensive ordnance included freefall bombs, unguided rockets, and torpedoes. Primary defensive elements were patrols of fighter and anti-submarine aircraft, each supported by search radar carried in specialized aircraft. Fighting ranges were now at least 250 miles, but aircraft not only extended combat range far over the

56. Louis A. Gebhard, *Evolution of Naval Radio-Electronics and Contributions of the Naval Research Laboratory* (Washington, D.C.: U.S. Government Printing Office, 1979), pp. 169–204, 206–208, 251–256, 299–307; Hezlet, *Electronics and Sea Power*, pp. 156–266; Norman Friedman, *Naval Radar* (Annapolis, Md.: Naval Institute Press, 1981); Fritz Trenkel, *Die Deutschen Funkmessverfahren bis 1945* (Stuttgart: Motorbuch, 1979), pp. 16–22, 159–193, 202.
57. Wagner, *American Combat Planes*, pp. 352–358, 408–420; Swanborough and Bowers, *United States Navy Aircraft*, pp. 176–181, 186–188, 305–309; Owen Thetford, *British Naval Aircraft Since 1912* (London: Putnam, 1962), pp. 29, 62–63, 96–99, 228–229, 318–319.

AIRCRAFT AND ELECTRONICS

250 MILES

STRIKE

EW

FIGHTER COVER

ASW

horizon: their speed allowed them to be concentrated or dispersed over a wide area. The critical dimension of battle became area instead of distance. Offensive and defensive capabilities could be defined in terms of concentric circles outlining the combat radii of given aircraft with given weapon loads. Electronics expanded combat conditions to include night and inclement weather. Ships provided bases for the aircraft and carried the second-line defenses of the fleet. It is notable that the technologies that transformed fleet combat from gunnery duels to air battles had little effect on the carriers themselves. The carrier of 1950 was essentially the same as the carrier of 1930. Conversely, the turbojet engine brought completely new technology to naval aircraft in 1948, but the effect on combat capabilities was small until a decade later.

Nuclear Weapons

Nuclear weapons obviously changed the means of waging war. At sea, the tremendous destructive capability of a single weapon meant that even with moderate accuracy, one shot had a high probability of disabling a target. More importantly, nuclear weapons changed the entire context of war. The risk of mutual annihilation now leaves the superpowers without reasonable prospects for political gain in a general war. Thus, nuclear weapons present new strategic problems and new tactical problems in the classical sense of strategy and tactics.

Nuclear weapons changed the context of warfare by raising deterrence to new levels of importance for strategy. The basic tenet of naval strategy in the Anglo–American tradition was that the key to utilizing the sea in wartime, while denying that use to the enemy, is neutralizing the enemy's main naval forces. Neutralization could come through blockade or by winning a series of small naval engagements, but the preferred approach was a single decisive battle.

The strategic problem raised by nuclear weapons is usually called linkage. If the adversary has nuclear weapons and the means to deliver them on your homeland, he is unlikely to suffer a decisive defeat, in this case at sea, without resorting to some kind of nuclear strike, either in retribution or to regain a military advantage. A strike to redress combat losses and regain the initiative would probably be against forces at their bases and the support facilities themselves. However, highly valued population, industrial, and cultural centers would inevitably suffer, either through collateral effects of the weapons or escalation in the face of attacks on home territory.

It has been argued that the navies of the two superpowers could, under certain conditions, engage one another without precipitating a general nuclear war. Certainly, naval operations can take place far from home territory and with negligible risk of collateral damage. If the stakes were high (but not too high) a naval confrontation in a remote area could develop into a clash of arms that remained localized. However, in the fast-moving events of war, a major defeat on land or at sea is just as likely to cause precipitous reaction as it is to lead to deliberative thinking. Furthermore, certain characteristics of modern naval forces strengthen rather than loosen the linkage between war at sea and general war. Naval strike aircraft are based in home territories, anti-submarine warfare can threaten the sea-based portion of nuclear deterrent forces, and satellites perform essential roles for both strategic and general purpose forces.

Technology did not immediately bring about a situation of mutual nuclear deterrence. Initial operational capability of the American nuclear force followed only two weeks after the practical demonstration of fission on July 16, 1945. But an American nuclear force was not actually deployed against the Soviet Union until May 1949, and more than a decade after the end of the war had passed before the Soviet Union clearly had the instruments of nuclear retaliation. In the summer of 1949, the Soviets both tested their first nuclear fission device and deployed their first long-range bomber, a copy of the American B-29. At least by 1950, then, they had the capability to retaliate against Western Europe for an American nuclear strike. However, it can be argued that the Russians thought that a nuclear threat to Europe would not deter the United States in a superpower confrontation. In November 1955, the Soviets first tested a thermonuclear weapon, and in July 1956, an intercontinental bomber was deployed to carry it.[58] Thus, by 1956, the Soviet Union could threaten the United States directly, and the Soviet leadership could feel more confident of deterring the use of nuclear weapons by the United States.

---

58. William Green, "The Billion Dollar Bomber," *Air Enthusiast*, No. 1 (October 1971), p. 265; Samuel Glasstone, *The Effects of Nuclear Weapons* (Washington, D.C.: U.S. Government Printing Office, 1962), p. 680.

At first, the American nuclear deterrent did not have an important link to combat at sea, because the Soviet navy was not capable of challenging, let alone defeating, Western carrier forces. In the first decade after World War II, the Anglo–American navies completely dominated the seas, and the United States had a virtual monopoly on the means of nuclear retaliation. By the time the Soviet navy offered any real threat to the battle fleets of the Western Alliance, a situation of mutual nuclear deterrence was well established. By an accident of history, Soviet naval forces were the first to realize any benefit from this situation, because they were inferior in capabilities and numbers for sustained conventional combat at sea.

In the tactical arena, nuclear weapons brought an unprecedented jump in destructive capability. This meant fewer weapons and less accurate delivery systems were required. The U.S. Navy was the first to develop nuclear weapons that could be delivered by carrier aircraft or shipboard launchers. The first tactical nuclear weapons developed specifically for naval use were intended for anti-submarine warfare. A special penetration bomb was designed to break through huge concrete bomb shelters like those built to protect German U-boat docks from Allied bombing during World War II. Nicknamed "Elsie," and officially designated the Mk 8, this bomb could also be used against ships. Tactical aircraft equipped to deliver it from carriers were first deployed in 1953. It was followed by the Mk 7 bomb and a depth charge called "Betty" and armed with an Mk 7 warhead.[59]

In 1956, the United States deployed the first carrier jet aircraft equipped to deliver nuclear weapons, the F9F-8B Cougar. Other important additions to the fleet's nuclear arsenal followed in the next five years. Nuclear warheads were first deployed in a torpedo (ASTOR, 1958), a surface-to-air missile (Talos, 1960), and a surface to underwater rocket (ASROC, 1961).[60] By 1961, then, the battle fleet had a complete suit of offensive and defensive nuclear weapons. The advance in capabilities was revolutionary, but the rate of change was evolutionary. At sixteen years from demonstration of fission to an actual operational capability, it had taken a remarkably long time to integrate nuclear weapons into the fleet for even the basic spectrum of missions.

---

59. National Atomic Museum, Albuquerque, files and displays, with special assistance from Richard L. Ray.
60. Ibid.

Nuclear technology brought a jump in weapon lethality, but it has not affected naval warfare as a straightforward improvement in combat capability. The advantages of tactical nuclear weapons are offset in the strategic context by the problem of linkage. The enhanced kill probability of combat nuclear weapons offers little advantage when using them means serious risk of a devastating general nuclear war. The leaders of both superpowers seem to realize this. There are indications that neither the American nor Soviet government would release tactical nuclear weapons for use under field command unless a general war was unavoidable.

The most significant effect of nuclear weapons on the battle fleet was to change the relative importance of its roles. After World War II, the rise of nuclear deterrence strategy and the lack of a rival battle fleet led the U.S. Navy to develop its fleet air forces for nuclear strikes against large land targets such as Soviet submarine bases, airfields, and industrial plants. The carrier task force became, first, a nuclear strike force, joining the Air Force Strategic Air Command in the role of deterrence in 1951.[61] It was, second, a mobile air force for regional conventional wars, and third, a means of dissuasion and a visible demonstration of national interests. In 1957, American submarines with cruise missiles went on their first deterrent patrols, and in 1960 ballistic missile submarines began to take over the role of sea-based nuclear deterrent from the carriers, which would be very vulnerable to nuclear attack.[62] The capability for sustained offense and defense, inherent in battle fleets but lacking in strike units such as missile submarines, became less important in the context of global nuclear war.

On the other hand, as the only type of naval force with the capability for both sustained offense and defense, the battle fleet remained a primary means of dissuasion and presence. It has been employed to dissuade the Soviets from intervening by sea in regional crises such as those in the Middle East and the Persian Gulf. U.S. carrier forces dissuaded China and Taiwan

61. John T. Hayward, "The Atom Bomb Goes to Sea," *The Hook*, No. 9 (Summer 1981), pp. 22–27; Norman Polmar, *Strategic Weapons An Introduction* (New York: Crane, Russak, 1975), pp. 19–20; U.S. Navy Department, *United States Naval Aviation 1910–1970* (Washington, D.C.. U.S. Government Printing Office, 1970), pp. 185, 187.
62. Norman Polmar, "Die ersten Marschflugkorper für den Einsatz in See," *Marine Rundschau*, No. 79 (Februar 1982), pp. 76–84. Specific dates for deterrent patrols given in *Dictionary of American Naval Fighting Ships* under names of individual submarines: *Tunny* (SSG-282), *Grayback* (SSG-574), *Growler* (SSG-577), *Halibut* (SSGN-587).

from attacking each other in 1950 and 1958, and Idi Amin's Uganda from attacking Kenya in 1976. In the role of demonstrating national interests, called "presence," a carrier battle group can put first-class air power into a region without base rights that require time to arrange and can compromise delicate political sensitivities in a region.

In sum, nuclear weapons represent a critical threshold for naval warfare, as they do for warfare in general. The significant result of this technological development was partly in fighting capabilities, but more in the context of naval strategy and the relative importance of battle fleets in war. It was a different world after 1956. As always, naval battles could mean general war, but now, general war between the superpowers could bring mutual annihilation. In this situation, the relative importance of battle fleets in war declined, especially when compared to naval strike forces such as ballistic missile submarines. In regional, nonnuclear conflict and in roles of dissuasion and presence, the highly visible and capable battle fleet has retained important functions. To fully appreciate the evolution of the battle fleet after 1956, this subtle shift from decisive combat to dissuasion through the *capability* for decisive combat must be kept in mind.

Guided Ordnance

The next transition raised the limits of fighting capabilities in a world without rival battle fleets. Guided standoff weapons, high-performance carrier aircraft, and computerized tactical data links were adopted first by the American and somewhat later by the British and French navies. Their chief rival at sea, the Soviet Union, developed only portions of these technological packages at first, applying them to extended coast defense rather than a sustained combat capability on the high seas.

The advent of guided ordnance led developments in three general areas that brought fundamentally new conditions to combat at sea. First, guided weapons replaced free-fall bombs, straight-running torpedoes, and guns as the primary types of ordnance used by the fleet. Dependence on electronics, acoustics, and electro-optics for guidance added a new element in weapon employment and made countermeasures a new form of defense. Second, jet aircraft matured and transformed naval aviation. Improvements in jet engine and airframe design introduced supersonic dash speeds to aerial combat, and allowed strike and fighter aircraft to cruise at more than twice the speed possible with reciprocating engines. Steam catapults, the angled flight deck, and the mirror landing system became necessary for carriers to launch and recover both high-performance and long-endurance aircraft. Third, ever more complex battle management was made possible with the introduction of digital computers and direct data links between many airborne and surface platforms.

Between 1956 and 1961, the U.S. Navy led the way in introducing guided ordnance as the essential weapons for all important functions of the battle fleet. Aircraft were armed with missiles and continued to be the main platforms for weapons and sensors used in long-range, first-line offense and defense. Armed principally with guided weapons, ships retained close-range, second-line defensive functions. Guided ordnance enhanced fighting capability by enabling an air or surface platform to deliver explosive warheads from considerable distances, often under conditions of poor visibility. These features allowed air attacks to be made from outside the effective range of defensive gunfire. Guided weapons also gave ships a defensive capability against high speed jet aircraft just when guns were becoming inadequate in range and accuracy. A fundamental change that accompanied the ascendance of guided ordnance was the addition of guidance links to the basic functioning of the weapon system. Guidance, added to fire control, increased the accuracy, range, and flexibility of weapons, but it opened the system to electronic and other forms of countermeasures.

During World War II, two types of guided ordnance were introduced to naval combat, although experimentation began as early as 1916. The homing torpedo was the first example of a guided weapon actually employed by a navy. In 1943 both German submarines and American aircraft with merchant convoys began using passive acoustic homing torpedoes.[63] The other guided weapon introduced during the war was the anti-ship missile. Most were radio-controlled missiles or glide bombs, and all were launched from land-based aircraft. These crude weapons were developed hastily and immediately sent to operational units. Some had initial success before countermeasures were developed. Notable examples are the German HS 293 and Fritz X weapons. Most, however, achieved little or never got beyond developmental testing.[64]

---

63. E.W. Jolie, *A Brief History of U.S Navy Torpedo Development* (Washington, D.C.: U.S. Government Printing Office, 1978), pp 36–39; Terry Hughes and John Costello, *The Battle of the Atlantic* (New York: Dial, 1977), pp. 282, 921–293

64. Bill Gunston, *The Illustrated Encyclopedia of the World's Rockets and Missiles* (New York: Crescent, 1979), pp. 105–108, 113, 118–121.

The year 1956 saw the first guided ordnance equip a battle fleet. For long-range air defense, carrier-fighter squadrons were deployed that year with Sparrow and Sidewinder air-to-air missiles. For close-range air defense, the Terrier surface-to-air missile joined the fleet aboard cruisers. The Tartar and Talos surface-to-air missiles became operational in 1960. The Tartar was small enough to be carried by destroyers, and armed 23 such ships by 1964. Although considered short-range by standards of the day, the Tartar was effective at 10 miles, farther than most heavy AA guns. The Talos could hit a high-speed jet aircraft at over 75 miles.[65]

Between 1956 and 1959, Soviet Tu 16 bombers were equipped with anti-ship missiles, but like their German predecessors, these naval aircraft were tied to land bases. Carrier strike aircraft were first deployed with air-to-surface missiles in 1959. The first was the AGM-12 Bullpup, with a 250-pound warhead and a range of 6 nautical miles. It was soon followed by the Bullpup B, carrying 1,000 pounds of explosive to 9 miles.[66] Thus, an aircraft could remain outside the range of a ship's anti-aircraft gun batteries and guide its ordnance with precision onto a moving target. This meant that without air cover, a ship now needed surface-to-air missiles for defense. Today, French carrier aircraft are equipped with the 39-mile Exocet AM39. U.S. carrier aircraft currently have four types of air-to-surface weapons available, including the 60-mile Harpoon.[67] In 1957, both the American and Soviet navies put the first operational cruise missiles on surface combatants, but neither the Regulus I nor the SS-N-1 had the range of carrier aircraft if employed against moving targets such as ships.[68]

The battle fleet carried a full spectrum of guided ordnance by 1961. In that year, the last component of the set was introduced when ships received a standoff ASW system in the form of ASROC. This is a short-range ballistic rocket, but it acts as the means of delivery for a torpedo with acoustical homing guidance. Subsequent surface-to-underwater systems, such as the

---

65. T. Blades, "USS Galveston: The First Talos Guided Missile Cruiser," *Warship*, No. 4 (1980), pp. 226–233; Norman Friedman, "The 3-T Program," *Warship*, No. 6 (1982), pp. 158–166, 181–185; "Guided Missile Cruisers 1959–67," *Dictionary of American Naval Fighting Ships*, Vol. 3, pp. 820–823.
66. Gunston, *Rockets and Missiles*, pp. 123, 134.
67. Ronald T. Pretty, ed., *Jane's Weapon Systems 1979–80* (London: Mcdonald and Jane's, 1979), pp. 142–144, 151–152; Gunston, *Rockets and Missiles*, pp 123–125, 131; R. Meller, "The Harpoon Missile System," *International Defense Review*, No. 8 (February 1975), pp. 61–66.
68. Gunston, *Rockets and Missiles*, pp. 79, 91.

Australian Ikara (operational 1964), the French Malafon (1965), and the Soviet SS-N-14 (1970), are guided missiles that carry homing torpedoes.[69]

Aircraft, still the primary ordnance carrier of the battle fleet, also changed in substantive ways. Advances in jet engines and airframe technology contributed to the new fighting regime of fleets. The change from piston to turbine engines came much more gradually than the substitution of guided weapons for guns. The first jet aircraft flew in Germany in 1939. The first jet did not land and take off from a carrier until 1948, and the first jet squadrons finally went to sea on operational deployments in 1950. Payload limitations of jets kept them in the fighter role, with propeller planes dominating both tactical strike and long-range nuclear strike missions for another eight years.

The A3D Skywarrior became the first operational jet capable of long-range nuclear strikes from carriers in 1957. The same year, the FJ-4B Fury joined the fleet as the first single-engine jet capable of carrying more ordnance than a propeller plane of similar size and weight. Strike elements of U.S. carrier air groups made a rapid transition to jets during the remaining two years of the decade. However, most air groups retained one squadron of propeller-driven AD Skyraiders until well into the sixties.[70]

The potential of high-performance jet aircraft could not be realized by navies until the aircraft carrier also benefited from new applications of technology. Techniques of shipboard launch and recovery represented a major obstacle to the introduction of supersonic aircraft. Aircraft performance was simply outstripping the carrier's ability to handle its planes safely. In 1955, HMS *Ark Royal* went into service with all three features essential for jet-age naval aviation. She had steam catapults that could accelerate heavier aircraft to the higher airspeeds required for swept-wing jets to become airborne. The old hydraulic catapults had about reached the physical limits of their capabilities. Until Colin C. Mitchell developed a practical steam catapult in 1950, it seemed that high-performance jets might be confined to shore bases. Second, the *Ark Royal* had an angled flight deck. This permitted safe recovery

69. Norman Friedman, "SS-N-14: A Third Round," *U S Naval Institute Proceedings*, Vol. 105 (June 1979), pp. 110–114; Pretty, *Jane's Weapon Systems 1979–80*, pp. 107–108; Gunston, *Rockets and Missiles*, pp. 256–257.
70. Detailed data on the composition of air groups for specific carriers on each deployment are given in *The Hook*, Vol. 6 (Spring 1978), pp 26–27, Vol. 6 (Summer 1978), pp. 26–27, Vol. 6 (Fall 1978), pp. 28–29; Vol. 6 (Winter 1978), p. 30; Vol. 7 (Fall 1979), pp. 26–27; Vol. 8 (Spring 1980), pp. 24–25; Vol. 8 (Summer 1980), pp. 30–31.

of jet aircraft, which must land at relatively high speed or stall. A straight flight deck meant a slow approach and cutting the engine upon landing. If the tailhook missed the arresting cables, only a heavy net barrier prevented collision with aircraft parked forward. An angled deck allowed landings safely above stalling speeds for swept-wing jets, because the aircraft could go around for another pass if it missed the arresting gear. Third, the *Ark* had a landing system that used lights and a mirror to show the pilot the proper glide slope for a safe landing as he made his approach at over 120 miles per hour.[71] In 1956, the United States deployed its first four carriers having these three vital features developed by the Royal Navy.[72]

Command and control, often considered within the broader context called battle management, was becoming a significant problem because of its growing complexity. The concept of a single type of all-purpose capital ship was now quite obsolete. From the time destroyers joined the blue water battle fleet, offensive and defensive functions had been dispersed in a number of different surface and later airborne platforms. Three-medium combat became four-medium combat with the widespread dependence on electronics during World War II. Weapon guidance added new capabilities and vulnerabilities to the employment of ordnance. Both guided missiles and high-performance jet aircraft significantly increased the pace of combat. All of these developments have made naval combat an extremely complex process.

Combat information and control centers evolved into a complex nexus of radar scopes, plotting boards, and communications equipment during World War II. But reading, recording, correlating, and relaying the data was done manually. In the age of jets and missiles, however, the situation would change too rapidly to plot on a board by hand with a grease pencil. A partial solution, and certainly an essential tool, came with automated data processing, display, and relay, made possible by the refinement of electronic digital computers. The first to be deployed by a fleet was the U.S. Navy's Naval Tactical Data System (NTDS), which went to sea in 1961 on the carrier *Oriskany* and the guided missile ships *King* and *Mahan*. The prototype for

---

71. Norman Friedman, *Carrier Air Power* (Annapolis, Md.: Naval Institute Press, 1981), pp. 93–95, 99–100, 106–107; Paul Beaver, *The British Aircraft Carrier* (Cambridge: Patrick Stephens, 1982), pp. 125–141.
72. Norman Friedman, "SCB-27: The Essex Class Reconstructions," *Warship*, Vol. 5 (1981), pp. 98–111; John S. Rowe and Samuel L. Morison, *The Ships and Aircraft of the U.S. Fleet*, 9th ed. (Annapolis, Md.: Naval Institute Press, 1971), pp. 4–5.

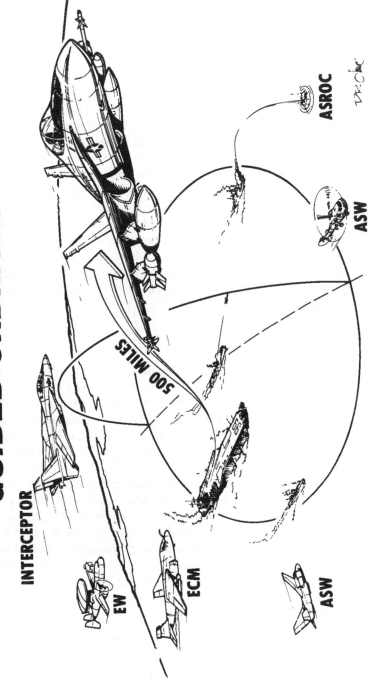

GUIDED ORDNANCE

INTERCEPTOR

EW

ECM

ASW

500 MILES

ASW

ASROC

Britain's Action Data Automation Weapon Systems (ADAWS) was installed on HMS *Eagle* shortly before the carrier was recommissioned in 1964.[73]

These systems collect, process, store, and present informatioin from various sensors on many platforms. The sensors can be electronic, acoustic, or optical, and the platforms are usually both ships and aircraft. All are tied together by electronic data links. The tactical data systems provide automated organization and video display of information for command and control. With this type of system and a competent team to operate it, a force commander has the services of rapid threat detection and assessment as well as systematic allocation of weapons to targets. In modern situations of multiple, high-speed missile and aircraft systems, both for employment by a force and as threats to it, automated data processing has become a basic element of naval combat.

Ocean Area C^3I

We are probably now in the midst of the next major transition. It has already brought ocean area coverage in command, control, and communication, as well as intelligence, navigation, and meteorology information. The same transition will make it possible to conduct tactical reconnaissance over areas as large as the North Atlantic or Indian Oceans by the end of the decade. The culmination of this series of developments may well bring tactical integration of ocean areas.

The major technological development that permitted these advances in capability is the earth-orbiting platform. After the Soviet Union placed the first man-made satellite in orbit around the earth in 1957, it was another seven years before an orbiting system intended specifically for a navy was in place. This system for navigation became operational in 1964 when the first of three Transit series was completed to become the Navy Navigation Satellite System (NNSS). A decade later, the Soviet navy may be said to have reached an initial operational capability with the first basic set of space-based support systems. By then, the Soviets had communications relay, routine

---

73. Raymond V.B. Blackman, ed., *Jane's Fighting Ships 1968–69* (New York: McGraw-Hill, 1968), p. 360; Harris R. Robinson, "C³: Progress and Developments," *U.S. Naval Institute Proceedings*, Vol. 108 (December 1982), pp. 114–117; R.D.S. Stubbs, "Development Trends in Data-Handling Systems," *International Defense Review*, Vol. 8 (February 1975), pp. 54–58; J Claisse, "Tactical Data Handling in the Royal Navy," *International Defense Review*, Vol. 4, No. 5 (1971), pp. 436–439.

monitoring of weather, navigation information, and intermittent radar surveillance of the oceans, all using satellites.[74] Ocean area tactical support of naval operations thus reached an initial operational capability in 1974. Ocean area systems now perform many more than these four basic functions, and further expansion of their capabilities is in progress. A brief review of these systems shows the broad spectrum of their functions and their growing importance.

The fleet benefited from weather satellites as early as 1960 when the first Tiros (Television and Infra-Red Observation Satellite) went into orbit. The first 10 of the Tiros series were both meteorological survey and rudimentary military reconnaissance satellites. The Soviet Union launched meteorological satellites in component tests as early as 1963 with Cosmos 14. The first operational Soviet weather satellite was Cosmos 122, launched in 1966, and the Meteor series gave the Soviets regular weather data from 1969. In 1972, the U.S. Navy tested the delivery of real-time weather data directly from satellites to ships at sea. Other related programs have added to the services provided to naval forces. Satellites monitor ice movements and solar activity as it relates to variations in the earth's magnetic field. In the future, the National Oceanic Satellite System will use radiometers to detect small changes in water temperature of the oceans, an important type of data in anti-submarine warfare (ASW) operations.[75]

Navigation aids in the form of earth satellites are developing into global positioning systems with accuracies down to meters. The Transit series of NNSS has been mentioned. Probably the first navigation satellites operational for the Soviets were Cosmos 385 and 422, placed in orbit in 1970 and 1971.[76] The U.S. Navy's Global Positioning System, also called NAVSTAR, is operational, but it will be expanded in this decade. The first six satellites were placed in orbit by booster, and another twelve will be orbited using the Space Shuttle. According to current plans, the system will provide latitude and

---

74. David C. Holmes, "NAVSTAR Global Positioning System: Navigation for the Future," *U.S. Naval Institute Proceedings*, Vol. 103 (April 1977), pp. 101–104; Reginald Turnill, *The Observer's Spaceflight Directory* (London: Frederick Warner, 1978), p. 184; SIPRI, *Outer Space—Battlefield of the Future?* (London: Taylor and Francis, 1978), pp. 135–136, 139–140.
75. Turnill, *Observer's Spaceflight Directory*, pp. 188–190; Barry Miller, "USAF, Navy Join in Weather Program," *Aviation Week and Space Technology*, December 3, 1973, pp. 52–55; SIPRI, *Outer Space*, pp. 145–157.
76. Walter B. Hendrickson, Jr., "Satellites and the Sea," *National Defense*, No. 67 (October 1982), pp. 26–28, 84; SIPRI, *Outer Space*, pp. 135–143; Gebhard, *Evolution of Naval Electronics*, pp. 407–408.

longitude anywhere on the globe by 1985 and altitude anywhere above the globe by 1987. Position accuracy is advertised as 16 meters in three dimensions and 0.1 feet per second in velocity.[77]

In the field of satellite communications, experimental and commercial units were followed by the first of the Soviet Molniya series in 1965, with seven operating by 1967. The first seven satellites of America's Initial Defense Satellite Communication System were orbited with a single launch in 1966. Improved systems have been placed in orbit since. One of the most significant for naval forces is the FLTSATCOM system, which has provided not only communications but also computer-to-computer data links since 1977.[78]

The last group of satellite systems to be developed is intended for ocean surveillance. Aside from photographic satellites, which are really intelligence gatherers, the surveillance platforms use radar, infrared, and passive electronic detectors. The Soviets orbited Cosmos 198, probably their first radar ocean reconnaissance satellite (RORSAT), in 1967. Their first paired RORSATs were operational in 1974. Cosmos 954, which came down in Canada in 1978, was a later version of a high-powered RORSAT. The White Cloud program put an array of passive electronic ocean reconnaissance satellites (EORSATs) into orbit in 1976. Other U.S. programs in the 1980s are Clipper Bow for high-resolution radar and Teal Ruby for infrared sensor patterns.[79]

Ocean area sensors are not confined to orbiting platforms. Growing numbers of acoustic systems are coming into operation. Probably the first such system was SOSUS (Seabed Sound Surveillance System), made up of passive hydrophone arrays fixed on the ocean floor. The first arrays were under test by 1948, but the system was probably not ready to support operational forces until the end of the 1950s or early 1960s. Open sources claim detection ranges of hundreds of miles and accuracy to fix a submarine's position within a radius of 30 to 50 nautical miles.[80]

77. "NAVSTAR GPS: Global Positioning System," *Aviation Week and Space Technology*, June 28, 1982, p. 16.
78. Turnill, *Observer's Spaceflight Directory*, pp. 182–184; SIPRI, *Outer Space*, pp. 105–109, 114–129.
79. Alan Hyman, "Ocean Surveillance from Land, Air and Space," *Naval Forces*, Vol. 3, No. 2 (1982), pp. 56–58; "The Soviet Military Space Program," *International Defense Review*, Vol. 15, No. 2 (1982), pp. 149–152; SIPRI, *Outer Space*, pp. 43–44, 91–92; "Expanded Ocean Surveillance Effort Set," *Aviation Week and Space Technology*, July 10, 1978, pp. 22–23; Philip J. Klass, "Soviets Push Ocean Surveillance," *Aviation Week and Space Technology*, September 10, 1973, pp. 12–13.
80. Norman Polmar, "The U.S. Navy: Sonars, Part 2," *U.S. Naval Institute Proceedings*, Vol. 107 (September 1981), pp. 135–136; Joel S. Wit, "Advances in Antisubmarine Warfare," *Scientific American*, February 1981, p. 32; Hyman, "Ocean Surveillance," pp. 58–59.

More recently, the U.S. Navy began development of a large, towed array sonar to complement the SOSUS system, called the Surveillance Towed Array Sensor System (SURTASS). The sensor for this system is a passive hydrophone array, several hundred feet long, which will be towed at the end of a 6,000-foot cable by a small ship. Data will undergo preliminary processing onboard the ship before being transmitted via satellite to a data processing center ashore. Twelve units were under construction or on order by 1982. The first is scheduled to become operational in 1983. A project to provide the fleet with rapidly deployable moored arrays, designated MSS or RDSS, is also underway.[81]

To date, these ocean area sensor and communications systems have had only support functions. As the big navies have become more and more dependent on this support, its importance has naturally increased. However, there are critical limitations which have prevented the transition from ocean area C^3I support to ocean area tactical integration. One problem is resolution versus area of coverage in radar and optical surveillance satellites. Broad area coverage tends to be low in resolution, and therefore target discrimination. High resolution tends to be confined in area of coverage. Before a high resolution sensor can be utilized effectively, it must be directed to within the vicinity of the target or considerable time will be consumed scanning merchant shipping and empty areas of the ocean. As one analyst put it, "in this situation, a word is worth a thousand pictures."

A second major problem is the time lag between data collection and its availability to a deployed weapon system. Unlike fixed land targets, naval platforms are constantly moving. At 25 knots, a ship can be anywhere in an area of 54 square miles after ten minutes, and 490 square miles after thirty. Images, acoustic returns, and electronic intercepts must be correlated with geographic references, processed for use by guidance systems, and transmitted to the unit that will use the information. In current systems, the time lag between detection by a broad area sensor and completion of the fire control solution allows a potential target to move well beyond missile guidance range. A missile's homing sensors simply cannot acquire a target much beyond its horizon without the assistance of a local sensor.

---

81. John D. Alden, "Tomorrow's Fleet," *U.S. Naval Institute Proceedings*, Vol. 107 (January 1981), p. 117; Vol. 108 (January 1982), p. 119; Norman Polmar, *The Ships and Aircraft of the U.S. Fleet*, 12th ed. (Annapolis, Md.: Naval Institute Press, 1981), p. 198; Couhat and Baker, *Combat Fleets, 1982/83*, pp. 801–802.

The solution will probably be advanced data processing. In consonance with a theme of this paper, system integration will precede new components. By correlating the general positions from more than one ocean area sensor with intercepted radar and radio signals, a ship can be identified and the estimate of her position refined. High-powered over-the-horizon radar, large airborne radar, and more advanced sensors in underwater and orbiting systems will certainly improve the accuracy surveillance for targeting purposes. However, the capability to process vast amounts of data will accelerate this transition. The recent development of super computers such as the CRAY-1 and the Control Data Cyber 205 make this capability possible in the immediate future.[82] With ocean area sensors, high-speed data processing, and instantaneous global data links, ocean area C^3I support will become ocean area tactical integration. A cruise missile utilizing this type of ocean area targeting grid can have an effective range of two or three thousand miles without any basic changes in engine or airframe technology.

Conclusions

With today's perspective we can reconsider how technological innovation has produced fundamental changes in the methods and conditions of warfare at sea. Salient improvements in operational capabilities provide the best measure of important change. Usually developments in technology are considered first and then their effects. In analyzing trends, changes in capabilities provide a better context. Within this context, the essential technologies can be identified. The means and rate of introducing these essential technologies then provide insights into the dynamics of each major change or discontinuity. This survey focused on conspicuous improvements in the fighting capabilities of battle fleets, but its conclusions can be applied to other aspects of naval warfare, such as commerce protection, amphibious operations, and mine warfare. The key is to identify the essential capabilities for the general mission, whatever that mission or function might be, and then to identify the technologies essential for those capabilities.

Based on the cases considered in this study, three general conclusions can be made about the dynamics of technological change in the naval sphere. One is that change is usually evolutionary but it can be dramatic. When a

---

82. Wit, "Advances," p. 33; Hyman, "Ocean Surveillance," p. 59, Norman Friedman, *Modern Warship Design and Development* (Greenwich, Conn.: Conway Maritime Press, 1979), p. 101.

OCEAN AREA TACTICAL INTEGRATION

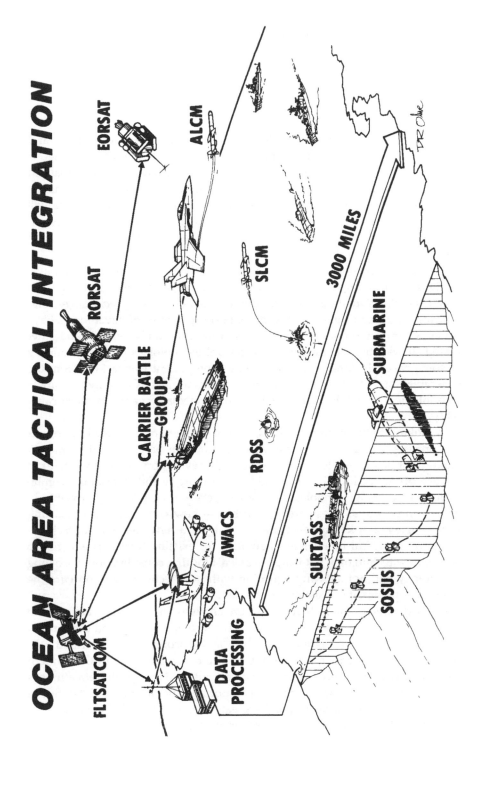

change in combat capability is truly dramatic, its effects are far-reaching, and the time required for the transition is relatively short. Important change is not always planned, and the full potential of such a change is seldom realized in advance. It does not necessarily occur in wartime. The rate of such change has no apparent correlation with how basic and important the change will be, and the rate of change has not increased as technology has progressed. However, when dramatic technological change has occurred, it has had profound effects on the balance of naval power.

Another conclusion is that *new* technology has not revolutionized naval warfare. Here an important distinction is made between new technology and new capabilities made possible by applications of existing technology. Significant changes in the military and political capabilities of naval forces have come when long-existing technologies were eventually refined and integrated. It was the final integration of several technologies that came quickly in some cases. In other cases an essential component was lacking from the ensemble, but by itself it would have been useless. Certainly no single technological "breakthrough" has brought immediate change in naval capability.

In looking to the future, the dynamics of change are most likely to reflect the process found in these case studies from the past. Instead of merely looking for some new technology that might revolutionize naval warfare, as we are prone to do today, it should prove more useful to examine combinations of existing technologies. Their effects will be felt first, probably in one of three ways. These are: 1) synthesis—new combinations of existing technologies, 2) a keystone—a missing link for a new ensemble of technologies, or 3) tactical innovation—new uses for existing forces. This is not to ignore new technology, but it argues at least for first consideration of existing technology and not allowing a fixation with new technology to obscure the important potential of existing technology.

A third conclusion is that important changes in capability are seldom reflected in obvious physical features or the size of warships and other naval weapon platforms. A significant feature in the ironclad was its rifled ordnance, in the dreadnought battleship its fire control system, in the modern aircraft carrier its airborne sensors and computers. None of these features is obvious in a photograph. The most important characteristic of the Soviet "Oscar" class submarine is not its great size, but the likelihood that its missiles will use space-based sensors for guidance.

Projections about the future tend to be general predictions of radical change or detailed rejection of anything but gradual change. There is a gap between prediction that is ultimately correct but happens too far in the future to be of use to planners and skepticism that precludes the possibility of rapid and fundamental change. Unfortunately, most analysis falls at these extremes. Taken together, these conclusions offer an alternative. It is the kind of perspective that can facilitate assessment of the current balance of naval power and can help to identify ways in which technology is most likely to change that balance.

By concentrating on dramatic improvements in mission capability, one can see more clearly the nature of fundamental changes in the way war has been conducted at sea, and predict more safely the essence of future revolutions. In this era of renewed great power naval rivalry, such an approach could be more useful to planners of policy, budgets, and advanced research.

Will Strategic Submarines Be Vulnerable?

In view of the increasing emphasis placed upon strategic submarines under the Reagan strategic program announced October 2, 1981, it is of interest to review the prospects for survivability of such submarines in the foreseeable future. This is particularly timely because the Scowcroft Commission has confirmed the U.S. inability to identify a survivable land-basing posture for the MX missile and because the Soviet Union will presumably soon be faced with the vulnerability of its own silo-based ICBM force, whether by reentry vehicles on U.S. ICBMs or from U.S. SLBMs.

Antisubmarine warfare (ASW) techniques and capabilities important for *strategic* purposes are quite different from those which can be employed in *tactical* antisubmarine operations. Strategic offensive submarines are able to carry out their mission—delivering nuclear weapons against the homeland of an opponent—while, at the same time, limiting their own vulnerability by utilizing evasive modes of deployment and operation. Tactical, or attack, submarines, on the other hand, must approach their target—warships, merchant ships, a chokepoint to be mined, or the like—to be successful; this limits their flexibility in operational decisions. Furthermore, the contest between tactical submarines and ASW forces may take place over months or years, involves no trailing of submarines but the kill of submarines essentially on sight, and could be modulated by either side to its own advantage. In a long war of attrition, for example, the naval forces of one side may be kept at home or in sanctuaries, so that the enemy's attack submarines would have no targets. The attack submarines themselves may be kept at home or out of danger if using them were deemed too hazardous because of their vulnerability. In contrast, to be effective and worth contemplating, ASW against strategic submarines would have to threaten to destroy almost all offensive submarines within a few days at most. Otherwise, ASW would be superfluous, since both U.S. and Soviet forces would be vulnerable over a period of months to repeated attacks on their accustomed ports.[1]

Richard L. Garwin is IBM Fellow at the Thomas J. Watson Research Center, Yorktown Heights, New York, Adjunct Professor of Physics at Columbia University, Adjunct Research Fellow at Harvard University, and Andrew D. White Professor at Cornell University.

1. This vulnerability over protracted periods is not unique to submarines. Obviously, nuclear attacks on the accustomed bases of strategic bombers (accompanied by fallout and attack on

International Security, Fall 1983 (Vol 8, No 2) 0162-2889/83/020052-16 $02 50/1

Here, I will not consider strategic ASW capabilities which could detect and destroy an opposing force only over a period of weeks, but will look instead at those capabilities which conceivably could pose a threat to strategic submarines over a period of one week or less. One can imagine wars of attrition against strategic submarines, but should such a nightmare actually occur, the logical counter would be to attack enemy military bases with nuclear weapons. In any event, such protracted war scenarios are not particularly relevant for evaluations of possible technological developments that would (newly) threaten the survivability of strategic submarines.

It should be noted at the outset that even the sudden destruction of a substantial fraction of the deployed strategic submarine fleet, taken alone, would not constitute a disabling blow against either U.S. or Soviet retaliatory capability, any more than the planned survival of only 50 percent of the land-based missiles would vitiate that system, or the inability of 50 percent of the strategic bomber fleet to take off or to penetrate Soviet air defense would negate the value of the air-breathing strategic component. Additionally, even potential future ASW capabilities which appear to threaten strategic submarine operations and deployments as such are currently practiced are not a peril if the postulated ASW technique or system could be substantially countered by modifying submarine operations, by countermeasures that could fool the detection system, or by reliable means of counterattacking the ASW system before it had destroyed a substantial fraction of the strategic submarine force.

Some advocate "moving the strategic force to sea" in order to reduce (or reduce the consequences of) actual or perceived vulnerability of land-based ICBMs. Although I believe that even a vulnerable Minuteman force is valuable and non-provocative, and that small, single-RV ICBMs have a good future even under some future SALT agreement, it is true that ICBM-range SLBMs (such as the Trident II [D-5] missile, the MX, or a small ICBM) can now be given accuracy equivalent to that specified for the land-based MX. We shall note later that the SSBN force can be controlled and communicated with about as well as can a land-based force, so the choice between land and sea may well be made on the bases of cost and vulnerability. The latter is the topic of this article.

---

other airfields) would lead to attrition of that force over a period of days or weeks. The landing of saboteurs or even fallout in the land-based missile fields would prevent access for maintenance and could degrade their capabilities over a period of weeks or months. All components of strategic offensive forces, thus, share a "use it or lose it" characteristic, over a period on the order of days or months.

As for cost, a recent publication[2] drawing upon work sponsored by the Navy and Defense Department estimated that a system of small submarines (SUM) "is at least $10 billion less expensive than the drag strip (multiple protective shelter basing of MX) for deploying and operating 850 survivable and effective warheads." It noted also the likelihood that the drag-strip deployment "would be even larger and more expensive or require an active and costly ballistic missile defense" actually to provide that number of survivable warheads. This basing mode was ultimately rejected in favor of placing the MX in existing (modified) Minuteman silos. Yet, in attempts to bypass the many problems with land-basing, SSBN systems of the future may contain also encapsulated MX missiles, carried horizontally outside the pressure hull of a small submarine, two or four on each side. Carrying true ICBMs (as opposed to shorter-range missiles), these submarines would find no advantage in moving far from their home ports in the continental U.S. and Alaska, so the Soviet ASW threat within a few hundred miles of U.S. coasts is relevant, as are the special means we might take to counter that threat at such great distance from the Soviet Union and so close to U.S. territory.

The Potential Effectiveness of Current ASW Technologies

For purposes of discussion, potential threats to strategic submarines may be grouped in three categories:

—Those in which deployed strategic submarines are kept within range of an attack weapon; this is known as "trailing."
—Those in which the attacker can narrow the area of uncertainty in which the submarine is deployed to one that is much smaller than the overall potential deployment zone, so that one or more individual search-and-kill platforms (e.g., aircraft) can be directed to a relatively small area to find the strategic submarine and attack it; this is known as "tracking."
—Those in which the entire deployment (or hiding) area must be searched at the beginning of hostilities and strategic submarines could be destroyed only as they were detected, localized, and attacked; this is know as "open-ocean search."

---

2. S.D. Drell and R.L. Garwin, "Basing the MX Missile: A Better Idea," *Technology Review*, May 1981.

THE TRAILING THREAT

The potential deployment zones of nuclear-powered ballistic-missile-launching strategic submarines (SSBNs) are limited—with only minor exceptions—only by the necessity to remain within missile range of their targets. In view of the long range of modern submarine-launched ballistic missiles, these potential deployment zones are very large and the problem of locating the SSBNs accordingly difficult.[3] One way of gaining information about the precise location of the target SSBN might be to trail it with nuclear-powered attack submarines (SSNs) from such short ranges that the target submarine would be vulnerable to attack with torpedoes or submarine-launched, rocket-carried nuclear warheads (SUBROC) if an order were given to destroy the SSBN fleet. Indeed, it would be feasible to trail SSBNs from surface ships equipped with an appropriate sonar. However, given the acute interest of an SSBN in knowing whether it (and its whole fleet of siblings) is held in trail, the availability of submarine-deployed towed arrays, the possible existence of "delousing facilities" in the open ocean,[4] and the like, it is inconceivable that a fleet-wide covert trailing operation could be long maintained. Thus, any trailing which occurred might as well be overt (which is much easier technically) and carried out at very short ranges in order to reduce the probability of loss of trail and to facilitate attack upon demand.

Such a trailing vehicle could use imaging sonar systems which project audible pulses (or utilize higher frequencies) to 1 nautical mile or less, which are then reflected strongly by the submarine so that the trailing could be aided by following the submarine image on the sonar. The trailer would thus have advance warning of impending maneuvers by its target and be able to maintain an advantageous position. The well-known variability of the ocean in refracting sound waves gives no protection against such short-range active

3. With Moscow as a target, ocean operating areas (in millions of square kilometers) achievable with various SLBM ranges are: 2,800 km—5.5 million; 4,600 km—19 million; 7,400 km—62 million; 11,100 km—180 million. In this article we have standardized on metric units, for those more comfortable with nautical miles and yards, 1 nmi = 1.85 km; 1 sq nmi = 3.43 sq km; 1 yd = 0.91 m; 1 kn = 1.85 km/hr.
4. The "towed array" is a cable hundreds of meters long with microphones spaced in its interior which can be towed by a submarine or a surface ship to provide a sensitive receiver of sounds generated by submarines or other trailing vessels. In addition to the sensitivity, the array determines the direction of origin of the sound to within one degree or so. The "delousing facility" might be a region of fixed detailed acoustic surveillance provided by the host country so that a submarine can traverse the area to learn whether it has a trailer attached. If the trailer chooses not to enter the area, it loses trail; alternatively, if it continues to trail the submarine, it is detected and subject to harassment or diversion.

sonar trailing, since the refraction of sound in the ocean never exceeds an angle of 15 degrees, and the trailing vehicle could remain (in adverse waters) at a distance from the quarry less than a few times its depth so that the dip of the sound ray exceeds the possible refraction angle. The (acoustic) vision might be distorted, but it would not be blocked.

Continuous active trailing would require at least three trailers per deployed SSBN and an appropriate kill weapon, but the most demanding requirement would be to reliably acquire trail in a time much shorter than the submarine deployment tours of two months or so. Although SSBNs operate from only a few ports, it can by no means be assured that overt or covert intelligence-gatherers outside that port could not detect and identify emerging SSBNs at relatively short range, and assign trailers to them. Presumably, this would not occur in peacetime within the 12-mile (22-km) limit, but unless operational countermeasures were taken, such restraint could not be assured, particularly during crises. Such operational countermeasures could include: potential target SSBNs creeping along their home coasts and unexpectedly dashing for open waters; their emerging in pairs or in the company of friendly SSNs so that the trailers would often choose the wrong submarine; the passing of the potential target submarine through regions of artificially high acoustical noise; and the establishment of zones near the SSBN base in which friendly naval forces would pose a physical hazard to the trailers if they followed the SSBN to this operating area. Trailing also might be deterred by the consequences for political relations between nations involved. Presumably, political tensions would rise sharply if an SSBN fleet were put under active trail.

But finally, if one side made the investment to acquire and operate effective trailing, it could still be countered if the SSBNs under trail ejected explosive charges (of limited lifetime) at times of their own choosing, which would destroy any trailers. It is probably the inevitability and effectiveness of this potential countermeasure, more than anything, which keeps both the U.S. and the Soviet Union from building a fleet of active sonar trailers.

THE TRACKING THREAT

Given the search and kill radius of a homing torpedo (1 km or more), or the range of a rocket-propelled nuclear warhead (20 km or more), the tracking threat differs from the trailing threat (i.e., requires a further step of localization) only if the dimensions of the tracking uncertainty area of a given SSBN exceed about 20 km, so that there is indeed a need to search for the SSBN in an uncertainty region exceeding 1000 square km. On the other hand, given an operating area for some tens of submarines on the order of 20 million square km, searching the entire operating area to destroy every submarine encountered would correspond to a search requirement on the order of 1 million square km per submarine. This reduction from 1 million to 1,000 square km search area is the origin of concern with the tracking threat. Thus, a system which in peacetime could track each deployed SSBN and localize it to an operating area of even 100,000 square km would allow localization and kill with perhaps 1/10th the hunter-killer force required for open-ocean search.

Such tracking systems must be characterized not only by their precision, but also by their "time-late." Even if such a system could gain precise knowledge of the target submarine's location, the information would be useful only if it could be made available to the weapon system to be used for the attack before the target had moved significantly. An acoustic system, for instance, would receive tracking information at a time-late enforced by the travel time of sound in the ocean (a speed of 1.5 km per second). Thus, acoustic systems with detection ranges of 1,000 km have a time-late on the order of 10 minutes, while those with detection ranges of 5,000 km would have a time-late on the order of one hour. During these times, submarines operating at a reasonable ten-knot patrol speed could have moved on the order of 3 or 20 km respectively. Since the submarine can move in any direction at a speed up to 18 km/hr (10 kn) or so, the uncertainty area after ten minutes is 30 square km; after one hour, 1,100 square km; and after two hours, 4,400 square km.

The only existing long-range system capable of tracking and localizing submarines in this way is advanced passive sonar. Arrays of listening devices may be deployed on the ocean bottom and the signals they detect reported by oceanic cable to processing stations on land. The noise radiated by submarines in the frequency range below a few hundred Hz[5] can travel thousands of km with little attenuation. In a noise-free ocean these signals can be used to provide a line of bearing to the source submarine. The detection of a submarine by several such arrays of listening devices (or alternatively its detection at many individual hydrophones spread throughout the ocean) could be used to localize the submarine to an accuracy which under the best of circumstances might be in the range of some tens of km.

Three problems limit these potential capabilities, however. First, the ocean is full of noise coming from many sources including natural origins and

---

5. One Hertz is one cycle of oscillation per second.

thousands of ships which sound more or less like submarines. The second difficulty stems from the fact that sound in the ocean is refracted strongly by the general reduction of water temperature and compressibility at depth.[6] Thus a submarine which is clearly audible at 700 km may be totally inaudible at 680 km. Third, the technical problem of keeping sensitive equipment working for long periods in a difficult ocean environment vastly reduces the capability one might otherwise assign to such listening systems.

If we assume for the moment the possibility of tracking some fraction of the deployed strategic submarine fleet at long range, what would be the resulting ASW threat?

Submarines could be attacked without further localization by *barraging* the uncertainty area with nuclear weapons delivered by land-based missiles. Because of the time-late of sound propagation and the notional 30-minute flight time of an ICBM (as well as the size of the original uncertainty area of the tracking system), a single warhead would provide only a small kill probability against the target submarine. Submarines exposed to such a threat would be well advised to patrol normally at a depth of about 100 meters, to minimize the lethality of a warhead of a given yield.[7] A 1-megaton warhead descending to optimum depth in the ocean would have a kill radius of about 5.6 km against a submarine at 100-meter depth, and would thus pose a threat to the survival of submarines within an area of about 100 square km. We have noted that perfect localization at acoustic range at a range of 5,000 km corresponds to a time-late of about 1 hour and an uncertainty area of about 1,100 square km. Adding 30 minutes flight time for an ICBM to deliver its warhead leads to the requirement to barrage some 2,300 square km, which would require 2,300/100 or some 23 single megaton warheads to destroy a single undecoyed SSBN detected at 5,000 km range on a perfectly accurate acoustic surveillance system.

6. This results in a complex sound velocity profile which in deep water bends sound rays so that they plunge repeatedly to depths of 3 km or so, returning near the surface at intervals of 50 km.

7. Because the shock wave pressure at the surface of an underwater nuclear explosion is zero (due to the addition of a reflected impulse), submarines are least vulnerable near the surface. Furthermore, in this way the strength of the submarine against high pressures is available for resisting the overpressure of the explosion, whereas if the submarine were operating at maximum depth, a relatively small additional explosive pressure could crush it. The submarine cannot operate *at* the surface because it would be detectable by vision, by the noise produced by cavitating propellers, and the like. Similarly, a nuclear warhead is most effective when it is detonated at an optimum depth; detonation at or close to the surface wastes the explosive energy by venting the bubble produced

Of course, many nuclear warheads might be employed against such a valuable target as a strategic submarine. If one imagines that a force incorporating 2,000 megatons yield is made available for this ASW activity, then the overall operating area which could be barraged would be about 200,000 square km. Given the deployment of perhaps 30 SSBNs, the tracking system and the resulting barrage would pose a serious threat to their survival only if the accuracy of tracking limited the uncertainty area per submarine to 7,000 square km or less (or if there were an intermediate stage of localization to this accuracy or better). Such a large number of warheads could barrage an uncertainty area generated by 2 hours delay after perfect localization of the submarines, or by a localization accuracy of 50 km (or a combination of the two). This is the magnitude of the threat, if unopposed. Should such a large system capability emerge, it could still be countered by jamming (to deny the detection of valid submarine targets); by decoys (to add to the valid submarine targets a sufficiently large number of apparently valid targets that the opponent could not destroy them all); or by destruction of the detection system.

Over the years, physicists, acousticians, oceanographers, and those interested either in imperiling submarines or in preserving them have learned of the complexity of the ocean. The long-range acoustic path is predictably obstructed by seamounts; the convergence-zone propagation of sound limits the reliability of detection; ocean noise may mask submarine acoustic signatures; submarines have local sensors to enable them to stay in particularly favorable near-surface water layers, those from which the sound cannot propagate to long distance; and submarines can operate in shallow water in which there is *no* good long-range propagation of sound.

These are the fundamental problems which would be faced by any long-range acoustic tracking system, even without the target submarine taking any countermeasures. For the detection of submarines traveling at high speeds (35-55 km/hr), the immense variability of the ocean is less of a problem because the sound radiated at such speed is dominated by the noise of the submarine's propeller. This source of noise becomes negligible at speeds below about 18 km/hr. Thus, SSBNs which are eminently detectable and vulnerable while traversing the broad oceans at very high speeds to reach their operating areas can slow to a discreet patrol speed at which their radiated noise is much reduced.

Long-range detection with intermediate localization to improve accuracy and carry out the attack would require the dispatching of weapon platforms to localize the target submarine. These vehicles could be supersonic aircraft like the Backfire, but they would take several hours to reach the search area. They could also be missile-delivered automatic sensors which would search within the original uncertainty area and then report back by radio to enable attack by other missiles. The localization could proceed by active acoustic means, by directional passive acoustic means, or by a pattern search for the disturbance of the earth's magnetic field created by the heavy steel hull of the submarine.[8]

Furthermore, an attack on strategic submarines with nuclear weapons would spoil the ocean basin for long-range acoustic detection for many hours because of the very intense sound produced by the nuclear explosion itself and its subsequent multiple reflections from the ocean boundaries. Moreover, there would be a substantial problem in categorizing submarines as enemy or friendly, or as SSBNs or SSNs, in partial or unreliable detections in which there might be only a line of bearing rather than a cross-fix by two or more detecting arrays.

There also are potential counters to long-range passive acoustic tracking, including the use of artificial noisemakers to increase the oceanic basin noise. Since even a noisy submarine radiates a total acoustic noise of only 0.1 watt, and quiet submarines in the range of 0.01 watt or less, it would be a trivial matter to provide a long-endurance noisemaker which could transmit a re-corded submarine signature for a period of hours or days. The provision of hundreds or thousands of such noisemakers could well eliminate the possi-bility of detecting submarines in the first place. Some dozens of noisemakers provided with a few-knot mobility could simulate SSBNs themselves, making attack on detected "SSBNs" unprofitable.

Launching the SSBN missiles during the flight time of the land-based missiles used to barrage its operating area would negate the purpose of the attack. The U.S. continually receives information on Soviet missile launches from infrared warning satellites. Unless these satellites were destroyed preemptively, itself a signal of an impending attack, there would be 10 minutes or more warning of a massive barrage attack against submarines. Depending on (or forcing) an ability to launch submarine-based missiles from under an attack is not without its own problems, chief among them being the instability it might introduce into the strategic posture if the submarine missiles have enough accuracy to be thought to imperil ICBM silos.

---

8. Most submarines can be detected by the magnetic disturbance to a range of 0 5 km, but not exceeding 1 km.

All in all, there seems to be little prospect of a long-range acoustic system which could hold in track a large fraction of an uncooperative SSBN fleet. The possible follow-up localization forces just noted do not at present exist, and in any case could be countered by noisemakers or other means. Further, there seems to be no known detection phenomenon other than acoustics capable of providing signals which could be used to track SSBNs at long distances. Prospects for the discovery of such techniques in the future are discussed subsequently.

THE AREA-SEARCH THREAT

If SSBNs cannot be trailed and killed on command, and cannot be tracked and localized and then destroyed all of a sudden, could their entire operating area be searched and the submarines destroyed as they were found? The detection of submarines by arrays of acoustic sensors can be ignored here, since that possibility is subsumed under the tracking threat. Logically, an area-search threat must involve something which is too expensive or too provocative to use all the time, but which would be capable of searching the entire deployment area in a few days or less and reliably detecting, identifying, and localizing SSBNs so that they may be destroyed. One can imagine the use of short-range sensors on numerous fast-moving platforms such as aircraft to search out a 20-million square km deployment area. If the sensor used were a magnetometer with a sweep width of 1 km, then a single aircraft operating at 400 km per hr would require 50,000 hours to sweep the area. Even a fleet of 100 aircraft would require 500 hours, and there would be no guarantee that either a random or a pattern search of the operating area would find all the submarines, even if the sensor were totally effective. This result is due to the relatively narrow sweep width, which would allow submarines in their normal operations to drift from an as-yet unswept into an already-swept area, and thus to be missed by further sweeping. The result is that 37 percent of the submarines would be undetected after 50,000 air-plane-hours of search, 14 percent after 100,000 hours, and so on.

Because of this latter problem, the only significant threat to deployed submarines would arise from active or passive sonar. But if long-range active sonar were used to sweep the operating area, it could be heard far beyond the range at which it could detect the target submarine, allowing the target to evade by operating near the surface or in other areas unfavorable for acoustic propagation, or by maintaining nose-on orientation with respect to the sonar (to reduce the echoing area). The use of passive sonar would give no such indication of sweeping activity. The most effective mobile passive sonar at present is a towed array of detectors which can be towed by surface ships or submarines. In either case, a vessel making a few knots could tow such an array several hundred meters long at appropriate depths, with the capability of detecting rather noisy submarines out to distances of a few hundred km. Sound propagates to that distance, however, only by repeated refraction through the ocean depth, with the previously noted inconveniences of convergence-zone detection. In any case, even a ship with a 300-km passive sonar range could detect submarines only in a neighboring area of 200,000 square km, so that about 100 such ships would be required in even favorable circumstances to obtain single line-of-bearing detections on some fraction of the submarines in the 20-million square km deployment area. Further, in order to exploit these detections, the detected submarines would have to be distinguished from surface ships, and each such contact would have to be explored and attacked successfully. This process could be undertaken by helicopters or fixed-wing aircraft, which were vectored by the ship towing the sonar array. Uncountered, a fleet of some hundred such towed arrays and aircraft could detect relatively noisy submarines and run them down within some hours. Still, because of the complexity of sound travel in the ocean, the result would be a gradual attrition of the strategic submarines, not the near-instantaneous destruction of the SSBN force.

The Effectiveness of Current ASW Technologies

The U.S. is widely credited with having deployed advanced fixed acoustic arrays. The U.S. also has tested and deployed towed arrays and advanced magnetometers. Tactical ASW benefits from the use of sensors and weapons mounted on helicopters and aircraft, as well as from advanced capabilities on surface warships. U.S. capabilities against Soviet submarines are greatly aided by the relatively high noise levels emitted by existing Soviet submarines, but even so the Soviet Union can probably maintain the security of most of its deployed SSBNs against a *preemptive* attack by operating them with moderate caution (e.g., at low speeds) and in ocean regions unfavorable for detection.

Current Soviet capabilities against U.S. SSBNs are believed to be virtually nil—a result of the low noise level emitted by U.S. SSBNs, the failure of the U.S.S.R. to deploy long-range acoustic sensors, and the more primitive state of Soviet computing and signal-processing technology.

U.S. defense leaders, commenting on the security of U.S. SSBNs, have steadfastly maintained that these weapons are highly invulnerable through the 1980s, but might be threatened by some capability not yet foreseen. Former Secretary of Defense Harold Brown in his annual report for Fiscal Year 1981 noted that the Soviet "VICTOR-class nuclear-powered attack submarine remains the most capable Soviet ASW platform. At present, neither it nor other currently deployed Soviet ASW platforms constitute a significant threat to our SSBNs." Secretary Brown in that same report notes that our strategic submarines patrol "virtually unchallenged in the vast ocean areas and present a multi-azimuth and so far untargetable retaliatory capability." He notes that the greater range of the Trident-I missile "considerably enhances survivability of the SSBN force, allowing these 12 Trident backfitted submarines to operate in much larger ocean areas while on-station, thus hedging against the possibility of a Soviet ASW breakthrough." But in the course of arguing in support of a land-based MX deployment, the Secretary and other Defense officials have suggested that there might be an ASW breakthrough which would result in the "oceans becoming transparent," or at least that submarines would be vulnerable to a breakthrough which *did* make the oceans transparent.

In 1980 Secretary Brown expressed confidence that aircraft in plain view of Soviet radars could, by the use of "STEALTH technology," become invisible. It is remarkable that he was at the same time alluding to the possiblity of some future unanticipated Soviet ASW breakthrough, without noting the likelihood of effective countermeasures in the ASW sphere. Commenting on these remarks in 1980, the commander of the U.S. submarine force, Vice-Admiral Charles H. Griffiths, said that the oceans were a great place to hide because "they're becoming more opaque as we understand more about them." Thus, U.S. navy sources in their public utterances give no support to the thought that "the oceans are becoming transparent."[9] The Scowcroft Commission Report of 1983 also shows no concern for strategically significant vulnerability of an evolving SSBN force.

Emerging Technologies

Fundamentally, the detection of submarines involves either a signal *generated* by the submarine itself and received by the detection system or the *reflection*

---

9. Admiral Griffiths, in September 1981, confirmed to the author that he continues to maintain these views and notes that he was quoted accurately and in context.

by the submarine of a signal from the detection system. The detection of generated signals corresponds to passive systems, such as the passive acoustic detection system previously mentioned. Other potential signals include the detection of radioactive products emitted by the submarine's nuclear reactor, the detection of light emitted by ocean fauna disturbed by the submarine, of gases emitted by the submarine, and the like. The use of reflected signals implies active systems which must be capable both of bringing energy *to* the submarine and receiving the reflected energy. Active sonars are the most commonly known source of such systems, but active magnetometers have also been considered, and the use of lasers to detect objects at depths in the ocean is much discussed.

Normal communications to U.S. SSBNs is via VLF radio transmitters employing large vulnerable towers and antenna systems. The radio waves at 5–16 kilohertz have a wavelength (at 10 kHz) of 30 km; like all waves, they are launched inefficiently from antennas much smaller than a wavelength. However, there is so much static (from distant lightning) in this band that even a small antenna like that on a pocket transistor radio is adequate to pick up enough signal (and static), so that a larger receiving antenna is no better.

VLF radio waves penetrate into seawater a meter or so, and the SSBN patrols at depth at low speed while holding an antenna within a meter or so of the ocean surface; it does this either by trailing a long buoyant cable or by towing a streamlined buoy containing an antenna. Either of these techniques works well but is an operational inconvenience. The VLF transmitters are very vulnerable to nuclear attack, but their function is satisfactorily assumed by TACAMO EC-130 aircraft equipped to transmit VLF signals to the SSBN fleet by a trailing antenna wire many km long. Indeed, the Reagan strategic program, announced October 2, 1981, will put VLF receivers on the bombers of the Strategic Air Command to improve the reliability and survivability of communications with the air-breathing strategic component. Modern technology shows a way to reduce the inconvenience associated with towing an antenna near the surface while the SSBN (whether SUM or Poseidon or Trident) patrols at depth to hide and survive. A "communications fish"[10] weighing 50 kg and powered by a lead-acid storage battery would swim at patrol speed just over the SSBN, a meter or two below the sea surface, receiving the VLF signal in the water and relaying it via megahertz

---

10. R.L. Garwin, "Fish Ragu (Fish, Radio-Receiving and Generally Useful)," JASON Technical Note JSN-81-64, August 1981.

acoustic link to the submarine below. Such high frequencies are strongly attenuated in the water, so that they would not add to the detectability of the SSBN. The storage batteries would propel the fish all day at a patrol speed of 7 km/hr, or for about an hour at 18 km/hr. A recharged fish would swim up from the submarine as required to relieve the duty fish for recharging. If no VLF signal were received, the SSBN or the fish could put up a microwave antenna for a few seconds (on schedule) to listen to communications from a special system of strategic communications satellites.

The much-commented-upon (but in this case exaggerated) difficulty of *communicating* with submarines implies a great difficulty of active detection. Electromagnetic waves are very strongly absorbed or reflected by seawater, so that radio frequencies from the high audio range (10 kilohertz and up) through the entire microwave range are reflected at the water's surface. Radio waves with frequencies in the low audio range can penetrate some tens of meters into the water, but the very long wave length (3,000 km for 100 Hz) could make it difficult to localize a submarine even if the disturbance caused by the submarine could be isolated from disturbances of waves, whales, ships, and the like, and from lightning-produced ambient noise. Not until the visible range does seawater transmit electromagnetic waves, and even here the absorption of light is extreme, allowing light to penetrate only some 100 meters in the clearest water. Nevertheless, there is much discussion these days of the use of "blue-green lasers" for one-way communication to and detection of submarines.[11]

Blue-green laser ASW would involve the use of satellites or aircraft on which the lasers would be mounted. These would be used to scan the ocean surface (penetrating to a depth of 100 meters or so) and detect disturbances in the received signals. Whether satellites or aircraft were used, clouds would

---

11. The color of clean ocean water above coral reefs or white bottom in general indicates to the eye that blue-green light penetrates most effectively. Even so, the penetration is limited, and the light as it penetrates is refracted by surface waves and ripples. The laser light would be sent to one possible submarine location after another in the operating area, scanning that area in the same way that a facsimile machine or a TV screen scans an object or an image. About 1,000 pulses of light per square km would be required, and the presence of a submarine at depth would be indicated by an increase in back-scattered light coming at a time corresponding to the depth of the submarine (for a "white" submarine), a reduction in light from a white or black submarine *beyond* the depth of the submarine, or from a similar disturbance in the signal. The *problems* are to obtain a laser of requisite characteristics, to provide adequate numbers of lasers and platforms (aircraft or satellite), to obtain a useful system in the presence of clouds, and (especially) to persuade the submarines to swim sufficiently close to the surface that they can be detected in this way.

totally vitiate any capability to detect submarines, although any communication function of such lasers could in principle survive passage through clouds. Detailed analysis, independent of progress in laser technology, shows that there is no possibility of strategically significant blue-green laser ASW because even the optimum laser color does not penetrate (in a round-trip) to the comfortable operating depth of existing submarines.

With the advent of fiberoptic transmission of information, the connection of acoustic sensors with information processing stations becomes both more convenient and less costly. Satellite radio relay could be used for similar purposes. Thus, one might imagine short-range direct-path passive submarine detection by bottom-mounted hydrophones covering the entire SSBN operating area. About 500,000 sensors on a 10-km grid could do the job reliably; they of course would have to be monitored in real time. This could be done automatically, but means would have to be taken to ensure that transmission of the acoustic data to the monitoring nation could not readily be impeded, and the whole system would have to be protected against jamming or other disruption. Such a system might be intended for steady operation, or it might be called into operation (aside from tests) only in the event of a crisis or an actual decision to destroy the opposing SSBN fleet. The latter mode might depend on radio transmission to satellites, although such satellites could be jammed by powerful ground-based transmitters. Such a dispersed array of short-range sensors might be countered by the use of jammers or decoys, or by attack on the sensors or their communication nets.

Against passive acoustic ASW, the technologies currently known for reducing radiated noise, for raising the ocean noise level in the region of submarine operations, and the provision of decoys to simulate submarine noise would seem to have the advantage over prospective developments in sensor technology and systems. Jamming and decoys seem also to be considerably cheaper and more rapidly deployable than vast arrays of sensors. Dragging the ocean bottom to cut long-range communication by cable or fiberoptics is an old art.

Conclusion

The oceans are so big that short-range detective devices would be needed by the hundreds of thousands to make the strategic submarine force vulnerable to attack. Long-range detection mechanisms of strategic significance are limited to acoustic detection and are readily countered.

Submarines will always be vulnerable in port, and in small numbers in the open ocean, but a nation with a force of dozens of submarines and significant ocean presence can keep the strategic submarines survivable when deployed. To do this will ultimately require countermeasures such as decoys or jammers, and a continuing awareness of technological developments and deployments.

It seems unlikely that either the U.S. or the U.S.S.R. could hope to achieve a capability for preemptive strike against the other's deployed SSBNs. Consequently an effective strategic doctrine must accommodate the continued capability of each nation to attack the other with SLBMs. Technological advances will likely give those missiles 100-meter accuracy, and communication technologies can provide reliable, timely links to the SLBM fleet. Thus, it makes sense to continue to depend on SSBNs as an important part of the strategic forces.

Although continued invulnerability of strategic submarines does not depend on arms control measures, there are some other benefits of arms control. Thus, a prohibition on the patrol of SSBNs closer than 1,000 km to each nation could ensure the time necessary to allow bombers to leave their bases before they were destroyed by a preemptive SLBM attack. Similarly, a ban on trailing submarines by active sonar would simply save the concern, funds, and forces which would otherwise be devoted to effective countering of the active trailing threat. If sanctuaries for the operation of SSBNs were negotiated, they would be of interest primarily to avoid the loss of SSBNs during limited warfare, in which SSNs might be hunted in the open ocean and SSBNs imperiled.

U.S. strategic doctrine still must reckon with the reality that sufficient resources expended for the destruction of any component of the strategic force could result in its gradual attrition, whether that be land-based missiles, air-breathing weapons such as air-launched cruise missiles, or strategic submarines. Whatever the possibility of such attrition warfare, the U.S. should have a capability to deter or to counter such a threat. Among strategic offensive forces thus far discussed, a fleet of strategic submarines is our greatest assurance of continued invulnerability.

The Submarine in Naval Warfare, 1901–2001

This article surveys the evolution of submarine technology, submarine capability, and strategy for the use of submarines. It traces change in the operational capabilities of submarines since their introduction, evaluates the past effectiveness of submarine forces in war, and suggests how their roles and capabilities are likely to develop in the future. It also addresses the current debate over the proper roles of submarines in naval strategy and discusses prevalent misconceptions about their past and present capabilities.

Submarines are fundamentally different from other warships. Because they function in the underwater medium, submarines tend, unlike surface ships and aircraft, to operate best in isolation; they require unique combinations of weapons and sensors; and they require tactics based on stealth and surprise. They are most capable in the role of hunter in hit-and-run attacks, in attrition warfare, and as platforms for single-salvo strikes ashore. They are least capable in missions that require prolonged exposure and the capability for sustained defense, such as sea control, naval presence, and projection of force ashore in a manner that requires more than a single salvo. Submarines further differ from surface and naval air forces in being most effective when dispersed rather than concentrated. Finally, submarines are different in that the strategies that give them their greatest warfighting potential do not conform to the classical Mahanian naval strategy of defeating the enemy by annihilating his main naval forces. Instead, whether they are employed in commerce warfare, as in the past, or for the delivery of nuclear weapons, submarines are the most effective means for a navy to circumvent classical battle and engage in direct anti-state warfare.

This article is drawn from research in long-range trends being conducted at the Los Alamos National Laboratory, which is operated by the University of California for the U.S. Department of Energy. The conclusions and opinions expressed herein are solely those of the author. I would like to thank Linda L. Riley and Thomas W. Dowler for their very helpful comments and suggestions, as well as Norman Friedman for providing material on British submarines

Karl Lautenschläger is a Staff Defense Analyst at the Los Alamos National Laboratory. He has been an Advanced Research Scholar at the Naval War College and a Visiting Faculty Member at the Fletcher School of Law and Diplomacy, and was a naval officer for five years with two combat deployments to the Tonkin Gulf.

International Security, Winter 1986-87 (Vol. 11, No. 3)
© 1986 by the President and Fellows of Harvard College and of the Massachusetts Institute of Technology.

Submarines have been in regular naval service only since 1901 and have been effective as warships only since about 1910. Yet during their relatively short history, developments in technology have given them the capabilities to perform six basic roles in naval warfare. By the outbreak of World War I, submarines were fully capable in three roles: coast defense, naval attrition, and commerce warfare. Their capacity to perform three additional missions— projection of power ashore, fleet engagement, and assured destruction— matured in the 1960s, after a long period of relative equilibrium in submarine technology that lasted well into World War II. All six remain the basic mission capabilities of submarines today. Current trends suggest three further developments in the near future: a new capability to perform strategic counterforce missions, a decline in the capacity to wage commerce warfare, and the possibility of a new capability in the form of decisive naval battle.

The history of how these capabilities were developed and used in war suggests five principal conclusions about submarine warfare. First, submarines possess no general immunity against countermeasures. Although they are difficult to find and largely immune to attack while cruising submerged, they become vulnerable once they disclose their presence by attacking. In fact, when actively employed in most combat missions, submarines are usually more vulnerable than other types of warships. This reflects the conflicting requirements of lethality and survivability in submarines and is a basic problem of submarine operations.

Second, navies have had difficulty solving the twin problems of properly integrating new submarine technologies into existing force structures and strategies and organizing timely measures to counter new submarine technology. Submarines have sometimes been used unwisely: their existing capabilities have not been fully utilized, while they have sometimes been prematurely assigned new roles before they were ready to carry them out. At the same time, states facing emerging submarine threats have sometimes been slow to respond effectively, leaving themselves vulnerable. Thus submarines have been overutilized, underutilized, and under-prepared-against. This reflects the general problem that military organizations face in adopting appropriate strategies for exploiting and countering fundamentally new technologies and capabilities.

Third, competing demands on available submarine forces during wartime have often prevented them from realizing their full potential through a concerted effort in one strategy. The capability to perform several types of missions brought a tendency to divide and allocate the force to perform all of them. With the singular exception of the U.S. submarine campaign in the Pacific during World War II, this problem has plagued the submarine forces of the world from their inception. Since 1960, this problem has been aggravated by the addition of essential roles, such as strategic nuclear deterrence and the related missions of hunting and protecting ballistic missile submarines, just as submarines are becoming more complex, expensive, difficult to build, and therefore fewer in number.

Fourth, submarine campaigns, against either naval forces or merchant shipping, are not a quick and simple route to victory. They are major undertakings of profound complexity. They extend over entire oceans and involve hundreds of submarines and anti-submarine vessels, and since they require prolonged effort to produce effect, there are ample opportunities for significant change in technology and tactics.

It is therefore delusive to draw simple analogies between past submarine campaigns and current naval problems. For example, it is often argued that the current Soviet submarine fleet has many times the commerce destruction potential of the World War II German submarine fleet, since "a few" German U-boats sank substantial numbers of merchant ships off the U.S. Atlantic coast in 1942, and the current Soviet submarine fleet is much larger than the pre-World War II German submarine fleet.[1] In fact, German U-boat strength in 1942 was about the same as Soviet submarine strength today, and Germany built more submarines in four years than the Soviet Union has built in the last forty.[2] Due to a variety of changed circumstances, Soviet potential to

---

1. For an example of how such mythology gains credence through repetition, see Lockheed Aircraft Corporation's two-page spread advertisements: "In 1942, we took a beating from a handful of enemy subs. Today we have to be ready for 377 unfriendly subs." U.S. Naval Institute *Proceedings*, Vol. 108, No. 10 (October 1982), pp. 20–21; or "If the enemy had 377 subs in 1942, the Battle of the Atlantic could have gone the other way." U.S. Naval Institute *Proceedings*, Vol. 108, No. 9 (September 1982), pp. 18–19.

2. Unclassified sources list 378 submarines for the Soviet navy in 1982, when the Lockheed advertisements cited above were published. Subtracting the 90 ballistic missile submarines, but counting all other types including training boats, the Soviets had 288. The numerical trend has been gradually downward since then. In 1942, the German navy had between 259 and 397 U-boats in commission for an average monthly strength of 330 submarines. Of these, deployed U-boats (*Frontboote*) rose from 101 in January to 168 in August 1942. Jean Labayle Couhat and A. David Baker III, *Combat Fleets of the World, 1982/83* (Annapolis: Naval Institute Press, 1982), pp. 581–582, 602–615; Bodo Herzog, *U-Boote im Einsatz 1939–1945* (Dorheim: Podzun, [1971]), p. 126; and Germany, Seekriegsleitung, "Ubootsverluste (Stand 24.8.42)," Anlage zu 1.Skl.Ib 1663/42, 29 August 1942, PG 31762F, National Archives Microfilm Series T1022, roll 3407. Based on many unclassified sources, a reasonably accurate estimate of Soviet submarine construction from 1946 through 1985, including those for export, is 724 new units. Between the beginning of May 1941 and the end of April 1945, the German navy commissioned 1,007 newly constructed U-boats.

wage commerce warfare is probably substantially smaller than that of Germany in World War II. The opposite conclusion derives from assessing a few rather than the several factors that determine the capability to wage commerce warfare.

Another basic flaw in many current analogies between the present situation and World War II lies in the common assumption that quick and decisive results are possible: that the U-boats were on the verge of victory and that next time we will not be so lucky. While the Battle of the Atlantic was a bitter struggle for the Allies with profound importance for the outcome of the war, the U-boats required prolonged effort to sink as many ships as they did, and they never came close to severing Allied sea lines of communication. Submarine campaigns, whether against commerce or warships, are inevitably protracted affairs. This fact has significance in a world in which the possibilities for nuclear escalation increase with the duration and intensity of conventional conflict.

Fifth and finally, this historical review suggests that the persistent debate over whether submarines or major surface units will have primacy in naval warfare is sterile and misdirected. Submarines and surface fleets are not alternatives to one another. They have parallel, complementary, and independent functions. Naval strategists do not face an "either/or" choice between surface forces and submarines, but rather the task of balancing these forces in a way that enhances the capacity of the whole navy to achieve overall mission goals.

With these problems in mind, this article will describe the basic characteristics of submarines, review the evolution of their basic roles and capabilities, and project current trends in submarine capabilities to the end of this century.

Characteristics of Submarines

The basic characteristics of submarines can be distilled into three generalizations. First, a submarine's effectiveness in war and deterrence depends on stealth, surprise, and a high probability of destroying its target on the first shot or salvo. Second, tactical reconnaissance and target acquisition pose persistent problems for submarine forces used against naval units and merchant shipping. Third, the most effective uses of submarine forces depend on unconventional strategies that differ from classical Anglo-American concepts of naval warfare.

The first set of characteristics—the importance of stealth, surprise, and a high single-salvo kill probability—applies to the use of submarines in warfare at sea, in striking targets ashore, and in deterrent or dissuasive roles by the threat of either.[3] Submarines hardly ever engage in combat, in the sense of sustained use of weapons for assault and defense. They generally carry only enough ready weapons for one or two salvos, and they have little means of defense. In today's naval warfare, they cannot shoot down a torpedo-carrying aircraft or an incoming missile, although they can use decoys to distract torpedoes. Aircraft operating in conjunction with surface ships usually have a greater sensor/weapon range than a submarine, and, with very few exceptions, submarines are highly vulnerable to a single hit.

On the other hand, submarines are very difficult to find when they are not launching torpedoes or missiles or using active sensors such as radar or pinging sonar. This poses a three-part problem for anti-submarine warfare (ASW) units, which must not only detect, but also locate the submarine and direct ordnance against it with precision.[4] As long as a submarine remains passive and quiet, it is seldom detected at all. Thus, the first element of a submarine's ability to survive is not its capability to defend itself or its resistance to damage as in a surface ship; it is its ability to remain undetected.

If a submarine is to be employed as a weapon system, it must make a transition from stealth to active use of its weapons. The employment of an active sensor and the launch of a weapon gives away its position and makes it vulnerable. Passive sensors are capable of detecting and locating a target, but precise data for launching a weapon usually come from active sensors, such as pinging sonar. Whether fire control sensors are active or passive, the launch of weapons is routinely a noisy affair, which discloses the presence of the submarine.[5]

In the attack situation, survival therefore depends upon surprise and a high single-salvo kill probability. Surprise allows the submarine to attack first, just as it compromises its stealth. A high probability of destroying the intended target with a single salvo of torpedoes or missiles allows success without requiring the submarine to remain in contact with enemy forces.

---

3. An excellent summary of many operational characteristics of submarines is provided by Norman Friedman, *Submarine Design and Development* (Annapolis: Naval Institute Press, 1984), pp. 9–16.

4. David R. Frieden, *Principles of Naval Weapons Systems* (Annapolis: Naval Institute Press, 1985), pp. 189–285.

5. The main exception is torpedoes that "swim" out rather than being launched out of a torpedo tube

Destruction with a single weapon would be ideal, but in practical terms, simultaneous firing of two or more is required for a high probability of a lethal hit. Then, the submarine can immediately commence evasive maneuvers and attempt to regain concealment in the ocean depths.

A second set of characteristics in submarines makes tactical reconnaissance and target acquisition major problems. By its very nature, a submarine is short-sighted and vulnerable when on or near the surface, and it is blind but has acute hearing when submerged. Optical and radar sensors have a very short range in submarines because they are close to the water and therefore the line-of-sight horizon is only a few miles. Passive detection of surface ships and aircraft from electronic emissions of their radar and radio equipment is possible at hundreds of miles under ideal atmospheric conditions; this can help a submarine in its search for targets. But submarines cannot obtain the precise and continually updated location necessary to fire a missile at a moving target by this method.

When submerged, a submarine is served by highly sensitive acoustic sensors, called sonar. These too are limited in range compared with electronic sensors that function in the atmosphere, and their performance varies considerably, depending on environmental conditions. Underwater ambient noise, surface disturbances, thermal layers, depth of water, the topography of the ocean floor, and the speed of the sensor platform all affect acoustic sensor range and accuracy. Many conditions, such as those related to thermal layers and higher speed, reduce sonar performance. However, the use of bottom bounce or the focusing of bending sound paths into convergence zones can extend today's detection ranges, in the latter case to 35 and 70 miles but with "blind" areas in between.[6] Sonar can be used as a passive listening device or an active transmitter and receiver of sonic signals, roughly comparable to radar. In general, passive sonar has greater range but is less satisfactory for fire control. While at shorter ranges (out to about 10 miles), active sonar provides more refined target location data, but can disclose the stalking submarine's presence.

Airborne or earth-orbiting platforms can provide submarines both with tactical reconnaissance to allow them to close for a torpedo attack or target location and with firing data necessary to fire missiles from long range. As these capabilities are developed, submarines will be able to sink moving ship

---

6. Edwin W. Shaar, Jr., "ASW and the Naval Officer Oceanographer," U.S. Naval Institute *Proceedings*, Vol. 104, No. 2 (February 1978), pp. 43–49

targets from hundreds of miles just as aircraft have been able to do since the middle of World War II. The persistent problem with any external reconnaissance or targeting system is that it has the same vulnerabilities as the systems that serve the surface fleet. As the evolving capabilities of submarines are surveyed in this paper, it will become clear that as submarines become more capable against first-line naval combatants, they assume vulnerabilities not associated with other submarine roles and missions. This might be called the submarine's capability/vulnerability paradox.

The first two sets of characteristics lead to a third and less familiar aspect of submarines: the best strategies for employing submarines do not conform to the classical principles of naval warfare. Thus, the evolution of submarine technology and tactics is also the evolution of unconventional strategies at sea.

A central part of classical, or conventional, naval strategy is the destruction of the enemy's naval forces in order to gain military objectives as a means to political objectives. The Anglo–American approach, as codified and advocated by Philip Colomb, Alfred T. Mahan, and Julian Corbett, is to concentrate forces for a few decisive battles to gain the military objective of controlling the seas, so that they could be used for military and commercial transportation, while denying that use to the enemy. The alternative version of classical naval strategy, historically adopted by the continental powers of Europe, is to deny an opponent the opportunity to engage in decisive battles for command of the sea while defending important coastal points and conducting attrition warfare against the opponent's naval and merchant shipping. Maintaining a fleet-in-being was a means of threatening the enemy with an inferior battle fleet and tying down a substantial portion of his forces, while avoiding major battles and the risk of defeat at sea. Attrition was a way of wearing the enemy down by attacking weak points. In this context, commerce raiding, or *guerre de course*, was used to harass the enemy, but it was not considered by itself to be a war-winning measure.[7]

From the adoption of fleet tactics for the seagoing gun platform in the early seventeenth century until the development of submarine warfare, fleet actions were the only way to gain control and thereby secure the use of the seas. Although dispersed attrition warfare and fleet-in-being strategies were often the only alternative for a continental power facing a stronger maritime nation, they could not bring a decision in a war at sea. They could only

---

7. Geoffrey Till, *Maritime Strategy and the Nuclear Age* (London: Macmillan, 1982), pp. 1–43.

contribute to the general war effort and avoid defeat by stalemate. Thus, for three centuries, major fleet actions were the most direct and efficient method to win a war at sea.

When first employed as regular units of the world's navies at the beginning of this century, submarines were just another instrument of classical naval strategy or a defensive alternative to it. They were intended for attrition warfare against the enemy's naval units, either on offensive sorties to the vicinity of the enemy's naval bases or as a kind of extended coast defense force. However, this soon changed. The development of submarine capabilities produced alternatives to the classical Mahanian approach to naval warfare.

Submarines have been employed in two basic kinds of unconventional strategy. Both are unconventional (that is, not classical) because the immediate objective is the enemy state rather than its armed forces. Nonmilitary objectives are attacked, and, if possible, the adversary's military forces are avoided altogether. The first kind was intended to cripple maritime commerce, which is an economic component of a state that lies exposed outside its borders on the sea lines of communication. Conceived first for surface torpedo boats in the late nineteenth century, it changed the function of commerce warfare from a means of harassing an opponent's flanks to major offensives against shipping that could decide the outcome of a war.[8] This type of strategy was attempted with submarines three times in two world wars with varying degrees of success.

The second kind of unconventional warfare at sea came when ballistic missiles armed with nuclear warheads were deployed aboard submarines at sea. This gave submarines the capability to destroy population centers, industrial capacity, or economic infrastructure of a state, and the threat of this kind of destruction is of course the basis of assured destruction strategies for deterring war between nuclear powers today.

Unconventional naval strategies using submarines provided alternatives to the classic approach in naval warfare but they have not replaced it. The strengths and weaknesses of these strategies are best explored as they evolved with the new capabilities brought to submarine forces by new ap-

---

8. Theodore Ropp, "Continental Doctrines of Sea Power," in Edward Mead Earle, ed., *Makers of Modern Strategy* (Princeton: Princeton University Press, 1943), pp. 446–456. See also Theodore Ropp, "Development of a Modern Navy: French Naval Policy, 1871–1904" (Ph.D. dissertation, Harvard University, 1937), pp. 33, 258–275.

plications of technology. Thus, we now turn to the salient developments in the operational capabilities of submarines.

Evolving Capabilities

During the twentieth century, technological developments have given submarines six generic capabilities of significance. These six capabilities—coast defense, naval attrition, commerce warfare, projection ashore, fleet engagement, and assured destruction—remain the basic roles of submarines in naval warfare today. However, the evolution of these capabilities was not a smooth process. In some cases, a basic new role was assigned before submarine capabilities were adequate to carry it out, with the requisite technology coming only later. In others, technology was refined sufficiently to provide new capabilities before services adopted a role to utilize them fully.

David Bushnell's barrel-like *Turtle* of 1776 was probably the first functional submarine. But 125 years would pass before a submarine was commissioned for service in a major navy. Continuous development began in 1860, with about fifty experimental prototypes being built in various countries during the rest of the century, and although most submerged and resurfaced successfully with men in them, none was a practical weapon system. Historical literature is replete with declarations that one experimental boat or another marks the advent of the modern submarine, but many technical problems had to be overcome first. These included ballast systems, underwater propulsion, surface and underwater endurance, air supply for the crew, underwater stability and control, suitable weapons, and the means for navigating while submerged.[9] In fact, the modern submarine was not born with one invention. It evolved gradually over many decades before it could become a truly effective naval craft.

The submarine actually became an effective warship in 1910, with the introduction of what was technically a seagoing submersible torpedo boat. This first submarine with all the basic technology to make it effective as a

---

9. Murray F. Sueter, *The Evolution of the Submarine Boat, Mine and Torpedo* (Portsmouth: J. Griffin, 1907), pp. 5–261; Hans–Joachim Lawrenz, *Die Entstehungsgeschichte der U-Boote* (München. J.F. Lehmanns, 1968); Richard Compton–Hall, *Submarine Boats: The Beginnings of Underwater Warfare* (New York: Arco, 1984); Wallace Hutcheon, Jr., *Robert Fulton: Pioneer of Undersea Warfare* (Annapolis: Naval Institute Press, 1981), pp. 31–50; and John M. Maber, "Nordenfelt Submarines," *Warship*, Vol. 8, No. 32 (1984), pp. 218–225.

naval weapon system was the Imperial German Navy's fifth *Unterseeboot*.[10] Commissioned for service in 1910 as U-9, the 485-ton submersible had a range of 3,200 nautical miles, giving it an operational radius of 1,000 miles with five days on station in an assigned patrol area. The boat was armed with six fairly reliable Schwartzkopf torpedoes, and was the first submarine to be equipped with a gyrocompass, which enabled it to navigate while submerged.[11] Although the German navy did not introduce diesel engines in its submarines until 1913 when U-19 was commissioned, the first U-boats had relatively safe Korting kerosene engines.[12]

Coast Defense

Submarines were formally given their first role in naval warfare before they acquired the capability to accomplish it. Several basic technical problems had yet to be solved when the major navies began putting small submarines into service for coast defense, nearly a decade before U-9 was commissioned. The submarines built with navy funds before 1901 were one-of-a-kind projects used only for experiments and limited training. But beginning in 1901, France put submarines into regular service, followed by Britain in 1902, the United States in 1903, and Russia in 1904.[13] All of these early submarines had operational radii of only a few hundred miles as well as major deficiencies in their design, and none had any means of navigating under water. Although first to introduce diesel engines in submarines, the French navy clung to either steam and electric or all-electric propulsion in spite of the inherent limitations of each because of the problems encountered with early diesels. The British, American, and Russian navies adopted dangerous and unreliable gasoline engines for the submarines in their first service flotillas.

---

10. Actually eight boats were built to the same basic design beginning with U-5, of which four were commissioned in 1910, although U-9 went into service three months before U-5 did. Four earlier boats were much less capable and largely experimental in nature.
11. Eberhard Rössler, *The U-boat: The Evolution and Technical History of German Submarines*, trans. Harold Erenberg (Annapolis: Naval Institute Press, 1981), pp. 23–27. There are several somewhat important translation errors in other sections of this English language edition.
12. The problem with the clouds of white exhaust from the kerosene engines was solved by stockpiling a more expensive grade of fuel for use only in wartime. Ibid., p. 33.
13. The first submarines commissioned in these four navies for regular service and not merely for experiment and training were: *Sirene* and *Triton* (France), December 1901; *Holland* No. 2 and *Holland* No. 4 (Great Britain), August 1902, *Adder*, later A2, and *Moccasin*, later A4 (United States), January 1903; and *Delfin* (Imperial Russia), June 1904.

The numbers required for an actual operational capability were not available immediately. In 1905, the French navy was the first to have the equivalent of a flotilla of sixteen submarines in service.[14] By the end of 1909, all four navies listed above had in fleet service the equivalent of one to four flotillas of sixteen submarines each. However, pure numbers and the official adoption of the submarine by the major navies could not compensate for critical limitations in operational capability. It was not until the second decade of commissioned service that the technological ensemble was completed and the component technologies were refined sufficiently to give the submarine reliability and fighting effectiveness.

Small submarines with short range have retained their usefulness in contemporary navies because they are relatively inexpensive. Although inadequate for a major naval power with an oceangoing fleet and overseas interests, the coast defense capabilities of submarines allow a few of them to make a valuable contribution to coastal security. Led by Sweden in 1904, countries with modest navies acquired small numbers of submarines for this role. The practice has continued, and for the past decade, Third World countries have been acquiring the latest types of diesel-electric submarines at an impressive rate. Today more than thirty small navies have from 3 to 15 modern submarines for coast defense. Many represent the best in diesel-electric submarines for range, speed, and armament, and are better than many nuclear submarines in quieting.[15]

Naval Attrition

As a type of submarine warfare, naval attrition is used here to indicate a strategy of wearing down enemy naval forces through gradual attrition of

---

14. There is nothing magical about the number 16. Surface and subsurface torpedo boats have traditionally been organized into flotillas of 6, 8, 12, or 16 units. To be useful in warfare, submarines need to be deployed in numbers, making a single large flotilla a reasonable baseline here for dating the beginnings of a combat capability. Different measures are used later in this paper for missile submarines.
15. Mark Hewish, Christopher Dawson, and Bob Dicker, "Diesel–Electric Submarines and Their Equipment," *International Defense Review*, Vol. 19, No. 5 (1986), Special Supplement; Jean Labayle Couhat and A.D. Baker III, eds., *Combat Fleets of the World, 1986–87* (Annapolis: Naval Institute Press, 1986); Christian Eliot, "Nuclear and Conventional Submarines," *Naval Forces*, Vol. 5, No. 1 (1984), pp. 60–72; Klaus Winkler, "Developments in the Design of Conventional Submarines," *Naval Forces*, Vol. 4, No. 6 (1983), pp. 50–58; Ulrich Gabler, "Further Development of Conventional Submarines," *Military Technology*, Vol. 7, No. 3 (1983), pp. 42–48; and F. Abels, "Developments in Conventional Submarine Design," *Naval Forces*, Vol. 5, No. 6 (1982), Special Supplement, pp. 61–65.

ancillary, obsolescent, and independently steaming warships. In contrast to commerce warfare, which is attrition of merchant shipping, it is directed specifically at combatant vessels. At this early stage in their evolution, submarines were not yet capable of engaging a first-line battle fleet, except through chance encounters that occasionally led to a single "hit and hide" attack. It would be several decades before new technologies would enable submarines to intercept and engage a concentration of high-speed surface combatants.

As diesel engines were refined and fuel capacities increased, submarines acquired a seagoing capability, and their roles expanded to include naval attrition. The typical seagoing submarine had a range of 3,200 miles on the surface at 8 knots using diesel engines and a range of 65 miles submerged at 5 knots on electric motors powered by batteries. The Royal Navy was the first to deploy seagoing boats in flotilla strength, with its first sixteen "D" and "E" class boats joining operational units between 1911 and 1914, although they lacked a basic capability for underwater navigation until 1914, when they were fitted with gyrocompasses.[16]

The early seagoing submarines introduced two forms of naval attrition, and in the midst of World War I they were employed in a third. British submarines were intended to cruise in the approaches of enemy naval bases and sink warships. This was an offensive anti-warship mission and explains why the Royal Navy called their seagoing submarines "overseas boats." A defensive version of this strategy was envisioned in the French and American navies for their first seagoing boats. They planned a kind of extended coast defense in which their submarines would cruise several hundred or a thousand miles from home bases and sink as much of the enemy fleet as possible before it could approach friendly shores. Anti-submarine warfare was established as a third form of naval attrition in 1917, when the Royal Navy made hunting U-boats the primary task of its submarine force.

Submarines engaged in naval attrition for the first time during World War I, when navies on both sides adopted this strategy. British submarines patrolled off German naval bases located on the North and Baltic seas, and German submarines planted mines and lay waiting to torpedo Allied war-

---

16. Great Britain, Ministry of Defence, Ship Department, *The Development of HM Submarines, From Holland No. 1 (1901) to Porpoise (1930)*, by A.N. Harrison, BR 3043, January 1979, pp. 4.1–4.25, 22.3–22.4, and Appendices 1 and 3. British submarine D-1, commissioned in 1909, was a seagoing diesel-electric prototype of shorter range than its near sisters and was employed mainly for trials for its first few years in commission.

ships in waters near their bases in the British Isles and the Mediterranean. German submarines were also used to keep the British fleet away from German shores, Austro–Hungarian submarines were deployed to keep French and Italian warships out of the Adriatic, and both attacked Allied warships supporting the landings at Gallipoli.[17]

Submarines were not particularly effective against battle fleets steaming in formation. Their greatest success against these forces during the European conflict of 1914–18 was in constraining their areas and modes of operation. Fleet commanders on both sides suffered considerable anxiety over submarines, in spite of the fact that their losses to submarine attack were minimal. No first-line capital ship on either side was sunk by a submarine. Only two British cruisers and one destroyer were torpedoed and sunk while steaming with a battle fleet, and of its modern warships, the German navy lost only two destroyers to British submarines.[18]

The general problem for submarines attempting to engage a battle fleet was that they could seldom get into firing position. Even at modest cruising speeds, the fleet was two to three times faster than a submarine running submerged. If a submarine surfaced, it would be vulnerable to the fleet's destroyer screen, and it would still have a speed disadvantage. Second, in order for submarines to conduct attrition warfare against first-line naval forces with any effect, they had to sink its dreadnought battleships. Even if a submarine could get into firing position, first-line capital ships were very difficult to sink because of their relatively stout construction, extensive hull compartmentation, and internal torpedo bulkheads. The six dreadnoughts torpedoed by submarines on both sides during the war were operating alone at slow speed, and they were repaired within a matter of several weeks.[19]

---

17 Arthur Hezlet, *The Submarine and Sea Power* (New York: Stein and Day, 1967), pp 24–42, 67–84; Arno Spindler, "The Value of the Submarine in Naval Warfare," U.S Naval Institute *Proceedings*, Vol. 52, No. 5 (May 1926), pp. 844–851; and Wladimir Aichelburg, *Die Unterseeboote Osterreich-Ungarns*, 2 vols. (Graz: Akademische Druck- und Verlagsanstalt, 1981), Vol. 1, pp. 75–76, 84–91, 121–123, Vol. 2, pp. 490–495.

18. HM cruisers *Nottingham* and *Falmouth* (19 August 1916), HM destroyer *Scott* (15 August 1918); Henry Newbolt, *Naval Operations*, History of the Great War Based on Official Documents, Vols. 4 and 5 (London: Longmans, Green, 1928, 1931), Vol. 4, pp. 35, 38, 45–46. German fleet torpedo boats (destroyers) *V 188* and *S 33* were sunk by British submarines on 6 October 1914 and 3 October 1918, respectively. Erich Gröner, *Die deutschen Kriegsschiffe 1815–1945*, 7 vols. (3rd ed., München: Bernard & Graefe, 1982–), Vol. 2, pp. 49, 54.

19. One French and five German capital ships were torpedoed by submarines. *Jean Bart*, torpedoed 21 December 1914 (103 days to repair); Robert Dumas and Jean Guiglini, *Les Cuirasses Frances de 23.500 Tonnes*, 2 vols. (Grenoble: 4 Seigneurs, 1980), Vol. 1, pp 54–55, 236–237. *Moltke*, 19 August 1915 (32 days); *Grosser Kurfürst*, 5 November 1916 (97 days); *Kronprinz*, 5 November

However, submarines sank many second-line and obsolete warships, because the older ships were poorly protected against underwater explosions and because they were employed on independent patrol missions and on slow speed operations in confined waters or close to shore. Still, overall results were not great compared to the losses from gunnery actions and mines.[20] During World War I, submarines thus contributed to offshore coast defense through dissuasion rather than sinking warships. They had little effect on battle fleets but modest success against second-line units in their attempts at attrition warfare against warships.

In the Second World War, submarines of both sides engaged in naval attrition in the Atlantic, Mediterranean, and Pacific theaters. Their performance against fleet units was much better than in the First World War.[21] In the Atlantic and Mediterranean, German U-boats sank 54 first-line surface warships, including 3 carriers and 2 battleships.[22] In the Pacific, American submarines sank 62 fleet units, including 5 carriers and a battleship.[23] Ironically, the Japanese navy, which emphasized the employment of submarines against warships, was not nearly as successful, sinking only 9 first-line warships, including one carrier.[24] However, early in the war, Japanese submarines helped curtail U.S. fleet operations by damaging 2 battleships and a carrier on two separate occasions. The most spectacular feat of a submarine operating against a fleet formation took place when the Japanese submarine

1916 (31 days); *Westfalen*, 18 August 1916 (46 days), *Moltke*, torpedoed 25 April 1918 while under tow after major machinery casualty and flooding (137 days); Hans H. Hildebrand, Albert Röhr, and Hans-Otto Steinmetz, *Die deutschen Kriegsschiffe*, 7 vols. (Herford: Koehler, 1979–1983), Vol. 3, p. 32, Vol. 4, pp. 55, 137–138, Vol. 6, p. 47. The British dreadnought *Audacious* was sunk indirectly by a submarine when it struck a mine laid by a German U-boat and sank 27 October 1914; Julian S. Corbett, *Naval Operations*, History of the Great War Based on Official Documents, Vols. 1–3 (London: Longmans, Green, 1920–1923), Vol. 1, pp. 249–251.

20. Great Britain, Royal Navy, *Navy Losses* (London: His Majesty's Stationery Office, 1919), pp. 3–6, 8; and Jean Labayle Couhat, *French Warships of World War I* (London: Ian Allan, 1974), pp. 290–294. See also pp. 12–122 for details of circumstances surrounding individual losses. Aldo Fraccaroli, *Italian Warships of World War I* (London: Ian Allan, 1970), pp. 13–71; and Gröner, *Die deutschen Kriegsschiffe 1815–1945*, Vol. 1, pp. 46, 78–80, 85, 128–140, Vol. 2, pp. 43–62.

21. Hezlet, *The Submarine and Sea Power*, pp. 124–136, 191–209.

22. Stephen W. Roskill, *The War at Sea 1939–1945*, 3 vols. (London: Her Majesty's Stationery Office, 1954–61), Vol. 3, Part 2, pp. 439–442, 448.

23. Anthony J. Watts and Brian G. Gordon, *The Imperial Japanese Navy* (Garden City: Doubleday, 1971), pp. 43–72, 127–164, 172–201, 257–295; and Hansgeorg Jentschura, Dieter Jung, and Peter Mickel, *Warships of the Imperial Japanese Navy 1869–1945*, trans. Antony Preston and J.D. Brown (Annapolis. Naval Institute Press, 1977), pp. 26–58, 80–87, 105–112, 141–153.

24. Jürgen Rohwer, "Die Erfolge der japanischen U-Boote 1941–1945," *Marine Rundschau*, Vol. 61, No. 4 (April 1964), pp. 86–88; and Mochitsura Hashimoto, *Sunk: The Story of the Japanese Submarine Fleet*, trans. E.J.M. Cole Grave (London: Cassell, 1954).

I-19 launched a single salvo of six torpedoes at the carrier *Wasp*, sinking the carrier, damaging a battleship, and sinking a destroyer.[25]

In wearing down the enemy's main fleets by attrition, the American and German submarine services in particular made major contributions to their countries' war at sea. However, the significant turning points in the struggle for control of the sea were still fleet engagements. The destruction of naval combatants in fleet actions was what blunted or sustained naval and amphibious offensives. Unlike the losses inflicted by submarines, the destruction of warships in fleet actions was concentrated in time and often caused by combinations of weapons. In the Pacific, American submarines destroyed or disabled over half of Japan's warship tonnage, but most of these ships were torpedoed in late 1944, after command of the sea had been decided in major naval battles off Midway, the Eastern Solomons, Guadalcanal, and in the Philippine Sea. In major fleet actions, the big killer of fleet units during World War II was aircraft. Submarines sank large numbers of warships, but their effect on enemy naval strength was through gradual attrition. By the end of the war, the submarine was thus a major contributor but not yet an arbiter in deciding who controlled the seas. For submarines, naval attrition had not yet become fleet engagement and certainly not decisive battle.

The first two types of naval attrition remain important uses of submarines today. In the South Atlantic war of 1982, for example, Argentina deployed its submarines for extended coast defense of the Falkland Islands, and Britain employed its in offensive anti-fleet operations that resulted in the sinking of the cruiser *General Belgrano*.[26] These roles are important for superpower navies as well. For example, just as the commanders of the British Grand Fleet had to be concerned about losses to U-boats in the North Sea during World War I, American carrier battle groups operating in the Gulf of Sidra today must be prepared to deal with Libya's small but relatively modern submarine force.

---

25. Ben W. Blee, "Whodunnit?," U.S. Naval Institute *Proceedings*, Vol. 108, No. 7 (July 1982), pp. 42–47.
26. Robert L. Scheina, "Where Were Those Argentine Subs?," U.S. Naval Institute *Proceedings*, Vol. 110, No. 3 (March 1984), pp. 114–120; Steven Gorton, "Thoughts on the Falkland Islands War," U.S. Naval Institute *Proceedings*, Vol. 108, No. 9 (September 1982), pp. 105–107; J.V.P. Goldrick, "Reflections on the Falklands," U.S. Naval Institute *Proceedings*, Vol. 109, No. 6 (June 1983), pp. 102–103; Carlos E. Zartmann, "An Old-Fashioned Modern War," U.S. Naval Institute *Proceedings*, Vol. 109, No. 2 (February 1983), p. 87; and John Byron, "The Submarine and the Falklands War," U.S. Naval Institute *Proceedings*, Vol. 108, No. 12 (December 1982), p. 43, Vol. 109, No. 4 (April 1983), pp. 11–12.

The third form of naval attrition, the use of submarines against submarines, has only recently become a significant wartime role of submarines. Although this was the primary mission of British submarines in the latter part of World War I, only 18 German U-boats, or 10 percent of the losses sustained by the German submarine force, were sunk by British submarines. German submarines sank 5 of their British counterparts, also representing only 10 percent of the losses sustained by the force. In World War II, these two submarine forces achieved even less against one another. The best performance was by American submarines, which sank 20 Japanese submarines, accounting for about 15 percent of the Imperial Navy's submarine losses. The main problem was target detection and location. In the First World War, submarines had to search for each other on the surface, because they had no means of underwater detection. In the Second World War, sonar provided submarines with an underwater sensor, but it had an effective range of only a few thousand yards under the best of conditions.

Today, submarines have become the most lethal anti-submarine systems under many conditions. Highly sensitive, long-range acoustic sensors give them the capability to detect and localize targets operating within a common underwater medium. Near an adversary's naval bases or under polar ice, submarines are essentially the only effective anti-submarine weapon and sensor platforms. Most significantly, the capability to hunt and destroy ballistic missile submarines now gives this third form of naval attrition importance as a threat to the submarine's role in nuclear deterrence through the capability of assured destruction. However, without significant advances in detection technology, neither superpower would probably destroy more than a few ballistic missile submarines, even in the most aggressive of attrition campaigns.

Commerce Warfare

The capability of submarines to destroy commerce on the high seas introduced a new form of naval warfare that did not depend on exercising control of the seas. Commerce destruction as a mode of naval warfare is difficult to describe, let alone assess accurately. Like most activities in war, the contributing factors are numerous and diverse and their overall effects cumulative.[27]

---

27. Several, but not nearly all, of these factors are explored in the development of a mathematical model for planning and assessing submarine warfare against merchant shipping in Robert

Some of today's outspoken commentators oversimplify the problem by searching for a few decisive factors, leading to erroneous conclusions about commerce warfare. But there is no question that the cumulative effects of the major submarine campaigns in both world wars have been significant for naval warfare and for the conduct of war as a whole.

Submarines acquired the capability to conduct commerce warfare in the last year before the First World War, as the German navy began to build up its submarine arm. The U-boats put into series production were comparable to the British "D" and "E" classes in speed and armament, but they had three times the range. Beginning with U-19, commissioned in 1913, diesel engines were used for surface propulsion in German submarines, but just as significant was the boat's range of 9,700 nautical miles. Except for specialized coastal types designed during the war to operate from bases seized in Flanders, German submarines built during the next five years had ranges of over 9,000 miles.[28] This oceangoing force could stay on patrol for four weeks at a radius of 2,500 miles from its bases. The German navy, alone among the navies of the world, had essentially, though not intentionally, skipped the coastal and seagoing stages in developing the operational capabilities of its submarines. Comparable cruising radius was introduced in American and Japanese submarines during the 1920s, in British submarines in the 1930s, and in Soviet submarines in the 1950s.

The oceangoing submersible torpedo boat would soon demonstrate formidable capabilities as a commerce destroyer, but in this case, the new capability was developed before the new role was adopted in a strategy. Before the outbreak of war in Europe, the German admiralty had no intention of employing its submarines for commerce warfare. The only prewar study of submarine requirements for a campaign against British merchant shipping was prepared by an obscure lieutenant named Ulrich-Eberhard Blum at the Submarine Inspectorate.[29] German war plans anticipated using submarines for coast defense off major ports and naval bases, for anti-warship patrols in the North Sea, and for naval attrition and reconnaissance in cooperation with

Eugene Kuenne, *The Attack Submarine: A Study in Strategy* (New Haven: Yale University Press, 1965).

28. Rössler, *The U-boat*, pp. 28–80, 328–333; and Groner, *Die deutschen Kriegsschiffe 1815–1945*, Vol. 3, pp. 28–62.

29. Philip K. Lundeberg, "The German Naval Critique of the U-Boat Campaign, 1915–1918," *Military Affairs*, Vol. 27, No. 3 (Fall 1963), pp. 106–107.

battle fleet operations in the North Sea. Not until the land war settled into a stalemate on the Western Front, did the German naval staff begin to seriously consider the possibility of using submarines to sever vital sea lines of communication to Great Britain.

Submarines have been used in three major campaigns against ocean commerce. German submarines sank shipping around Great Britain from 1914 to 1918, but a concerted effort to sever sea lines of communication only came in 1917–18. Germany attempted the same strategy in the Second World War, with the major surges in effort coming in 1941 and 1943. The third submarine campaign was waged by the United States against Japan, beginning in 1941 and continued for the duration of the Pacific war.

Each of these campaigns underwent a series of evolutions. There was no single key to the successes or setbacks on either side. Not only was there a series of initiatives and countermeasures in each case; the campaigns also comprised thousands of diverse and individual operations, each with its own set of tactical conditions and technical factors.

THE FIRST WORLD WAR

The German submarine campaign against Great Britain during the First World War demonstrated both that the submarine could be a lethal commerce destroyer and that it could be defeated, but there were only false starts and intermittent efforts for the first thirty months of the war.[30] During the first six months, U-boat commanders, on their own initiative, sank only 10 merchant ships totaling about 20,000 tons. Most of these were ordered to stop by surfaced U-boats according to international prize rules and sunk with scuttling charges in their holds after the crew pulled away in lifeboats.

In February 1915, Germany declared the first of two submarine blockades of the British Isles during the Great War. Because the early German submarines carried only six torpedoes, more ships could be sunk if the U-boats continued to operate under international prize rules. However, a number of

---

30. Albert Gayer, "Summary of German Submarine Operations in Various Theaters of War from 1914 to 1918," U.S. Naval Institute *Proceedings*, Vol. 52, No. 4 (April 1926), pp. 621–659. The official history of the U-boat campaign from the German naval archives is Arno Spindler, *Der Handelskrieg mit U-Booten*, 5 vols. (Berlin: E.S. Mittler und Sohn, 1932–34, 1941, 1966). Summaries of war patrols of nearly every U-boat are provided in U.S. National Archives and Records Service, *U-Boats and T-Boats 1914–1918*, prepared by Harry E. Rilley and Johanna M. Wagner, Guides to Microfilmed Records of the German Navy, 1850–1945, No. 1 (Washington: U.S. Government Printing Office, 1984), pp. 1–138.

ships, including passenger liners, were torpedoed without warning, bringing protests from neutral countries and causing the German admiralty to issue more and more restrictions (today called "rules of engagement"). At the same time, Britain gradually armed its merchant ships, and deployed heavily gunned decoys called Q-ships, making it increasingly dangerous for U-boats to attempt to operate according to international prize rules. This first attempt at concerted submarine warfare never gained momentum. The blockade was formally ended just six months after it had begun, under pressure from the neutral United States.

Although a few U-boats continued a restricted campaign against maritime commerce around the British Isles, most of the North Sea boats were sent to the Mediterranean. They inflicted substantial losses on shipping in the Mediterranean during much of the war, but this never threatened the survival of France or Italy and in some respects it was a diversion of effort, because the shipping vital to England was on the North Atlantic sea lanes. Restrictions on areas of operation and tactics were not the only obstacles to success. Until the numbers of submarines on patrol could be increased, their effect in any theater would be limited. As more U-boats were built and the newer boats went to sea with 8 to 12 torpedoes, the monthly rate of sinkings began to rise. In September 1916, the monthly total went above 200,000 tons for the first time.

Finally, in February 1917, Germany began a second unrestricted submarine campaign, hoping to knock Britain out of the war before the United States could mobilize to take part in the Allied war effort. The German navy had had oceangoing submarines since 1913, but it was not until this time that there were adequate numbers of these submarines to wage effective commerce warfare. The next six months were the worst for British shipping in either world war. German U-boats were able to sustain an average of 614,000 tons sunk per month for the entire period, and most of the sinkings were inflicted where they had the greatest effect: close to the British Isles.[31] The

---

31. Excellent statistical summaries of the World War I German submarine campaign are in Bodo Herzog, *60 Jahre Deutsche U-boote 1906–1966* (München: J.F. Lehmanns, 1968), pp. 67–129; Arthur J. Marder, *From the Dreadnought to Scapa Flow*, 5 vols. (London: Oxford University Press, 1961–1970), Vol. 5, pp. 110–120; Newbolt, *Naval Operations*, Vol. 5, pp. 387–429; C. Earnest Fayle, *Seaborne Trade*, History of the Great War Based on Official Documents, 3 vols. (London: John Murray, 1920–1924), Vol. 3, pp. 465–479; and Great Britain, Royal Navy, *Merchant Shipping (Losses)* (London: His Majesty's Stationery Office, 1919), pp. 162–164.

title of Admiral John Jellicoe's book about this period, *The Crisis of the Naval War*, aptly summarizes the British perspective on what was happening.[32]

Although loss rates were alarming, German submarines were never able to impose a blockade on Britain. Allied and neutral shipping not only delivered vital foodstuffs and raw materials, but supplied the war effort on the Western Front from overseas. No single countermeasure defeated the German submarine campaign of 1917, although reintroduction of the ancient convoy system is often cited. The combination of Allied shipping control, the availability of neutral shipping, the merchant convoy, special anti-submarine weapons, thousands of mines laid in the approaches to German submarine bases, and a vigorous construction program to replace shipping losses all contributed in significant ways.[33] Some brief examples show how these factors worked together.

A fundamental cause of Germany's failure to establish a submarine blockade was the availability of neutral shipping for the Allied war effort. The German high command calculated that Britain would be knocked out of the war if 600,000 tons of British shipping were sunk every month for six months, but this assumed that neutrals would be coerced into keeping their ships in port by the ruthlessness of the U-boat offensive. In the event, diplomacy put neutral shipping back to sea after some initial hesitation and, in effect, raised the requirement for victory to a rate of about 900,000 tons per month.[34] Even at its peak strength of 172 boats, Imperial Germany's U-boat arm could not hope to accomplish this.

Further scrutiny of how the U-boats were defeated shows that anti-submarine warfare was not like the more traditional methods of naval combat. Although there was a frantic search for technological antidotes to the submarine, the only useful weapon developed during the war was the depth charge, and it was hardly more successful in sinking U-boats than gunfire or ramming and far less successful than mines. Of the 320 U-boats that sortied during the war, 178 were lost, including 134 to anti-submarine measures. But loss rates for the U-boats were in fact higher in late 1916, long

32. John Jellicoe, *The Crisis of the Naval War* (London: Cassell, 1920). See also Marder, *Dreadnought to Scapa Flow*, Vol. 4, pp. 49–292; and John Jellicoe, *The Submarine Peril* (London: Cassell, 1934).
33. Hezlet, *The Submarine and Sea Power*, pp. 93–107; Fayle, *Seaborne Trade*, Vol. 3, pp. 454–458; James Arthur Salter, *Allied Shipping Control* (Oxford: The Clarendon Press, 1921); and Patrick Beesly, *Room 40: British Naval Intelligence 1914–1918* (London: Hamish Hamilton, 1982), pp. 253–270.
34. Hezlet, *The Submarine and Sea Power*, pp. 85–86, 90–92.

before the convoy system was introduced and when they were operating more against warships. There was something more fundamental at work here than weapons or convoy tactics.

In this new and very different kind of naval warfare, the important measure of success was not how many submarines were sunk, but how many merchant ships reached their destination. The defeat of the U-boats in 1917–18 was not due so much to actual losses, as to the submarines' growing difficulty in finding targets and an inability to get past the convoy escort to sink them when they were found. Centralized shipping control allowed the merchantmen to be routed around areas where U-boats were known to be operating, and even more significantly, convoys greatly reduce the opportunities for visual contact. Once in contact, the submarine was not so often sunk as it was forced to stay submerged below periscope depth while the convoy passed in safety.[35] Thus, the first major submarine campaign against ocean commerce was defeated not in the usual way of sinking enemy ships, but by preventing the enemy from sinking one's own ships.

THE ATLANTIC: ROUND TWO

The German navy waged its second submarine campaign against British commerce, beginning in the summer of 1940. The campaign developed into five major phases, each representing a different set of tactics, new countermeasures, and a shift in operating areas.[36] The objective was to sink as much Allied tonnage as possible for smallest losses to the U-boat force in what was called "tonnage warfare." The task for the U-boats had grown considerably since World War I. The British Empire and the United States alone had 33 million tons of shipping in 1939, and during the war they would build another 42 million tons. To have an effect on merchant fleets of this size, there would indeed have to be tonnage warfare. But seen in retrospect, tonnage warfare was carried out as a series of shifts toward and away from the ultimate objective of severing the sea lines of communication to Great Britain.

Today, countless allusions are made to this campaign as a model for what might happen in a third world war. However, most of these fail to take into account the many complex factors determining its course and final outcome.

---

35. Marder, *Dreadnought to Scapa Flow*, Vol. 4, pp. 285–286, Vol. 5, pp. 88–104.
36. Statistics and outlines of the various stages in the campaign are given in Herzog, *U-Boote im Einsatz, 1939–1945*, pp. 59, 85–86, 125–127, 187–188, 225; and Willem Hackmann, *Seek and Strike: Sonar, Anti-Submarine Warfare and the Royal Navy 1914–1954* (London: Her Majesty's Stationery Office, 1984), pp. 235–237, 239.

For reasons of both its popularity as a model and the complexity of its execution, the second Atlantic submarine campaign against commerce bears at least brief description.[37]

The first nine months of the European war saw only preliminaries to the U-boat campaign. The German navy had not prepared for a submarine offensive, because the U-boat campaign of 1917–18 was seen as a failure and a mistake, and because the admirals dominating the naval staff preferred a fleet of heavy surface units. The staff had gradually adopted a strategy of commerce destruction from the late 1920s, but the means were to be a kind of combined arms approach, using surface action groups, independent cruiser raiders, aircraft, and submarines. Lack of preparation before the war meant that an average of only 6 submarines operated in the Atlantic for the first several months of the war. Such small numbers were capable of little more than harassment, and in March 1940, all available U-boats were recalled to take part in the German invasion of Norway.

The first phase of the German submarine offensive against shipping really began in late May 1940, when the U-boats were redeployed to the western approaches of British seaports. In August, they began operating from bases on the Atlantic coast of German-occupied France, which significantly reduced transit time to patrol areas. Individual U-boats operated on the surface so that they did much to alleviate the perennial problems with tactical reconnaissance faced by all submarine forces; but it would also become a major vulnerability.

Numbers were a problem for both the offense and defense at this stage. In November 1940, Dönitz ordered the U-boats to coordinate their attacks on convoys in what he called wolf pack tactics, but since the monthly average of U-boats at sea was only 10 in this phase, the rate of sinkings stayed at around 200,000 tons per month, in spite of the fact that losses to ASW were insignificant. On the defensive side, a shortage of suitable warships forced the Royal Navy to escort merchant convoys for only a few hundred miles to and from British ports. This was also within the range of land-based aircraft,

---

37. Basic histories of the Campaign are Karl Donitz, *Memoirs Ten Years and Twenty Days*, trans. R.H. Stevens (Cleveland. World, 1959), and Roskill, *The War at Sea 1939–1945*. Detailed statistics are provided in Herzog, *60 Jahre Deutsche U-Boote 1906–1966*, pp. 209–296; and Jurgen Rohwer, *Axis Submarine Successes 1939–1945*, trans. John A. Broadwin (Annapolis: Naval Institute Press, 1983). Summaries of war patrols of nearly every U-boat are provided in U.S. National Archives and Records Service, *Records Relating to U-Boat Warfare, 1939–1945*, prepared by Timothy Mulligan, Johanna M. Wagner, and Mary Ann Coyle, Guides to Microfilmed Records of the German Navy, 1850–1945, No. 2 (Washington: U.S. Government Printing Office, 1985), pp. 23–198.

which actually sank no U-boats, but worried their commanding officers and kept them from attacking in daylight.

In the second phase of the campaign, beginning in April 1941, the U-boats moved their main area of operations farther westward, to the central Atlantic. German submarine strength rose steadily from 120 to 200 boats over the next six months and losses remained low, but rate of sinkings fell off during the summer as ASW measures improved. More escorts became available, and they were able to accompany merchant convoys all the way across the Atlantic. The cover of darkness was being penetrated by the middle of 1941, when escorts were equipped with radar that could locate a surfaced submarine out to about 3 miles, beyond effective torpedo range. The Admiralty was able to reroute convoys away from known concentrations of enemy submarines using direction finding on radio transmissions from the U-boats to Dönitz's command center. British naval intelligence also began breaking German naval codes and was able to learn much of what Dönitz intended for his submarines from the content as well as the numbers and origin of radio messages. In September, the German naval command began diverting its submarines to the Mediterranean to attack British naval forces which were threatening to sever German and Italian supply lines to North Africa. In November, as more U-boats were being sent to the Mediterranean and loss rates in the North Atlantic began to climb, Dönitz temporarily suspended U-boat attacks on convoys.

U.S. entry into World War II led to the third phase in the German submarine campaign in January 1942. For Dönitz's concept of tonnage warfare, this was an opportunity to attack unprotected shipping with little initial risk of losses. Unprotected, the shipping in American waters was easy prey for the U-boats, and the rate of sinkings rose rapidly. From March until November 1942, U-boat strength rose from 284 to 379, but only 6 to 12 boats were in American waters at any one time, because of long transit times, the demand for boats in other operating areas, and the fact that several months were required to train each crew for the scores of new boats. Yet German submarines sank an average of 500,000 tons of shipping per month in the Atlantic during this period, most of which was west of 50°W in a great crescent from Newfoundland to the Amazon delta. About one-third of this tonnage was American.

When the U.S. Navy introduced convoys and relays of coastal escorts in a kind of "bucket brigade" approach, the U-boats were ordered to move their area of operations southward. By late summer, their main area of success was the Caribbean. As Dönitz shifted the areas of operation farther south to avoid each new patrol and convoy area, the U-boats moved farther from their primary objective. Every Allied merchantman sunk added to the cumulative effort, but the United States was not dependent on the seas for survival and Britain was. For the limited number of U-boats available to put direct pressure on Britain, they had to attack ships steaming to and from the British Isles. Toward the end of 1942, as the coastal convoy system was extended, the numbers of American escorts increased, and the coverage of air patrols expanded, sinkings in American waters declined significantly.[38]

In the fourth phase, as the campaign in American waters lost momentum, the German navy renewed its offensive against the north Atlantic convoys from about October. By January 1943, it had over 400 U-boats in commission, of which over half were operational. Night wolf pack tactics were used in the mid-ocean area out of range of land-based aircraft. The worst month for the Allies was November 1942, when 743,000 tons of shipping were sunk by German submarines in all theaters. The culmination of the struggle came in March 1943 with a series of big convoy battles. In one case, 49 U-boats attacked two convoys totaling 88 merchantmen with only 14 escorts, sinking 21 ships for the loss of one submarine. After March, U-boat successes fell off rapidly and their own losses began to rise as Allied ASW measures were improved and expanded. By May, the offensive had been defeated, and at the end of August, Dönitz withdrew his submarines from the central Atlantic.[39]

In the fifth and last phase, the U-boats were dispersed to search for weak points in the Allied sea lines of communication, but from August 1943 until June 1944, when the U-boats were recalled to participate in the defense of "fortress Europe," the monthly tonnage sunk exceeded 100,000 on only one occasion.

A number of technical and tactical developments defeated the U-boats.[40] The gap in air cover was closed with the use of small escort aircraft carriers.

---

38. Samuel Eliot Morison, *History of United States Naval Operations in World War II*, 15 vols. (Boston: Little, Brown, 1947–1962), Vol. 1, pp. 114–418.
39. Jürgen Rohwer, *The Critical Convoy Battles of March 1943* (Annapolis: Naval Institute Press, 1977).
40. Some of these are outlined in detail in Hackmann, *Seek and Strike*, pp. 233–323; Alfred Price, *Aircraft versus Submarine* (Annapolis: Naval Institute Press, 1973), pp. 43–228; Peter Hodges and Norman Friedman, *Destroyer Weapons of World War 2* (Annapolis: Naval Institute Press, 1979), pp. 56–60, 131–140; and Alastair Mitchell, "The Development of Radar in the Royal Navy 1935–1945," *Warship*, Vol. 4, No. 13 (January 1980), pp. 2–14; No. 14 (April 1980), pp. 117–134.

Operations research introduced statistical analysis in solving the complex problems of submarine warfare to the great benefit of the Allies.[41] One conclusion of operations research was that large convoys lost fewer ships than small ones, and this allowed the most economical use of escort ships. With more efficient use of the growing number of escorts, the Allies were also able to form ASW support groups to reinforce convoys under heavy attack. High frequency radar, carried by ships and aircraft, was effective in locating U-boats at night, and its electronic pulses were not picked up by German warning receivers. Finally, British signal intelligence now provided sufficient information for 60 percent of the May 1942 to May 1943 convoys to be routed clear of U-boat patrols, again showing that avoiding submarines could be as important as sinking them in this kind of warfare.[42]

It has been a bitter struggle fought on a grand scale. The Germany navy commissioned a total of 1,171 submarines between 1935 and 1945. Of these, 940 sortied and 784 were lost, 593 to Allied ASW measures. But the U-boat offensives of the Second World War had not been as threatening as the campaign of 1917. For their heavy losses, the U-boats sank 14 million tons or about 17 percent of the 84 million tons of shipping available to the Allies during the war. These figures are significant when making comparisons between this historical case and the current or future Soviet submarine threat to the sea communications of the Western Alliance.

The second submarine campaign against Great Britain demonstrated a number of important things about commerce warfare. There would always be competing military requirements imposed on a submarine force, making it difficult to muster the numbers required for a concerted campaign. The withdrawal of U-boats for the Norway invasion in 1940, to conduct naval attrition in the Mediterranean in 1941, and to help defend against the Allied landings on the continent in 1944 are clear examples of conflicting priorities. Where merchant shipping was sunk could be as important as aggregate tonnage sunk, as shown by the relief of pressure on British shipping when the U-boats were sent to American waters in 1942. Outside sources of tactical reconnaissance helped submarines find targets, but represented sources of vulnerability, particularly to signal intelligence. In the course of hundreds of

---

41. C.H. Waddington, *OR in World War II: Operational Research Against the U-Boat* (London: Elek Science, 1973); and Keith R. Tidman, *The Operations Evaluation Group: A History of Naval Operations Analysis* (Annapolis: Naval Institute Press, 1984), pp. 17–94.
42. Patrick Beesly, *Very Special Intelligence: The Story of the Admiralty's Operational Intelligence Centre 1939–1945* (Garden City: Doubleday, 1978), p. 192. See also pp. 63–75, 92–122, 160–211.

encounters between U-boats and ASW forces, some advances in ASW technology were countered merely with changes in tactics, while others denied submarines the ability to attack and survive even when technical countermeasures were developed.

THE PACIFIC WAR

The right combination of circumstances allowed the U.S. Navy to wage a devastating submarine campaign against Japan during World War II. The most basic factor was Japan's vulnerability to commerce warfare. Over three-quarters of the country's requirements for seventeen basic raw materials and significant percentages of other raw materials and foodstuffs came from overseas. Compared to the shipping available to the Allies, the Japanese merchant marine was relatively small, having 1,600 ships totaling 6 million tons on hand when the war began. The Japanese merchant marine was working to capacity before the war and was sensitive even to small losses. The island nation had a limited shipbuilding capacity to replace losses. Adding ships built and captured during the war, the U.S. submarine force was attacking total maritime assets of only 3,100 ships of 10 million tons.[43]

At the time of the Japanese attack on Pearl Harbor, the U.S. Navy had 51 submarines stationed at forward bases in the Pacific. During the next four years, 249 U.S. submarines would conduct about 1,500 sorties against Japanese shipping, with the operational force in the theater never exceeding 156 "fleet boats." They sank half of the merchant tonnage available to Japan during the war. Another quarter of this tonnage was sunk by carrier and land-based aircraft, and 8 percent was sunk by mines. The combined result was to eliminate the vital services of the Japanese merchant marine.

U.S. submarines were able to operate deep in enemy waters from the first days of hostilities, even though Allied surface and land forces were losing engagements and being forced to retreat. Although Japan controlled the western Pacific for the first two years of the war, U.S. submarines were able to maintain pressure on Japanese merchant shipping, generally increasing their rate of sinkings until the last months of the war, when ships no longer ventured out of port. During 1944, carrier task forces made sweeps into

43. United States, Strategic Bombing Survey, *The War Against Japanese Transportation 1941–1945,* Pacific War, Report 54 (Washington: U.S. Government Printing Office, 1947), pp. 1–2, 13–20, 32, 53–54, 116–118. See also U.S., Strategic Bombing Survey, *Japanese Merchant Shipbuilding,* Pacific War, Report 48 (Washington: U.S. Government Printing Office, 1947).

Japanese home waters, sinking large numbers of ships. Land-based aircraft accounted for only a few ships sunk each month, but were able to maintain this modest rate of attrition for the entire war. Mines had their effect in the last months of the conflict, when Army Air Force B-29s could deliver them from island bases within range of the Japanese homeland.[44]

Submarines were by far the most important factor in the destruction of the Japanese merchant marine, yet their losses were very low, particularly compared to the casualties suffered by the German U-boat arm during the same period. Only 31 American submarines were lost to Japanese ASW measures and probably another 8 to mines.[45] The reasons for the low efficiency of Japanese anti-submarine countermeasures were institutional, doctrinal, and technological.

Between the wars, the Japanese admirals planned strategy based on decisive battle, and tactics were developed accordingly. As a result of this emphasis, they ignored commerce protection almost completely in both building programs and fleet training. There were few escort ships in the Japanese navy in 1941, and large scale construction of this type of vessel was not undertaken until late in the war. Until April of 1942, the Japanese navy had no unit assigned to convoy escort. The training situation was much the same. Officers questioned after the war said that before 1942 they had never seen exercises involving defense against submarine attacks on merchant shipping.

During the war, most Japanese navy personnel did not wish to be assigned to convoy escort duty, but preferred instead the more glamorous offensive operations of the Combined Fleet. Those who manned the escort ships were unaggressive, poorly trained, and inadequately equipped for their mission. They were prone to accept the slightest evidence that a submarine had been sunk, thus giving up the attack too early and in many cases allowing it to escape.[46] Coupled with these factors were technical difficulties. Even as escort commands were established and expanded, the shortage of such basic items

44. Strategic Bombing Survey, *War Against Japanese Transportation*, pp. 2–8, 34–48, 114–134; Clay Blair, Jr., *Silent Victory: The U.S. Submarine War Against Japan* (Philadelphia: J.B. Lippincott, 1975); and Theodore Roscoe, *United States Submarine Operations in World War II* (Annapolis: Naval Institute Press, 1949).
45. John D. Alden, *The Fleet Submarine in the U.S. Navy* (Annapolis: Naval Institute Press, 1979), pp. 249–266; and W.J. Holmes, *U.S. Submarine Losses in World War II* (Washington: U.S. Government Printing Office, 1946). Detailed accounts of each submarine's loss are given in Blair, *Silent Victory*; and Roscoe, *United States Submarine Operations in World War II*.
46. Toshiyuki Yokoi, "Thoughts on Japan's Naval Defeat," U.S. Naval Institute *Proceedings*, Vol. 86, No. 10 (October 1960), pp. 68–75; and Y. Horie, "The Failure of Japanese Convoy Escort," U.S. Naval Institute *Proceedings*, Vol. 82 (October 1956), pp. 1072–1081.

as depth charges remained serious. Many ASW ships were not equipped with sonar. Japanese airborne radar was inadequate for detecting surfaced submarines at night. Although great confidence was placed in magnetic anomaly detection (MAD) gear installed in aircraft, probably no more than five American submarines were sunk by Japanese aircraft at sea.[47] In terms of cost exchange, the American submarine campaign against Japan was probably the closest thing to an offensive against negligible ASW opposition.

The submarine campaigns in both world wars represented fundamental innovation in the conduct of naval warfare. The objective of each campaign was to sink merchant shipping and if possible avoid engaging the adversary's naval forces. Instead of strategic and tactical concentration of naval forces for major fleet actions in a classical approach, there was both strategic and tactical dispersal. The submarine force was dispersed strategically to cover wide areas crossed by major shipping routes, and in World War I there was no tactical concentration of submarines. In World War II, German and American submarines would form small tactical concentrations using wolf pack tactics, but strategic dispersal was still a key to success. The counter to the submarine campaign was also strategic dispersal and tactical concentration. Since no decisive battle could be fought, naval forces used in antisubmarine operations were distributed in small groups, but concentrated tactically to protect convoys and to patrol the near approaches to major ports.

Commerce warfare remains an important role of submarines today, although the situation, as it is now evolving, is different than it was in World War II, as I will note below. To briefly summarize relevant conclusions from the three historical cases reviewed here, proper assessment of current and future commerce warfare scenarios using submarines must consider at least nine essential factors. These are: 1) the numbers, individual tonnage, and aggregate tonnage of merchant ships available to the target state; 2) the capacity of the target state's shipyards to expand its merchant marine and replace losses; 3) the availability of allied and neutral shipping to the target state; 4) the number of submarines in the attacking force; 5) competing mission demands that would be placed on the attacker's submarine force; 6) the attacker's shipyard capacity to expand its submarine force and replace combat losses; 7) the vulnerability of the target state's economy to serious

---

47. Atsushi Oi, "Why Japan's Anti-Submarine Warfare Failed," U.S. Naval Institute *Proceedings*, Vol. 78, No. 6 (June 1952), pp. 587–601; and U.S. Strategic Bombing Survey, *Pacific War*, Report 72: *Interrogations of Japanese Officials*, OPNAV-P-03-100, pp. 161, 196, 228, 441, 485.

Table 1. Submarine Campaigns in Two World Wars

	Germany vs. Great Britain 1914–1918	Germany vs. Great Britain 1939–1945	United States vs. Japan 1941–1945
Submarines			
In Commission	374	1,171	311
Sortied	320	940	249
Lost	178	784	48
Lost to ASW	134	593	31
Total Sorties	3,274	?	1,569
Merchant Shipping			
Tonnage Available (millions)			
Before War	43.1	41.4	6.0
Built during War	10.8	42.5	3.3
Captured	2.4	0.7	0.8
TOTAL	56.3	84.6	10.1
Tonnage sunk (millions)			
by Submarines	11.2	14.7	4.9
by Mines	1.1	1.4	0.4
by Surface Warships	0.6	1.6	–
by Aircraft	–	2.9	2.5
TOTAL	12.9	20.6	7.8
Percentage Losses			
Submarine Force	47.6	67.0	15.4
Merchant Fleet to Sub	19.9	17.4	48.5

loss of its shipping capacity; 8) the relative effectiveness of submarine and anti-submarine capabilities; and 9) the geographical relationship between vital shipping lanes, submarine bases and lines of transit, and sustained deployment areas of ASW forces.

Technological Equilibrium

For the submarine, the period between 1913 and 1943 saw little significant development either in new applications of technology or in new capabilities. After slow initial development of more than a century, rapid synthesis of technologies in the last decade before the First World War had produced three basic capabilities in submarines. After that, submarine technology settled down to three decades of stable equilibrium. The one new development was the advent of acoustics in naval warfare, but with a few exceptions, the capabilities of submarines and their weapons stayed about the same.

Table 2 compares the salient capabilities of U-27, the most advanced submarine type completed before World War I, with the German Type VIIC U-boat and the American *Gato* class "fleet boat," the standard submarines used in the Atlantic and Pacific commerce warfare campaigns of World War II. In range and speed, both surfaced and submerged, performance is virtually the same. Torpedo range and speed are also similar. The main areas of improvement for submarines were operating depth, made possible by stouter hull construction, and better armament in terms of the numbers of torpedoes carried and the lethality of their warheads.[48]

Significant new developments in submarine technology began to emerge in 1944, when the German navy transformed the submersible torpedo boat

Table 2. Submarine Capabilities in Two World Wars

	U 27—1914	Type VIIC—1940	Gato—1941
Displacement (surfaced):	664 tons	750 tons	2,025 tons
Range (surfaced): (submerged):	9,800 nm @ 8 kn 85 nm @ 5 kn	8,500 nm @ 10 kn 80 nm @ 4 kn	11,000 nm @ 10 kn 96 nm @ 2 kn
Speed (surfaced): (submerged):	17 knots 10 knots	17 knots 8 knots	20 knots 9 knots
Diving Time:	45–80 seconds	30 seconds	30–50 seconds
Operating Depth:	160 feet	330 feet	300 feet
Torpedo Tubes:	4 20-inch	5 21-inch	10 21-inch
Torpedoes (carried): (range):	6 9,000 yd @ 27 kn	14 8,200 yd @ 30 kn	24 9,000 yd @ 31 kn 4,500 yd @ 46 kn
(warhead):	360 lbs TNT	617 lbs TNT	643 lbs HBX

48. Submarine development in the interwar period is summarized in Friedman, *Submarine Design and Development*, pp. 37–43; and in Ermino Bagnasco, *Submarines of World War Two* (Annapolis: Naval Institute Press, 1977), pp. 24–28. See also Great Britain, Ministry of Defence, Ship Department, *The Development of HM Submarines*, p. 12.1–29.4; Alden, *The Fleet Submarine in the U.S. Navy*, pp. 10–102; and Rossler, *The U-Boat*, pp. 88–119.

into a true submarine. Then, after the Second World War, submarines acquired in quick succession three fundamentally new capabilities of projection ashore, fleet engagement, and assured destruction. The first submarines with the entire technological ensemble necessary for each of these new capabilities completed tests and trials in 1957, 1958, and 1960, respectively. Integrated units with minimum numbers of submarines necessary to exercise each basic new capability were in service between 1960 and 1967.

Projection Ashore

In today's terminology, the projection of naval power ashore refers to the ability of naval forces to strike targets inland with manned aircraft or guided missiles and the ability to conduct amphibious operations. The concept emphasizes striking or seizing objectives a substantial distance inland, but it can be said that shore bombardment by gun-armed ships is a modest form of this capability. Submarines acquired a capability to strike targets a few hundred miles inland in 1957, when they were first deployed operationally with an armament of cruise missiles.

Small unguided rockets had been launched from the deck of a submerged U-boat in unofficial tests at Peenemünde during the summer of 1942, and two years later the German air force launched the first of several thousand guided missiles known as the V-1 at London and Antwerp.[49] In 1947, a U.S. version of the V-1, called the Loon, was the first cruise missile test-fired from a submarine. The series of tests that followed contributed to development of the Regulus cruise missile. The 500 nautical mile range of this missile enabled submarines to hit targets far inland, but it was the missile's nuclear warhead that gave it much more than the nuisance capability of the V-1s. The German missiles had been launched in great numbers, but submarines could only carry two to four missiles each. Therefore, the development of smaller (3,600-pound) nuclear warheads that could be carried by a missile was essential to give submarines the new capability. The U.S. Navy maintained a unit of four Regulus missiles (aboard either one or two submarines) on deterrent patrol in the Pacific from 1957 until 1964, when the mission was taken over by ballistic missile submarines.[50]

---

49. Jak P. Mallmann Showell, *U-Boats Under the Swastika· An Introduction to German Submarines 1935–1945* (New York: Arco, 1973), p. 114.
50. Norman Polmar, *The American Submarine,* 2nd ed. (Annapolis: Nautical and Aviation, 1983),

The Soviet navy had its own cruise missile program in which the first version of the SS-N-3 "Shaddock," a land-attack cruise missile with inertial guidance and a range about equal to the Regulus, was put aboard converted submarines beginning about 1959. By 1963, the Soviet navy had 17 submarines carrying 68 land-attack missiles, while the U.S. Navy stopped its cruise missile program with 5 submarines capable of carrying 17 Regulus missiles.[51] The mission of the Soviet force was most likely to provide theater nuclear strikes in support of land operations in the Baltic and Pacific areas, but the nuclear-powered cruise missile submarines of the "Echo I" class, introduced in 1960, could certainly have struck major U.S. population centers along both coasts.

In some ways, the submarine was an ideal platform to carry and launch cruise missiles against inland objectives. Since the target is fixed, no active guidance would be required. The submarine could launch its missiles and retire quietly while pre-programmed or inertial guidance took them to their targets.[52] Although a new capability and a notable technical achievement, the ability of submarines to project power ashore with cruise missiles, was at first only a modest addition to naval power. At the time it was introduced, the submarine-launched cruise missile was overshadowed in the U.S. Navy by the carrier, whose jet aircraft could not only deliver many nuclear weapons at more than twice the range of a cruise missile but also effectively deliver conventional munitions, compensating for their lower explosive power by flying multiple missions.[53] The early cruise missiles carried by both American and Soviet submarines had to be launched from the surface, leaving the submarine most vulnerable during the critical part of its mission. All of these factors reduced the initial significance of the new capability, and illustrate the difficulties that can arise in matching technology with strategy. But most

pp. 101–107; Norman Polmar, "Die ersten Marschflugkorper fur den Einsatz in See," *Marine Rundschau*, Vol. 79, No. 2 (February 1982), pp. 76–84; and Viktor Frampton, "Ask Infoser," *Warship International*, Vol. 22, No. 1 (1985), p. 104.

51. Norman Friedman and Przemyslaw Budzbon, "Soviet Union," in *Conway's All the World's Fighting Ships 1947–1982*, ed. Randal Gray, 2 vols. (Annapolis: Naval Institute Press, 1983), Vol. 2, pp. 493, 495; Siegfried Breyer and Norman Polmar, *Guide to the Soviet Navy*, 2nd ed. (Annapolis: Naval Institute Press, 1977), pp. 128–131, 156; and Norman Polmar, *Guide to the Soviet Navy*, 3rd ed. (Annapolis: Naval Institute Press, 1983), pp. 103–104, 109, 363.

52. For greater accuracy, the Regulus system used guidance by radio commands from a separate radar picket submarine.

53. Some argue that cruise missiles continue to be overshadowed by tactical aircraft capabilities as in Robert Nutwell, "'Silver Bullets' and Coups de Grace," *U.S. Naval Institute Proceedings*, Vol. 110, No. 6 (June 1984), pp. 73–79.

important was the introduction of ballistic missiles in submarines, which eclipsed the cruise missile submarine in both superpower navies within a few years.

Recent developments in various technologies are giving the submarine-launched cruise missile new importance in the role of projection ashore. With the introduction of compact cruise missiles the size of a torpedo, a submarine can carry larger numbers of land-attack missiles, and these can be launched from underwater. New guidance technology, such as terrain contour matching (TERCOM), provides cruise missiles with the accuracy necessary to strike tactical targets with conventional or low-yield nuclear munitions. Furthermore, as land-based nuclear delivery systems become more vulnerable to attack because of improving accuracy in short and medium range ballistic missiles, the submarine begins to look more attractive as a secure launch platform for medium-range tactical and theater nuclear systems.[54] The submarine's capability for projection ashore, in existence since 1957, will at last become a significant theater strike capability when compact cruise missiles are deployed aboard submarines in significant numbers by the end of this decade.

Fleet Engagement

In 1959, submarines first acquired the long-sought capability to operate with or against a battle fleet. The unsolved problem from the beginning had been speed. The first oceangoing submarines completed from 1913 were called fleet submarines (*Flotten-Uboote*), indicating the German navy's intention to employ them as part of the battle fleet. The limitations of available propulsion plants were soon evident, and as tactics were established, German emphasis was on coordinated but separate operations. The British Admiralty, however, established the requirement in 1912 for submarines with sufficient surface speed and sea-keeping qualities to accompany the fleet under all conditions. For the next twelve years, the Royal Navy pursued the elusive goal of high surface speed, building 28 large submarines in an attempt to meet the 1912

---

54. Richard K. Betts, ed., *Cruise Missiles: Technology, Strategy, Politics* (Washington, D.C.: Brookings, 1981), pp. 48, 83–91, 99–100, 388–393, 526–528; Miles A. Libbey III, "Tomahawk," U.S. Naval Institute *Proceedings*, Vol. 110, No. 5 (May 1984), pp. 150–163; and J. Philip Geddes, "The Sea Launched Cruise Missile," *International Defense Review*, Vol. 9, No. 2 (April 1976), pp. 198–202.

requirements. Most of these were steam powered, and all were unsatisfactory.[55]

A major step was taken at the end of World War II when new ensembles of technology were developed to give submarines greater speed submerged than on the surface. In 1937, the Japanese navy built a single experimental submarine capable of just over 21 knots submerged, and a modest program produced three production boats in 1945; but the major developments took place during World War II in Germany.[56] Propulsion designs proceeded along two lines: a hydrogen peroxide turbine that needed no outside air supply and a diesel-electric system with powerful electric motors and a high capacity battery outfit. The second system was adopted for mass-produced operational units of the large, long range Type XXI "electroboat" and the much smaller Type XXIII coastal boat. Both had streamlined hull designs for better underwater speed and a snorkel for running the diesels and charging the batteries without surfacing. The best American submarines in the Pacific campaign were capable of 20 knots on the surface but only 9 while submerged. While the Type XXI could do only 16 knots using diesels on the surface, it was capable of 17 knots on electric propulsion while submerged and its underwater endurance was several times greater than that of any submarine built to that time. With high-capacity batteries and the snorkel, submarines at last also had propulsion for long range operations while submerged. The surface torpedo boat that could submerge had become a submarine torpedo boat that did not need to surface. When Germany surrendered in May 1945, there were 120 Type XXI and 62 Type XXIII U-boats in commission, but only a few were ready for war patrols.[57]

The electroboat became the new standard for submarine performance with the completion of six U.S. *Tang* class submarines in 1951–52. In the meantime, the great expense of the high-capacity battery plants led to the slightly less capable Guppy (greater underwater propulsion program) conversions of 50 U.S. fleet boats between 1947 and 1951.[58] The Soviet navy did not fully exploit the advantages of electroboat technology at first. Its 236 postwar "Whiskey"

---

55. Richard Compton-Hall, *Submarine Warfare. Monsters and Midgets* (Poole, Dorset: Blandford Press, 1985), pp. 18–30, 43–50.
56. Dorr Carpenter and Norman Polmar, *Submarines of the Imperial Japanese Navy* (Annapolis: Naval Institute Press, 1986), pp. 100, 116–117.
57. Rossler, *The U-Boat*, pp. 168–187, 198–204, 208–210, 214–247; Eberhard Rossler, *U-BootTyp XXI*, 3rd ed. (Munchen: Bernard und Graefe, 1980).
58. Friedman, *Submarine Design and Development*, pp. 53–66, and Norman Friedman, "Project Guppy," *Warship*, Vol. 3, No. 9 (1979), pp. 38–44.

boats were only modest improvements over the U.S. fleet boats of World War II, but with 14 instead of 24 torpedoes and less endurance, indicating an intention to use them for extended coast defense rather than for commerce destruction. Of the more than 300 Soviet submarines completed between 1950 and 1957, only 20 "Zulu" class boats were true oceangoing submarines, comparable to the German Type XXI in performance.[59] Today, the basic design concepts developed by the German navy are found in all diesel-electric submarines, the Type XXI being the model for oceangoing units and the Type XXIII the pattern for coast defense submarines for the last four decades.[60]

Although more than an evolutionary development, the electroboat did not introduce a basic new capability for submarines. It improved an existing capability, making submarines much more effective against troop and merchant convoys, but the advent of the fast carrier task force during the Second World War left even these very fast submarines with insufficient speed to engage the battle fleet in its new form. Furthermore, ASW aircraft, equipped with radar, sono-buoys, MAD gear, and homing torpedoes represented formidable threats to submarines snorkeling or maneuvering at periscope depth to attack.[61]

The basic new capability to engage a first-line battle fleet came with the introduction of nuclear propulsion and a hull form optimized for underwater speed. The U.S. Navy's *Albacore*, completed as a "hydrodynamic test vehicle" in 1953, was the first submarine with a hull that was *optimized* for underwater speed. The USS *Nautilus*, completed in 1955, was the first to have a nuclear propulsion plant, and the *Skipjack*, first to combine the two features, was operational in 1959.[62] Submarines now had the capability for sustained high underwater speed. They could cruise for weeks at 20 to 25 knots like carrier task forces, and they could match the carrier force's 30-knot speed in combat

59. Although often called a Soviet version of the Type XXI U-boat, the "Whiskey" class submarine is 35 percent smaller and not comparable in speed submerged, in range both surfaced and submerged, and in armament. Compare data on "Whiskey" and "Zulu" in Couhat and Baker, *Combat Fleets of the World*, 1986–87, pp. 508–509; Polmar, *Guide to the Soviet Navy*, 3rd ed., pp. 116–118; and John Jordan, "Soviet Attack Submarines," *Jane's Defense Weekly*, September 22, 1984, pp. 500–502, with Rössler, *U-BootTyp XXI*, pp. 149–151. The most common designations for Soviet submarines built since World War II are assigned by Western intelligence using the U.S. Navy phonetic alphabet. These designations are given here in quotes.
60. Ulrich Gabler, *Unterseebootau*, 3rd ed. (München: Bernard und Graefe, 1986).
61. Norman Friedman, *Carrier Air Power* (Annapolis: Naval Institute Press, 1981), pp. 122–129.
62. Raymond V.B. Blackman, ed., *Jane's Fighting Ships*, 1968–69 (New York: McGraw–Hill, 1968), pp. 398, 401, 403.

operations. The buildup of nuclear submarine forces was gradual, the United States completing a nominal flotilla of 16 high performance nuclear attack submarines (SSNs) in 1966 and the Soviet Union matching this force a year later.[63]

Until this time, the only weapon available to submarines for attacking surface ships was the torpedo, although the effectiveness of this weapon was improved considerably with the introduction of acoustic homing by the German and American navies in 1943 and wire guidance by the U.S. Navy in 1946. The Soviet navy introduced a rudimentary stand-off capability with the modification of its land-attack cruise missile, the SS-N-3 "Shaddock," as an anti-ship weapon. Between 1961 and 1969, 45 "Echo II" and "Juliett" submarines were built to launch these missiles while surfaced against NATO aircraft carriers from about 25 miles using radar guidance from the submarine. With the aid of a Tu 20 "Bear D" radar aircraft, introduced in about 1967, the "Shaddock" missile could hit a moving ship at 250 miles. In 1967, the Soviet navy also commissioned its first "Charlie I" class submarine, which can fire eight SS-N-7 "Siren" missiles while submerged at ships 35 miles away, although acoustic conditions probably reduce this range considerably.[64]

Submarines were capable of engaging a force of surface combatants and causing serious losses, but parallel developments in ASW systems prevented them from acquiring the capability for decisive battle. Concern that the Soviet navy would produce Type XXI submarines in large numbers led the U.S. Navy to develop significantly better ASW capabilities for its carrier task forces. Fixed-wing ASW aircraft had operated from slower escort aircraft carriers to protect merchant convoys since the middle of World War II, and from 1953 they were complemented by helicopters with dipping sonar. Operating from much larger, faster carriers after 1954, both types of aircraft could protect the fleet as well. In 1961, the U.S. Navy introduced variable depth sonar in some destroyers and armed many others with the first true stand-off ASW weapon (called ASROC for antisubmarine rocket) with a range

63. Ibid., pp. 397–398; Budzbon and Friedman, *Conway's All the World's Fighting Ships 1947–1982*, Vol. 2, pp. 494, 497; and Polmar, *Guide to the Soviet Navy*, 3rd ed., pp. 108, 110. The Soviets built fourteen "November" class submarines, which did not benefit from the *Albacore*-type hull, but attained high speeds with a nuclear plant twice as powerful as that of the *Skipjack*. The first of these was commissioned in August 1958 but was probably not operational until the next year. Subsequent Soviet attack submarines had *Albacore* hulls.

64 Polmar, *Guide to the Soviet Navy*, 3rd ed., pp. 98–102, 333–334, 363; and Siegfried Breyer and Armin Wetterhan, *Handbuch der Warschauer–Pakt–Flotten* (Koblenz: Bernard & Graefe, 1983–85), sections 004.04, 007.02, 007.03.

ten times the previous maximum of 1000 yards. These surface and airborne systems precluded missile submarines from closing to 25 miles on the surface, and carrier-based fighter aircraft easily neutralized the capabilities of the "Bear D" radar aircraft. Thus, in spite of great potential range in submarine-launched anti-ship missiles, the nuclear-powered submarine and torpedo or short-range missile have been, until recently, the most effective combination against a first class battle fleet, and the requirement to fire from short range made stealth as important as speed.[65]

Assured Destruction

The advent of the submarine-launched ballistic missile and the deployment of these nuclear delivery systems in large numbers enabled submarines to make a significant contribution to nuclear deterrence through the strategy of assured destruction. In this case, technological developments provided new capabilities that combined with the submarine's inherent advantages at just the right time to solve an emerging strategic problem.

The emerging problem was how to maintain a secure deterrent force and thereby a low risk of nuclear war.[66] Immediately after the advent of nuclear weapons at the end of World War II, the Western powers saw American nuclear forces as a means to deter Soviet agression in Europe and prevent a major war. The Soviet Union developed its own nuclear weapons and had deployed a bomber force with intercontinental range by the end of 1956. At about this time, both superpowers were also adding thermonuclear weapons to their arsenals. The tremendous destructive power of each thermonuclear weapon and the intercontinental range of new bomber aircraft meant that from the late 1950s, each superpower could deal the other a direct and devastating blow. Thus began the age of mutual nuclear deterrence. However, the stability of this situation appeared to be threatened by the emergence of intercontinental ballistic missiles (ICBMs), which, when deployed, would give each side the capability to destroy the other's bombers on the

---

65. Norman Friedman, *Modern Warship Design and Development* (Greenwich: Conway Maritime Press, 1979), pp. 121–135; and Norman Friedman, *U.S. Naval Weapons* (Annapolis. Naval Institute Press, 1982), pp. 99–140, 256–269, 272–273.
66. Summaries of these developments can be found in Jerome H. Kahan, *Security in the Nuclear Age: Developing U.S Strategic Arms Policy* (Washington, D.C.: Brookings, 1975), pp. 9–98, David Alan Rosenberg, "The Origins of Overkill: Nuclear Weapons and American Strategy, 1945–1960," *International Security*, Vol. 7, No. 4 (Spring 1983), pp. 3–71; and Lawrence Freedman, *The Evolution of Nuclear Strategy* (New York. St. Martin's, 1981), pp. 22–68, 76–90, 134–171, 227–256.

ground at their bases in a preemptive first strike. In order to avoid a strategy that would require hair-trigger response in a crisis situation, the nuclear deterrent forces had to be able to survive a nuclear attack and still strike their targets.

A strategy of assured destruction meant giving nuclear forces a second-strike capability in order to reduce the risks of nuclear war in a crisis situation. The first step was to put part of the bomber force on 15-minute ground alert, so that it could become airborne (but not necessarily proceed to its targets) upon warning of an attack. A few years after ICBMs were introduced, they were put into hardened silos, which gave them the capability to survive a first strike, a capability only recently in doubt with the deployment of many ICBMs with multiple warheads of very high accuracy.

Even before ICBMs were based in hardened silos, submarines acquired a capability for assured destruction that has remained intact in spite of many technological developments. In the United States, the Polaris program produced a submarine that carried sixteen intermediate-range (1,200 nautical miles) ballistic missiles. The first of these submarines went on patrol in late 1960. Successive models of the missile brought increases in range to 1,500 and then 2,500 nautical miles, and by 1967 the U.S. Navy had 41 Polaris ballistic missile submarines (SSBNs) in commission.[67] With two-thirds of this force at sea, there were sufficient numbers of submarines on station at all times to launch missiles against the 300 largest cities in the Soviet Union. With 27 of the 41 boats on station and some degradation for system failures, the Polaris submarine missile force alone could deliver the 400 megaton equivalents assumed necessary for unacceptable damage of Soviet industry under the McNamara definition of assured destruction.[68]

Further improvements in assured destruction capability came with the introduction of technologies that brought multiple independently targetable reentry vehicles (MIRVs) in the American Poseidon C-3 missile in 1971 and intercontinental range in the Soviet SS-N-8 missile in 1974.[69] At the same

---

67. Harvey M. Sapolsky, *The Polaris System Development; Bureaucratic and Programmatic Success in Government* (Cambridge: Harvard University Press, 1972); and Norman Polmar, *The Ships and Aircraft of the U.S. Fleet*, 12th ed. (Annapolis: Naval Institute Press, 1981), pp. 20–23, 336–337. Detailed data on commissionings, conversions, and first deterrent patrols of each boat are given alphabetically by ship's name in United States Navy Department, *Dictionary of American Naval Fighting Ships* (Washington: U.S. Government Printing Office, 1959–1981).
68. Alain C. Enthoven and K. Wayne Smith, *How Much is Enough? Shaping the Defense Program, 1961–1969* (New York: Harper & Row, 1971), especially pp. 174–178, 207–208; and William W. Kaufman, *The McNamara Strategy* (New York: Harper & Row, 1964).
69. The evolution of the Soviet SLBM force is described in Robert P. Berman and John C. Baker,

time, developments in ASW technology do not appear to threaten the survivability of ballistic missile submarines in the foreseeable future.[70]

The Future

Future capabilities of submarines in naval warfare will be determined by new developments in technology and ongoing trends in building programs as they change or sustain existing force structures. In the near-term future, submarines will acquire a new capability and see the decline of one existing capability and the enhancement of another.

STRATEGIC COUNTERFORCE

The high accuracy of the D-5 Trident II submarine-launched ballistic missile will bring a new capability to submarines when it is deployed in 1989 by enabling them to destroy precise, hardened targets such as missile silos.[71] If building and retrofitting schedules proceed according to current projections, the U.S. Navy will have a 20-ship force by the year 2000 and be able to maintain two-thirds of it on station. The Soviet ICBM force, if maintained in its current configuration of 1,398 fixed silo launchers, could be destroyed by the Trident force in a first strike.[72]

The capability to destroy small, hardened targets with nuclear warheads is a function of warhead yield and the accuracy of the delivery system. This kind of accuracy is becoming possible through improvements in navigation and guidance technology. Since submarine-launched ballistic missile (SLBM)

Soviet Strategic Forces: Requirements and Responses (Washington, D.C.: Brookings, 1982), pp. 55–59, 62–65, 93–96, 106–108.

70. Richard L. Garwin, "Will Strategic Submarines Be Vulnerable?," *International Security*, Vol. 8, No. 2 (Fall 1983), pp. 52–67; and Donald C. Daniel, "Antisubmarine Warfare in the Nuclear Age," *Orbis*, Vol. 28, No. 3 (Fall 1984), pp. 527–552.

71. D. Douglas Dalgleish and Larry Schweikart, "Trident and the Triad," U.S. Naval Institute *Proceedings*, Vol. 112, No. 6 (June 1986), p. 76; and Roger F. Bacon, "Strategic Employment Concepts," *The Submarine Review*, Vol. 2, No. 3 (October 1984), pp. 4–9.

72. Unclassified estimates of missile performance vary. Against a nominal Soviet ICBM silo hardened to 3,000 psi, the C-4 Trident I would have an 8 percent single shot kill probability (SSKP), assuming a 1,500-foot CEP and a yield of 100 KT for each of its warheads. The D-5 Trident II would have a 77 percent SSKP, assuming accuracy as good as a 600-foot CEP and a nominal yield of 500 KT for each warhead. If each D-5 missile carried 8 warheads in a delivery system with these hypothetical characteristics and the number of submarines on station was increased from 13 to just 15, the force would have a 94 percent chance of disabling each of the 1,398 Soviet ICBMs in its silo. Data from International Institute for Strategic Studies, *The Military Balance*, 1985–86, p. 158; Couhat and Baker, *Combat Fleets of the World*, 1986–87, pp. 593, 616; and General Electric, Defense Electronics Division, *Missile Effectiveness Calculator*, 1965.

systems deliver long-range missiles from a moving underwater platform, the launch position as a reference point for the guidance system has been less precise than for an ICBM, which is launched from a fixed silo. Early SLBM systems had accuracies of from 3,000 to 6,000 feet, circular error probable (CEP).[73] More precise submarine navigation systems and the introduction of a stellar system that takes at least one star sighting to refine the missile's trajectory in the post-boost phase reportedly reduced the CEP of the C-4 Trident I missile to 1,500 feet, in spite of an increase in range to 4,000 miles, over three times that of the Polaris A-1 of 1960.[74] Improved stellar-corrected inertial guidance in the D-5 Trident II could decrease the CEP further even with a range goal of 6,000 miles.[75]

The new capability in submarines will increase strategic targeting options, and it will mean broader nuclear warfighting capabilities. It may or may not enhance the deterrent effect of the U.S. submarine force. Deterrence can be primarily the result of better warfighting capability, but this is not necessarily the case, particularly in the realm of strategic nuclear war, where what deters is subject to debate and impossible to measure. By opening a "window of vulnerability" on the Soviet land-based ICBM force, a hard target kill capability bestowed on submarines may well undermine stable mutual deterrence between the superpowers.

Counter-ICBM capability in submarines will have profound strategic effects far beyond a better warfighting capability that may or may not mean a more effective deterrent. By giving the essentially immune SLBM force the capability to destroy an adversary's land-based ICBM force in a single first strike, we raise a more serious version of the vulnerability problem that many in the late 1970s sought to eliminate from the Minuteman force by replacing it with a mobile version of the MX. The strategic implications of deploying a highly accurate version of the D-5 SLBM will become clear when the Soviets give their own SLBM force the capability to destroy the U.S. land-based ICBM force. If land-based ICBMs are to be retained after the introduction of highly accurate SLBM systems, both superpowers will inevitably develop

---

73. Circular error probable is usually based on test data indicating that 50 percent of the bombs from an aircraft delivery system or warheads from a missile delivery system will fall within a circle having the radius given as CEP.
74. Bill Gunston, *The Illustrated Encyclopedia of the World's Rockets and Missiles* (New York: Cresent, 1979), pp. 92–95; and Thomas B. Cochran, William M. Arkin, and Milton M. Hoenig, *Nuclear Weapons Databook*, Vol. 1, *U.S. Nuclear Forces and Capabilities* (Cambridge, Mass.: Ballinger, 1984), pp. 69, 74, 134–143.
75. Cochran, Arkin, and Hoenig, *Nuclear Weapons Databook*, Vol. 1, pp. 144–146.

countermeasures to make land-based missiles more survivable, and this will probably lead both sides to deploy exclusively mobile ICBM forces and ballistic missile defenses to protect them.

COMMERCE WARFARE

A second major capability being affected by developments in technology and force structures is the ability to wage commerce warfare using submarines. Current trends point to declining capability and little likelihood for a commerce warfare campaign of the magnitude experienced in World War II. In the unlikely event of future war between the superpowers, the obvious naval scenario would be a Soviet submarine campaign against Western maritime nations, which depend upon shipping to deliver essential raw materials and to move troops and military supplies. It is this scenario that will best serve to illustrate the current trend.

Most basic of several trends that make commerce warfare a thing of the past is the size of the objective: the overwhelming numbers of merchant ships operated by the Western maritime nations. The merchant marines of the Western Alliance currently number 38,000 ships totaling 183 million tons. This represents four times the shipping available at the beginning of World War II. The carrying capacity available to the Western Alliance is much greater if neutral powers are induced to keep their ships at sea. Liberia and Panama alone have 7,500 ships totaling 100 million tons, most of which is Alliance shipping registered under neutral flags of convenience.[76] The availability of neutral shipping cannot be assumed in a future war, but since it was a critical factor in the World War I submarine campaign, the Soviets cannot ignore it in their assessments of a future submarine campaign against shipping.

A second reason to question future capabilities for waging an effective commerce warfare campaign is numbers of available submarines. Sophisticated submarines cannot be built rapidly in large numbers, and the new capabilities have brought essential new missions, leaving fewer numbers to wage commerce warfare. Submarines are much more capable today than they were in the two world wars, but in commerce warfare, numbers are as important as capabilities. Even the most advanced submarine in the world can only be in one place at one time. It must expend at least one missile or torpedo to sink a ship, and with the size of today's merchant ships and past

---

76. Numbers of merchant ships and aggregate tonnage are given by country in Couhat and Baker, *Combat Fleets of the World*, 1986–87. Data is taken from *Lloyds Register of Shipping*, 1984.

experience as a guide, more than one weapon will be required to sink each one. After expending a typical load of 24 torpedoes, the submarine must return to base for more. Stalking targets, attacking each one, transits to and from base, reprovisioning and maintenance all take time. Therefore numbers are essential if the campaign is to have effect. To seriously threaten the survival of the maritime nations by destroying a good portion of their shipping, the experience of three campaigns in two world wars indicates that hundreds of submarines would be necessary just to start an effective campaign, and that monthly production rates would have to be in the dozens. Yet the trend today is to build ever larger and more sophisticated submarines, and as complexity has increased, force levels and building rates have declined significantly.

A projection of the makeup of Soviet submarine forces in 1995 shows serious limitations in numbers required for a commerce warfare capability if current trends continue. Not only are Soviet building rates down to about 7 or 8 boats a year compared with between 60 and 80 in the late 1950s, but the force has taken on several competing but essential missions.[77] Eliminating submarines that will be over 30 years old from the force and assuming an optimistic building rate of 10 new boats per year, the Soviet navy will have about 240 submarines in the mid-1990s. Reflecting both the current makeup of the Soviet submarine force and the most likely trends in its development, these 240 submarines will probably be assigned as follows. About 60 will be ballistic missile submarines, with a modest complement of 40 SSNs to protect them against American SSNs, although more are likely to be assigned to this mission. Another 20 submarines will probably be armed with land-attack cruise missiles, as replacements for the current theater strategic forces, such as the "Golf II" SSBs deployed in the Baltic. Countering a nominal 15 U.S. and French carrier battle groups with just 4 cruise missile submarines each would require another 60 submarines, and Soviet prudence would dictate that more be assigned to this mission.[78] Even these optimistic assumptions

---

77. Building rates were derived from completion dates for all Soviet submarines built since 1945. Sources are Budzbon and Friedman, *Conway's All the World's Fighting Ships 1947–1982*, Vol. 2, pp. 468, 492–499; Polmar, *Guide to the Soviet Navy*, 3rd ed., pp. 84–123; Couhat and Baker, *Combat Fleets of the World*, 1980–81, pp. 540–552; 1982–83, pp. 602–615; 1984–85, pp. 695–711; 1986–87, pp. 498–510.

78. Norman Polmar with Norman Friedman, "Their Missions and Tactics," U.S. Naval Institute *Proceedings*, Vol. 108, No. 10 (October 1982), pp. 34–44; Paul J. Murphy, ed., *Naval Power in Soviet Policy* (Washington: U.S. Government Printing Office, 1978), pp. 78–84, 112–117, 155–168; and Milan Vego, "Their SSGs/SSGNs," U.S. Naval Institute *Proceedings*, Vol. 108, No. 10 (October 1982), pp. 60–68.

leave the Soviet navy with only 60 SSN and diesel-electric boats to wage a campaign against *either* the ballistic missile submarines *or* the merchant shipping of the Western Alliance, and this takes no account of subs in each category that will be in transit to operational areas, used for training, and in the yard for refit. Given the notorious reputation Soviet submarines have for breakdowns and low availability, these last factors are significant.

In order to wage a campaign that would seriously threaten Western sea lines of communication, the Soviets would need a total of 450 to 500 submarines if the other commitments listed above are to be met as well. Regardless of the impressive achievements of the Soviet shipbuilding industry over the past two decades, there is little evidence that the Soviets will be able to produce such a force or to make good the losses they would suffer in what would be an intensive struggle.

Anti-submarine warfare has also improved significantly since World War II. Numerical strength in a submarine force is no longer sufficient. To wage a major campaign against shipping today, a submarine force needs both numbers and the best capabilities in each unit. Yet the kind of technical sophistication required to overcome advanced ASW techniques must come at the price of numbers, because again, the more sophisticated a submarine, the greater its cost in resources and manpower and the longer its building time.

There are several developments that not only enhance ASW capability but change the nature of such a campaign compared to its historical antecedents.[79] The advent of seabed sensor arrays makes barrier ASW an effective complement to localized ASW built around the convoy system. Another development, the towed sonar array, not only provides a complementary surveillance system for seabed sensors, in a slightly different configuration, it allows surface ships to acquire tactical data directly from the operating medium of the submarine, within thermal layers rather than through them. A third development is the ASW helicopter, which extends sensor coverage and represents a weapon delivery platform that moves three times as fast as any submarine. Since helicopters can easily operate from the decks of large merchantmen as well ASW escorts, future arming of merchant ships would be with helicopters, instead of deck guns as in the two world wars.

---

79. Joel S. Wit, "Advances in Antisubmarine Warfare," *Scientific American*, Vol. 244, No. 2 (February 1981), pp. 31–41; B.W. Lythall, "The Future of Submarine Detection," *Naval Forces*, Vol. 2, No. 2 (1981), pp. 41–49; and Norman Friedman, "The Evolution of Towed Array Sonar Systems," *Naval Forces*, Vol. 4, No. 5 (1983), pp. 76–81.

Finally, there is the problem of just how the Soviets would be able to wage a major submarine campaign against Western commerce for many months without an escalation to general nuclear war. Such a struggle is frequently postulated, but to be prolonged, the assumption must be that it could become bitter without the use of a single nuclear weapon. Setting aside political calculations on whether the West would engage in such a campaign without resort to nuclear weapons, there are at least two direct linkages between antisubmarine warfare and general nuclear war. First, Soviet dependence on radar and electronic ocean reconnaissance satellites (RORSAT and EORSAT) for submarine operations means that the Western navies would inevitably try to destroy these space platforms. However, the employment of anti-satellite weapons also threatens early warning satellites that are an integral part of strategic nuclear forces. Second, aggressive ASW operations by American SSNs in the Norwegian Sea and Arctic Ocean to preempt Soviet attacks on shipping would also threaten Soviet ballistic missile submarines, an essential element of strategic nuclear forces. The sinking of Soviet SSBNs in the course of ASW operations to preempt an anti-shipping campaign could appear to be deliberate attrition of Soviet SSBNs as a preliminary step to a strategic nuclear offensive against the Soviet Union. This would provide a strong incentive for the Soviets to escalate immediately to nuclear war.[80]

Assuming the war remains conventional, the Soviets would make better use of their limited numbers of submarines by attacking troop and supply convoys attempting to reinforce NATO forces rather than waging protracted warfare against ocean commerce. But given the risks of general nuclear war, the Soviets would gain even more by attacking the channel ports with bombers and intermediate-range ballistic missiles instead.

These trends reduce the potential impact of using submarines against commerce, but they will not make it disappear. We might say that commerce warfare capability in submarines is regressing to commerce harassment. In a major war, the Soviet navy would undoubtedly send some submarines to attack military and merchant convoys as part of a general war of attrition. Given the fortunately low likelihood of war between the superpowers, a more probable scenario is a maverick Third World country using its small force of submarines to strike at its enemies, large or small, by sinking some

---

80. Barry R. Posen, "Inadvertent Nuclear War? Escalation and NATO's Northern Flank," *International Security*, Vol. 7, No. 2 (Fall 1982), pp. 28–54; and Desmond Ball, "Nuclear War at Sea," *International Security*, Vol. 10, No. 3 (Winter 1985–86), pp. 16–21, 22–23.

of their shipping. Both of these cases indicate the need for considerable ASW capability in Western navies. While recognizing a decline in capabilities and probabilities of a major commerce warfare campaign, Western ASW forces are essential to prevent the Soviets from having a "free ride" in a limited attrition campaign and to counter third power assaults on vital shipping, such as tanker traffic from the Persian Gulf.

DECISIVE BATTLE?

As submarine forces are losing the capability to wage commerce warfare, they are gradually gaining in the capability to engage first-line naval forces. Once able to destroy the modern equivalent of a surface battle fleet, submarines will have acquired their eighth basic capability. This represents much more than wearing down enemy naval forces through gradual attrition of ancillary, obsolescent, and independently steaming warships, and it gives submarines more combat potential than the essentially hit-and-run tactics of fleet engagement that came with the introduction of high-performance nuclear submarines in 1959. It is the advent of classical battle capability for the undersea arm of navies. It is also the integration of submarines into some kind of fleet-type targeting, command, and control system, with all of the associated tactical problems long avoided by keeping submarines dispersed and independent.

At the beginning of the 1980s, Soviet and American submarines began to be armed with anti-ship cruise missiles that could be launched from underwater and hit moving ship targets hundreds of miles away.[81] This means that they can hide in hundreds of thousands of cubic miles of ocean until the instant of weapon launch. Many of the requisite technologies are common to the new generation of compact, land-attack cruise missiles recently coming to maturity. As more submarines are armed with long-range, anti-ship missiles, they will acquire the ordnance delivery capability to destroy a large combat formation of surface warships.

However, against maneuvering targets, the added capability of real-time tactical reconnaissance and targeting information is essential. In this respect, the new generation of cruise missile submarines will be subject to the capa-

---

81. Michael McCGwire, "The Tomahawk and General Purpose Naval Forces," in Betts, *Cruise Missiles*, pp. 231–247; and John Jordan, "'Oscar': A Change in Soviet Naval Policy," *Jane's Defence Weekly*, May 24, 1986, pp. 942–947.

bility/vulnerability paradox.[82] The obvious platforms for the required sensors are aircraft and low-orbit satellites. Both are vulnerable to attack by any fleet with modern carrier-based aircraft. A high-performance fighter with a small anti-satellite (ASAT) missile has been shown to be effective, making ocean reconnaissance satellites probably more vulnerable to carriers than carriers are to weapons systems served by these satellites.[83] Aircraft can also provide target acquisition and tracking data, but surveillance aircraft are defenseless against carrier-based fighters. If the reconnaissance and targeting aircraft are protected by fighters, then we have come full circle to the requirement for carriers to counter carriers, and submarines are just one part of an integrated battle situation. Thus, although the submarines themselves will be extremely difficult to counter with fleet ASW defenses at missile launch range, they must have outside support, not only from tactical reconnaissance platforms but also for the protection of those platforms. At least for the anti-warship mission, it appears that as submarines move closer to the capability for decisive battle, they will also have to become more integrated with and dependent upon fleet surface and air units.

Conclusions

After 125 years of technical experimentation, submarines joined the navies of the world as warships, and in the next 60 years they evolved from unimportant ancillary craft into a central element of national security. A rapid synthesis of technologies during the last decade before World War I gave submarines the basic capabilities of coast defense, naval attrition, and commerce warfare. There followed three decades of sustained equilibrium in submarine technology and basic capabilities. World War II brought the first and only successful submarine campaign against merchant shipping and precipitated many new technical developments. Another period of relatively rapid technological synthesis gave submarines three more basic capabilities between 1957 and 1967: projection ashore, fleet engagement, and assured

---

82. Norman Friedman, "C³ War at Sea," U.S. Naval Institute *Proceedings*, Vol. 103, No. 5 (May 1977), pp. 124–141; and R.B. Laning, "Air Support for Submarine War," *The Submarine Review*, Vol. 3, No. 3 (October 1985), pp. 77–81.

83. "Defense Dept. Readies Asat Weapon for Third Test Firing in Space," *Aviation Week & Space Technology*, September 2, 1985, pp. 20–21; and "Defense Dept. Plans Next Test Firing of Air-Launched Asat System," *Aviation Week and Space Technology*, September 23, 1985, pp. 20–21. The effectiveness of satellite ocean surveillance is also probably overestimated. See Frank Cranston, "USN Carrier 'Disappeared' for Two Weeks," *Jane's Defence Weekly*, July 26, 1986, p. 112.

destruction. The near-term future will bring a seventh new capability, strategic counterforce. It will continue a decline in the capability to wage major commerce warfare campaigns, and it will enhance the submarine's new capability to engage and destroy first-line fleet units. Decisive battle therefore could possibly become an eighth major capability for submarines in coming years.

Both the successes and the problems of submarines are directly related to their separateness. Their effectiveness as a weapon platform and their survival in the face of countermeasures depend upon operating in a separate medium, and their most effective employment, whether as a means of naval attrition, a commerce destroyer, or a nuclear deterrent, has been while operating essentially in isolation from surface forces. Although naval establishments persist in their attempts to combine the roles and missions of submarines with those of the surface fleet, submarines have had no need to integrate. For their entire history, they have operated best in parallel to but separate from the surface fleet. The most bothersome aspect of this separateness is that it seems to challenge the basic tenets of Anglo–American naval doctrine, because the most successful submarine strategies do not conform to the classical model of naval warfare.

None of this need represent the challenge to established doctrine that is often assumed by proponents and opponents alike. In adding a new operating medium, new modes of operation, and new strategic concepts, submarines are not an alternative but an addition to the more traditional instruments of naval power. If submarine strategies and force structures are developed in this context, they can contribute even more effectively to the exercise of naval power in the future.

Stopping the Sea-Based Counterforce Threat

Harold A. Feiveson
and John Duffield

The ballistic missile submarine has long been thought to have an important counterforce role in the execution of nuclear war. This is especially true of potential Soviet first strikes against the United States. Soviet submarine-launched ballistic missiles (SLBMs) fired close to U.S. shores and using trajectories that minimize the time of flight have appeared well suited to attacks on U.S. strategic bomber bases and on many critical components of the command, control, and communication network.[1]

The importance to U.S. strategic thinking of this potential counterforce role for SLBMs was highlighted most recently by the President's Commission on Strategic Forces (The Scowcroft Commission). The Commission based one of its central conclusions on the contention that U.S. strategic bombers on ground alert could be attacked successfully *only* by SLBMs.

. . . [I]f Soviet war planners should decide to attack our bomber and submarine bases and our ICBM silos with simultaneous detonations—by delaying missile launches from close-in submarines so that such missiles would arrive at our bomber bases at the same time the Soviet ICBM warheads (with their longer time of flight) would arrive at our ICBM silos—then a very high proportion of our alert bombers would have escaped before their bases were struck. . . . If the Soviets, on the other hand, chose rather to launch their ICBM and SLBM attacks at the same moment . . . there would be a

---

An early version of this paper was presented at hearings hosted by the Federation of American Scientists in the Dirksen Senate Office Building, December 13–14, 1982. Excerpts were published in the *F.A.S. Public Interest Report*, January 1983.

Harold A. Feiveson is a Senior Research Analyst at the Center for Energy and Environmental Studies at Princeton University. John Duffield is a graduate student at the Woodrow Wilson School, Princeton University.

---

1. Alton H. Quanbeck and Archie L. Wood, *Modernizing the Strategic Bomber Force* (Washington, D.C.: Brookings, 1976); Roger D. Speed, *Strategic Deterrence in the 1980s* (Stanford, Calif.: Stanford University, Hoover Institution Press, 1979); Herbert Scoville, Jr., and David G. Hoag, "Ballistic Missile Submarines as Counterforce Weapons," in Kosta Tsipis, Anne H. Cahn, and Bernard T. Feld, eds., *The Future of the Sea-Based Deterrent* (Cambridge, Mass.: M.I.T. Press, 1973).

International Security, Summer 1984 (Vol. 9, No. 1) 0162-2889/84/010187-16 $02.50/1
© 1984 by the President and Fellows of Harvard College and of the Massachusetts Institute of Technology.

period of over a quarter of an hour after nuclear detonations had occurred on US bomber bases but before our ICBMs had been struck.[2]

By this argument, there is no strategic "window of vulnerability" that needed to be urgently addressed by the U.S. so long as ICBM silos could be attacked effectively only by enemy ICBMs. This reassurance, however, will evaporate once the Soviets develop the capability to strike *both* missile silos and bomber bases simultaneously with SLBMs.[3]

Lethality of SLBMs

Unfortunately, such a capability is now within the reach of the U.S.S.R. This is a result of the sudden and explosive growth in the number of submarine-based warheads during roughly the last decade and of projected improvements in SLBM accuracy. The number of Soviet SLBM warheads tripled from around 850 in 1977 to about 2500 today. This expansion will continue if the Soviets continue to replace their older SLBMs with the highly MIRVed SS-N-20. (See Figures 1 and 2.) In the case of the U.S., the number of independently targeted warheads carried by the submarine force (including submarines in port as well as those at sea) grew from less than 700 in 1972, when the first multiple independently targetable reentry vehicles (MIRVs) were deployed, to over 5000 today. (See Figure 3.) The current U.S. modernization plan is to deploy 20 Trident submarines, each with a complement of 24 Trident II (D-5) missiles (and with each missile carrying perhaps 8–10 warheads), by the end of the century. (See Figure 4.)

Despite this recent history of impressive growth on both sides, the submarine force posed no substantial threat to missile silos in either the United States or the Soviet Union as long as SLBM accuracies remained poor. The U.S. Poseidon and Trident I warheads are estimated to have a circular error probable (CEP) of 450 meters (m) at full range and the Soviet SS-N-18 a CEP of 550 m.[4] No estimates are publicly available for the latest Soviet MIRVed SLBM, the SS-N-20. At such CEPs, a 100 kiloton (kt) Trident I warhead would have less than a 10 percent chance of destroying a Soviet silo hardened to

---

2. *Report* of the President's Commission on Strategic Forces (Washington, D.C.: U.S. Government Printing Office, April 1983), pp. 7–8.
3. Jeremy Stone, "Missiles: Will We Ever Learn?," *The Los Angeles Times*, May 12, 1983, Part 2, p. 7.
4. A.A. Tinajero, *US/USSR Strategic Offensive Weapons: Projected Inventories Based on Carter Policies* (Washington, D.C.: Congressional Research Service, September 30, 1981), pp. 127, 129.

Figure 1 Soviet SLBM Warheads—Projected

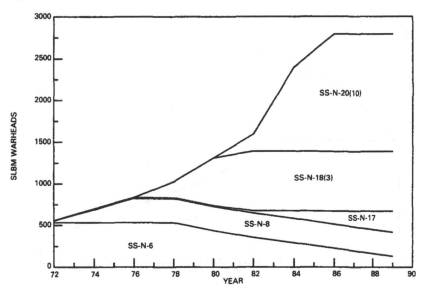

Sources: The International Institute for Strategic Studies (IISS), *The Military Balance
1982–1983* (London, 1983); The Stockholm International Peace Research
Institute, *World Armaments and Disarmament: SIPRI Yearbook 1981* (Lon-
don: Taylor and Francis, 1981); A.A. Tinajero (fn. 4, p. 75). The data assume
Soviet forces consistent with SALT I and II and also assume an average
SS-N-18 MIRV loading of 3 warheads.

2000 psi (pounds per square inch) or greater, assuming even perfect reliability. A 200 kt SS-N-18 warhead would have about the same probability of
destroying a similarly hardened U.S. silo. To be able to threaten missile silos,
the CEPs of these missiles would have to be significantly reduced, to approximately 200 m or less.

The U.S. Navy, under its Improved Accuracy Program, now contemplates
a CEP of about 200 m or less for the Trident I and on the order of 120 m for
the Trident II. It is only prudent to assume that a new highly accurate Soviet
missile could also be deployed in roughly the same time frame.

There is no question that, once these accuracy improvements have been
achieved, the ballistic missile submarine force will be seen by defense planners on both sides to constitute a very formidable counterforce threat against
the full spectrum of fixed land-based targets, including missile silos and

Figure 2   Soviet SLBM Force—Projected

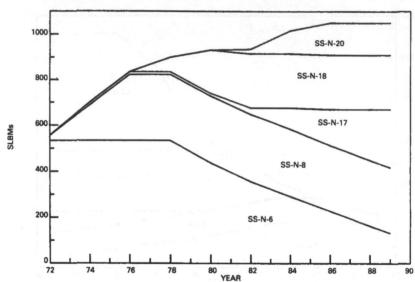

Sources: The Military Balance 1982–1983; SIPRI Yearbook 1981; A.A. Tinajero (fn. 4, p. 61).

hardened command and control facilities. This threat might be tempered for a time by the operational complexities of command and control and the firing coordination that would be required of a submarine counterforce attack, and—in the case of the threat to the U.S.—by the difficulty the Soviets have had in keeping large numbers of ballistic missile submarines on station.[5] However, none of these problems are clearly insurmountable.

Why Try to Slow the Development of Hard Target Kill Capability?

In one sense, the development of such a submarine-based counterforce threat should not matter very much (even if it made simultaneous attacks on bombers and missiles theoretically possible). For one reason, as SLBM ranges

5. Quanbeck and Wood, *Modernizing the Strategic Bomber Force;* Joel S. Wit, "American SLBM: Counterforce Options and Strategic Implications," *Survival*, Vol. 24, No. 4 (July/August 1982), pp. 163–174.

Figure 3   U.S. SLBM Warheads—Current Plans

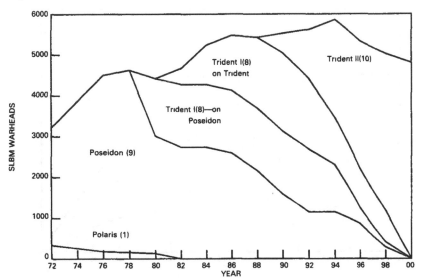

Sources: *The Military Balance 1982–1983; SIPRI Yearbook 1981;* Thomas Cochran,
William Arkin, and Milton Hoenig, *Nuclear Weapons Databook, Vol. I, U.S.
Nuclear Forces and Capabilities* (Cambridge: Ballinger, 1984), pp. 134–146;
Strategic Force Modernization (fn. 7, pp. 179–182).

increase, deployment trends of the SSBNs by both sides may be to greater
offshore distances where the forces can be better protected. This would imply
SLBM flight times comparable to those of ICBMs. In general, neither the
Soviets nor the Americans are likely to contemplate a first strike simply on
the basis that submarines patrolling close to the adversary's borders could
raise the theoretical possibility of success.

The dangers rather are in how such a threat will influence the worst case
planners on each side. For it is to be expected that a submarine-based counterforce threat will encourage steps towards launch-on-warning, reliance
upon preprogrammed routines, and delegations of launch authority to larger
numbers of decision-makers, measures which could substantially increase
the risks of accident and inadvertent escalation in a crisis. Finally, increased
vulnerability of missile silos and the command and control network would
encourage the deployment of still more weapons systems in an endless quest
for more survivability.

Figure 4   U.S. SLBM Forces—Current Plans

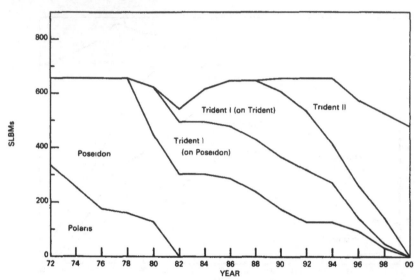

Sources: *The Military Balance 1982–1983; SIPRI Yearbook 1981; Nuclear Weapons Databook,* pp. 134–146; *Strategic Force Modernization* (fn. 7, pp. 179–182).

In the face of these prospects, several analysts have over the years suggested various arms control agreements to constrain SLBM development and deployment.[6] However, there has not been the slightest movement in this direction.

One reason for this failure is that SLBM limitations have always been unpopular within the broad defense community, an antagonism which goes well beyond a general lack of interest in arms control. In this case, the Navy, by agreeing to SLBM restrictions, would be accepting limitations on its operations to the benefit of the Air Force, while advocates of an assured destruction strategy would have to accept limitations on the leg of the triad that for so long has appeared both most secure as a second strike retaliatory force and also the least provocative.

---

6. See for example Quanbeck and Wood, *Modernizing the Strategic Bomber Force,* pp. 58–60.

Furthermore, many strategists do not believe that even very accurate SLBMs would prove destabilizing in a crisis since, unlike fixed ICBMs, they would be relatively untargettable.

Such lack of concern may prove to be shortsighted. Although a highly accurate SLBM force itself might not be vulnerable, it would place the fixed missiles of the adversary in a "use-them-or-lose-them" posture, compelling the adversary, as surely as with an ICBM threat, to consider strategies of launch-on-warning and preemptive attack, albeit only against other land-based targets. In the U.S. this danger can be seen clearly enough. As a Soviet submarine threat develops, the assurance that U.S. bomber and ICBM forces could not be attacked simultaneously would disappear and those concerned abut the vulnerability of U.S. nuclear forces would insist even more strenuously on a full panoply of remedial actions.

Of course, many defense planners in the U.S. are less worried about a Soviet sea-based hard target capability than they are intrigued by the emerging U.S. capability to be incorporated into the Trident II missile. Whatever are the merits and hazards of expanding U.S. counterforce capabilities in this manner, it is not likely to lead to a reduction in the Soviet counterforce threat to the U.S. and would probably increase it. Although conceivably it could encourage the Soviets in the long run to reduce reliance on their heavy fixed land-based missiles that now threaten U.S. ICBMs, it is more probable that an immediate impact would be to encourage an expanded Soviet sea-based force which itself would have hard target kill capability.

In fact, modifying submarine forces to give them enhanced counterforce potential has many of the same dangerous seductions that in the past led to the development of the MIRV and cruise missile. These developments have all appeared to give the U.S. an immediate (if only temporary) advantage, and from a suitably narrow perspective, they have even appeared advantageous for the purposes of arms control. Just as today the Trident II is rationalized by some observers on the grounds that it is less dangerous than the MX, the MIRV program once seemed to some to be the price that had to be paid to get the ABM treaty. And the cruise missile was initially justified as the less costly and less provocative alternative to the advanced manned bomber.

Is It Too Late to Stop a Counterforce SLBM Threat?

Of course, there is no foolproof way to prevent SLBMs from achieving improved accuracy, even if arms control measures were adopted. One reason is that accuracy improvements are inevitable even without substantial modifications of the missiles themselves. Many of the inaccuracies currently associated with SLBMs are due to uncertainties in the location, velocity, and orientation of the submarine launch platform, and in the earth's gravitational field near the submarine's position. These uncertainties can be substantially reduced through refinements in the submarine's navigation system and through improved gravitational mapping. Such efforts constitute a major part of the current "Improved Accuracy Program."[7]

Similarly, when the satellite-based Global Positioning System becomes available, the accuracy of currently deployed SLBMs could probably be increased with only minor changes in the missiles themselves (although there may be no current plans to do so). According to Richard Garwin, "It is likely that the accuracy of the Trident-I missile can be improved by a factor of four by the use of a modified missile guidance system capable of receiving signals from the NAVSTAR satellite while the missile is in powered-flight."[8] Such a modification would decrease the CEP of the Trident I to about 100 m, giving each reliably delivered Trident I warhead a kill probability of 86 percent against targets hardened to 2000 psi.

A Freeze on Deployment and Flight-Test Limitations

The full realization of high SLBM accuracy, however, would require a number of flight-tests. These include tests during development and full-system tests under operational conditions. The Trident I missile, for example, as of 1981, had been fired about 35 times from submarines as a part of its operational test program *in addition* to test firings during the missile's development. And the Navy plans about 40 operational test missile firings of the Trident II after initial deployment of this missile.[9] The Soviets, also, have traditionally undertaken large numbers of flight-tests of new missile systems. One estimate is

7. *Strategic Force Modernization Programs*, Hearings before the Subcommittee on Strategic and Theater Nuclear Forces of the Committee on Armed Forces, United States Senate, October–November 1981, pp. 198–199.
8. Richard L. Garwin, "Why Don't Those Who Assert the Fact of Soviet Superiority Support Feasible Moves to Redress the Balance Soon?," a contribution to the Symposium on Nuclear Superiority, *Harvard International Review*, March 20, 1980, p. 2.
9. *Strategic Force Modernization Programs*, pp. 199–200.

that they have typically required at least 15 flight-tests to develop a new missile.[10]

A ban on the deployment and flight-testing of new SLBMs and limitations on the flight-testing of existing SLBMs might therefore be expected to slow significantly the growth of confidence in SLBM forces as counterforce systems. These measures have the attraction for both the United States and the Soviet Union that, while they would place quite severe constraints on the development of a sea-based first-strike potential, they would not significantly affect the survivability of these forces as an effective deterrent. They also have a potential political advantage in the United States in that, by representing a clearly definable component of a nuclear weapons freeze, they could gather significant public support.

A FREEZE ON DEPLOYMENT: The SALT II Treaty prohibited the development of more than one new ICBM for each side but did not attempt to limit new SLBM types. This distinction may have reflected a greater concern about ICBMs as a destabilizing threat. In any case, without a prohibition on new types, it would be virtually impossible to contrive measures that could deny high accuracy capabilities to SLBMs.

In addition to a ban on new types of missiles, a freeze or. limit on the number and mix of existing SLBM types would help limit the growth of a sea-based counterforce threat. Such a freeze would forbid the replacement of older missile types (such as the SS-N-6, SS-N-8, or Poseidon) by existing modern types (such as the Trident I, SS-N-18, and SS-N-20). Some of the newer missiles have greater throwweight for a given range, which permits greater fractionation of warheads. It may also be somewhat easier over time to incorporate accuracy improvements in the larger newer missiles. Verification of a freeze that did not permit the replacement of one existing type of SLBM by another might require certain cooperative arrangements between the U.S. and the Soviet Union, including stipulations that would limit the size of launch tubes on the modern ballistic missile submarines.

A freeze instituted at the end of any specified calendar year would halt deployment at the levels projected in Figures 1 through 4 for that year, numbers which may then be compared to the expected path of deployments in the absence of a freeze.

It is worth remarking that an SLBM freeze has one particularly attractive

---

10. Strobe Talbott, *Endgame: The Inside Story of SALT II* (New York: Harper and Row, 1979), p. 55.

feature: the age distributions of U.S. and Soviet SLBMs are roughly the same. About 250 of the missiles on each side were less than 5 years old in 1983, and most were less than 10–11 years old. Construction of replacement submarines (such as the U.S. Trident and the Soviet Delta III) would be permitted under the freeze which we are discussing, although sharp limits on flight-testing of SLBMs might also discourage new submarine types, which the superpowers would want to test under operational conditions.

FLIGHT-TEST RESTRICTION: Since a certain number of confidence tests on each side will be required each year for existing missile types, a complete flight-test ban on SLBMs may not be a practical goal. The number of confidence tests could, however, be quite low. In the arms reduction proposal taken to Moscow in March 1977 by Secretary of State Vance, the U.S. suggested a flight-test limitation restricting confidence tests to two per year for each modern (ICBM) missile-type already deployed.[11] Such a low limit on the number of flight-tests permitted would place severe constraints on the development of high accuracy SLBMs. It seems likely that to establish high confidence in a new guidance system providing high accuracy would require a large number of tests.

We have not analyzed the verifiability of flight-test bans in any detail. However, given that the carefully elaborated Salt II Treaty included prohibitions on several classes of both SLBM and ICBM flight-testing, we think it likely that significant flight-testing of SLBMs could be monitored by national technical means.[12]

LIMITATIONS ON DEPRESSED TRAJECTORY AND TERMINAL GUIDANCE: To optimize the effectiveness of a surprise attack, SLBMs would, if possible, be fired along depressed trajectories in order to minimize flight times. Figure 5 shows the time of flight that would be required at various great circle distances for minimum energy ballistic trajectories and for missiles at less than their maximum range.

No point in the continental United States is located more than 1000 nautical miles (nm) from the open sea, so all targets would be within a range of about 1200 nm of the Soviet SSBNs patrolling within 200 nm or so of shore. At this distance, an SLBM fired on a minimum energy trajectory would take about 16 minutes to arrive at target, adding roughly two minutes for the slower

---

11. Ibid., pp. 55, 60.
12. Although the U.S. has raised some questions relevant to verification of the SALT II limitations on flight-testing, these involve essentially technical issues of interpretation of treaty provisions. See Federation of American Scientists, *Public Interest Report*, Vol. 37, No. 3 (March 1984), pp. 9–12, 14–16.

Figure 5 Flight Time vs. Range
For minimum energy and depressed trajectories (ballistic flight times)

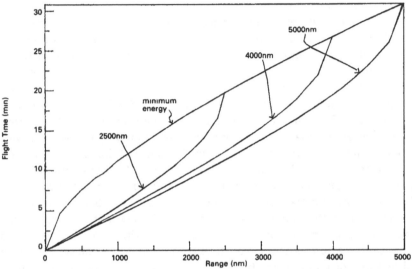

Note: The top line shows flight times at various ranges for minimum energy trajectory. The bottom three lines show the reductions in flight time which are
achievable with a missile of given maximum range using a depressed trajectory. For ranges greater than 1000 nm add 2 minutes to ballistic flight time to
allow for slower flight after launch and during reentry.

speeds at liftoff and reentry.[13] Using the highly depressed trajectories achievable by missiles with a 2500 nm or greater maximum range could cut this
time to only about 8 minutes.

The straightforward way to discourage the development of a capability to
employ depressed trajectories would be to ban flight-tests in this mode. At
one point in the SALT II negotiations, the U.S. proposed such a ban (with
the Soviets showing at least some interest in the idea). However, the ban
was not included in the final treaty.[14]

Evidently, neither the U.S. nor the U.S.S.R. has yet flight-tested an SLBM
in a depressed trajectory. U.S. Defense Department spokesmen asserted in
1975 that such tests could be observed through national technical means of
verification and that, even if the Soviets had at that time some fledgling

13. Speed, *Strategic Deterrence in the 1980s*, p. 144.
14. Talbott, *Endgame*, pp. 207–209.

capability to shape SLBM trajectories, two years of additional testing to develop an operational capability would have been required.[15] No such test series has been reported since.

A ban on flight-tests of maneuverable reentry vehicles would constitute another important barrier to the development of high accuracy SLBMs. At the low reentry angles implied by depressed trajectories, it will be virtually impossible to achieve high accuracy without some sort of terminal maneuver.

STANDOFF ZONES: A complementary way to reduce the threat posed by submarines patrolling off the coasts of the U.S. (and the U.S.S.R.) would be to require that SSBNs maintain more than a certain minimum distance from shore, an idea which has (most recently) been advocated by Richard Garwin. Even when depressed trajectories are used, each additional two hundred miles of distance a missile must travel to its target adds approximately one minute to the flight time. Without depressed trajectories, the additional time needed to cover the extra distance would be still greater. (See Figure 5.)

Verification of a standoff zone would be troublesome but not impossible. Although there is no way at present to monitor with total confidence an ocean zone that would possibly have to extend hundreds of miles from shore and thousands of miles along the coast, neither could a submarine assume that it could brazenly violate a standoff zone without detection and response. A violation would be unambiguous if detected.

In a crisis, presumably, the parties to the agreement would ignore the standoff zones if it suited their interest to do so. It seems, however, more likely that if a tense situation arose, each side would try to use its adherence to a standoff agreement to reassure the other. During a confrontation, each superpower would wish to show resolve, but it would also wish to protect its own forces and otherwise ensure that the adversary has no grounds or capacity to preempt. To penetrate close to the enemy's shore would increase one's own vulnerability; it certainly would not be reassuring—and the only resolve that it could convey would be the resolve to strike first.

Effects on Vulnerability

Since the described freeze on SLBMs would still permit the replacement of older submarines by quieter, faster, and otherwise improved types, it would

---

15. Quanbeck and Wood, *Modernizing the Strategic Bomber Force*, p. 59.

not significantly reduce the survivability of the SLBM force in this respect.[16] However, a freeze could affect the vulnerability of the subs by prohibiting the deployment of missiles with intrinsically longer range. The range of SLBMs determines the area of ocean from which submarines on patrol can immediately threaten enemy targets, and the greater the patrol area, presumably the more the difficulties faced by anti-submarine warfare (ASW) efforts.

The nominal ranges of modern types of SLBMs are already quite long, however. By the end of 1983, the Soviets had nearly 600 SLBMs carrying over 2500 warheads capable of striking Washington, D.C., New York, and other cities of the northeast United States from Soviet home waters. Similarly, the U.S. had 264 Trident I SLBMs carrying over 2100 warheads able to strike Moscow and other major targets in western Russia from U.S. coastal waters. Furthermore even these nominal ranges (and that of the shorter range Poseidon) could be increased substantially by off-loading some warheads on a MIRVed missile.[17]

Figure 6 shows, for example, that the range of the Trident I could be boosted from 4000 nm to 6000 nm by off-loading approximately three warheads. This would result in nearly a tripling of the ocean area from which an SLBM could reach Moscow.[18] Even with this warhead reduction, the Trident I component of the submarine force would contain well over 1000 warheads.

Thus far, however, the U.S. Navy has shown virtually no interest in increasing the range of SLBMs. Indeed, the Navy has configured the much more powerful Trident II to have the same nominal range as the Trident I. According to Admiral William A. Williams, the Director of Strategic and Theater Nuclear Warfare, the Navy is "not investing in additional range [with the Trident II] but in additional payload with more warheads per missile. . . . [W]e are not advocating the D-5 [Trident II] because of its greater range. The C-4 [Trident I] has a very comfortable range."[19]

Matters for the Soviets are perhaps less clear given the more advanced state of U.S. ASW capabilities and the smaller number of missile submarines

---

16. Allowed replacement of ballistic missile submarines is also consistent with most versions of the nuclear weapons freeze which have been advocated in the U.S.

17. Since a MIRV bus makes up about one-half of the total missile payload, a given fractional reduction in total payload would imply roughly twice as great a fractional reduction in reentry vehicle weight and in number of warheads, assuming no change in the bus.

18. Richard L. Garwin, "How Much Can New Technologies Improve Strategic Anti-Submarine Warfare?," Aspen Study Group on the U.S. Strategic Posture, November 10, 1982.

19. *Strategic Force Modernization Programs*, p. 187.

Figure 6   Range vs. Payload—U.S. SLBMs

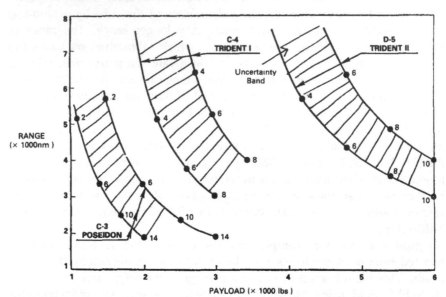

Sources: The uncertainty bands are due to slightly conflicting data given in the open literature. For the D-5, the band is determined by the assumption that a 6000 lb. payload will permit ranges of alternatively 4000 nm (Thomas Cochran, William Arkin, and Milton Hoenig, *Nuclear Weapons Databook: Vol. I, U.S. Nuclear Forces and Capabilities*, p. 145) and 3000 nm (IISS, *The Military Balance 1981–1982*, p. 110). For the C-4, the band derives from the worst case assumptions of *The Military Balance* (p. 110) and the comparison between the D-5 and the C-4 given in *Strategic Force Modernization* (See fn. 7, p. 168). For the C-3, the band derives from *The Military Balance* (pp. 104, 110). The maximum numbers of warheads shown are taken from *The Military Balance;* and we assume that the bus takes up approximately 50 percent of the full payload.

which the Soviets have typically chosen to keep at sea. But they, too, have large numbers of SLBMs with ranges of over 4000 nm. And the Soviets could off-load to increase the range of these missiles should such a course of action seem prudent. (See Figure 7.)

Reduction in Numbers of SLBMs

The simplest way to *reduce* the counterforce threat posed by submarine-launched ballistic missiles would be to cut their numbers. Such reductions

Figure 7   Range vs. Payload—Soviet SLBMs

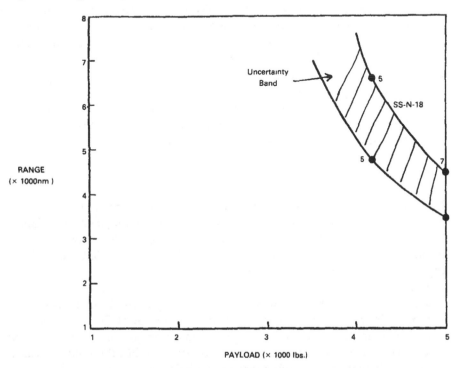

Sources: The Military Balance 1981–1982, p. 105. The uncertainty band derives from the assumption that at maximum payload of 5000 lbs. the range of the SS-N-18 is alternatively 4500 nm or this range reduced by 25 percent.

could make the submarine force more vulnerable by reducing the number of submarines that would have to be tracked and destroyed in a preemptive strike if the reductions were effected sub by sub rather than launcher by launcher. However, if reductions were accomplished by deactivating launch tubes in each submarine, SLBM reductions could be distributed among the SSBNs in ways to ensure that the number of submarines remained roughly constant even in the presence of sharp declines in the number of missiles.

There are, in fact, several ways by which the number of launchers each submarine carries might be reduced. The easiest approach would be simply to fill unwanted launch tubes with concrete. Currently several Polaris submarines are being converted to nuclear attack submarines in this manner.[20] Another approach would involve actually cutting missile tubes out of a submarine. Evidently, eight Soviet Yankee class SSBNs have had their launch tubes removed in this manner.[21] In any case, it is likely that a satisfactory way to eliminate missile tubes, launcher by launcher rather than sub by sub, could easily be devised once there was a will to do so.

Alternative ways to reduce the number of SLBM warheads without diminishing the survivability of the ballistic missile submarine force would be to move towards single warhead SLBMs, much as has been suggested for land-based missiles, and/or towards new smaller submarines.

Conclusion

If the United States and the Soviet Union are really worried about the survivability of their strategic forces, a freeze on the deployment and flight-testing of new SLBMs and sharp limitations on the flight-testing of existing SLBMs would have a significant impact on the level of the SLBM counterforce capability that could be developed.

That survivability can be enhanced in this way has been clear for many years. Analysts who have studied the vulnerability of the U.S. bomber force, for example, have almost always focused on the Soviet SLBM threat, and many of them have been led to suggest some form of limitation on SLBM forces. What has been lacking so far is not analysis, but the political will to seek such limitations through serious negotiation.

---

20. Captain John Moore, ed., *Jane's Fighting Ships, 1982–1983* (London: Jane's Publishing Company, 1982), p. 604.
21. Ibid., p. 462.

Part III:
Naval Operations—
Controlling the Risks

Nuclear War at Sea | *Desmond Ball*

$\mathbf{T}$he subject of nuclear warfare at sea, and the difficulties of controlling escalation of conflict at sea, has so far drawn very little attention from the strategic community.[1] This is despite the fact that more than a third of the nuclear weapons of the U.S. and the Soviet Union are deployed on sea-based platforms; the control of these weapons by central national authorities is physically loosest; and the doctrines and operational procedures associated with sea-based nuclear weapons are subject to less well-defined thresholds and, in some cases, are quite provocative. Moreover, there are good reasons for believing that the first use of nuclear weapons could take place at sea, and for concern that the escalation dynamics of nuclear warfare in this theater are far less constrained than those that would attend nuclear operations on land.

There are several reasons for this lack of attention. The nuclear weapons based at sea—both tactical and strategic—are less "visible." The details of their deployment and their supporting infrastructure are generally closed to public scrutiny. And the employment doctrines for these weapons have never been officially presented in a form that would permit informed critiques to be readily explicated.

Some of the considerations that require serious attention are as follows:
—Accidents at sea.
—The attractiveness of ships as nuclear targets.
—The launch autonomy of naval commanders.
—Problems raised by dual-capable systems and platforms.
—Anti-submarine warfare (ASW) strategy.

This paper was prepared for a conference held by the Institute on Global Conflict and Cooperation (IGCC) at the University of California, Los Angeles (UCLA) on April 26, 1985. An earlier version was used as the basis for a seminar at the Center for International Affairs (CFIA) at Harvard University on January 16, 1985.

Desmond Ball is Head of the Strategic and Defence Studies Centre, The Australian National University.

---

1. There are two particularly noteworthy exceptions to this observation. See Bruce G. Blair, "Arms Control Implications of Anti-Submarine Warfare (ASW) Programs," in U.S. House of Representatives, Committee on International Relations, *Evaluation of Fiscal Year 1979 Arms Control Impact Statements: Toward More Informed Congressional Participation in National Security Policymaking* (Washington, D.C.: U.S. Government Printing Office, 1978), pp. 103–119; and Barry R. Posen, "Inadvertent Nuclear War?: Escalation and NATO's Northern Flank," *International Security*, Vol. 7, No. 2 (Fall 1982), pp. 28–54.

—The vulnerability of the ASW and command, control, communications, and intelligence (C^3I) support structures and the incentives for preemption that derive from this.

—The U.S. Navy's doctrine for the conduct of offensive operations in forward areas.

—U.S. Navy doctrine for the employment of tactical nuclear weapons.

—Soviet doctrine for war at sea.

—The lack of U.S. contingency planning concerning the escalation dynamics of naval conflict, and the U.S. Navy's resistance to such planning.

Accidents at Sea

The sea is the only area where nuclear weapon platforms of the U.S. and the Soviet Union actually come into physical contact. Accidents have attended four sorts of naval activities: 1) covert submarine operations of the Holystone sort; 2) more routine monitoring activities; 3) games of "chicken"; 4) harassment for tactical military purposes.

THE HOLYSTONE UNDERSEA SURVEILLANCE PROGRAM

One of the most interesting, not to say dangerous and provocative, submarine surveillance and intelligence operations is a program code-named at various times Holystone, Pinnacle, Bollard,[2] and, most recently, Barnacle. This program, which is controlled from Office M-34 at the Atlantic Fleet Command headquarters at Norfolk, Virginia, has involved the use of at least four SSN-637 or Sturgeon-class attack submarines specially outfitted with sophisticated electronic equipment operated by special units from the NSA, which gather intelligence either inside or just outside Soviet territorial waters. The program was authorized during the Eisenhower Administration and placed under the direct control of the Chief of Naval Operations. In the mid-1970s, the schedule of Holystone missions was approved every month by the 40 Committee.[3]

---

2. Seymour M. Hersh, "Submarines of US Stage Spy Missions Inside Soviet Waters," *The New York Times*, May 25, 1975, pp. 1, 42; and Seymour M. Hersh, "A False Navy Report Alleged in Sub Crash," *The New York Times*, July 6, 1975, pp. 1, 26.

3. *The New York Times*, May 25, 1975, p. 42; and George B. Kistiakowsky, *A Scientist at the White House: The Private Diary of President Eisenhower's Special Assistant for Science and Technology* (Cambridge: Harvard University Press, 1976), p. 153.

The Holystone submarines are reported to engage in a wide range of intelligence operations, including close-up photography of the undersides of Soviet submarines and other vessels; plugging into Soviet underwater communication cables to intercept high-level military and other communications considered too important to be sent by radio or other less secure means; electronic observation of Soviet SLBM tests to monitor the various computer checks and other signals that precede test launchings; and the recording of "voice autographs" of Soviet submarines, which consist of tape recordings of the noises made by submarine engines and other equipment.[4]

The Holystone submarines generally operated within the 12 mile (19.3 kilometer) territorial limit claimed by the Soviet Union, and often within the 3 mile (4.8 kilometer) limit. For example, one Holystone submarine is reported to have collided with an E-class submarine in Vladivostok Harbor in the mid-1960s when photographing the underside of the Soviet vessel.[5] On one occasion in November 1969, the USS *Gato* is reported to have operated as close as one mile (1.6 kilometers) off the Soviet coast; later on the same patrol the *Gato* collided with a Soviet submarine 15–25 miles (24–40 kilometers) off the entrance to the White Sea, in the Barents Sea off northern U.S.S.R.[6] And another Holystone collision occurred in May 1974 in Soviet waters off the port of Petropavlovsk on the Kamchatka Peninsula.[7] Holystone collisions also occurred on at least six other occasions between 1961 and 1975.[8]

Each of these collisions—together with more than a hundred other Holystone intrusions that were probably detected by Soviet forces but that they were unable to locate[9]—could have been the catalyst for a chain of events that might have run from a localized engagement involving the intruding submarine to a full-scale nuclear exchange. The general orders for Holystone missions reportedly state that, if threatened, the submarines "have authority to use weapons."[10] In the November 1969 incident, the weapons officer of the *Gato* prepared to arm a SUBROC (UUM-44A) anti-submarine rocket (which carries a 1–5 kt W55 nuclear warhead) and three smaller nuclear

---

4. *The New York Times*, May 25, 1975, p. 42.
5. Ibid.
6. *The New York Times*, July 6, 1975, pp. 1, 26.
7. Ibid., p. 26.
8. *The New York Times*, January 20, 1976, p. 1.
9. Ibid., p. 4.
10. *The New York Times*, July 6, 1975, p. 26.

torpedoes, and the submarine was "maneuvered in preparation for combat."[11] According to one report:

Only one authentication—either from the ship's captain or her executive officer—was needed to prepare the torpedoes for launching.[12]

MONITORING ACTIVITIES

Accidents frequently occur during weapons field tests and naval exercises when monitoring vessels approach too close to the operations. On March 8, 1963, for example, a Soviet trawler came under fire from U.S. naval vessels some 70 miles east of Norfolk, Virginia, when it entered an exercise area.[13] More recently, on March 21, 1984, a Soviet Victor-1 nuclear-powered attack submarine was disabled when it collided with the USS *Kitty Hawk*, which was conducting maneuvers off the coast of South Korea during Exercise Team Spirit '84.[14]

GAMES OF "CHICKEN"

Numerous accidents have occurred when U.S. and Soviet ships have engaged in games of "chicken," in which each captain attempts to gauge the resolve of the other. "Chicken" has been described by Admiral Elmo Zumwalt as:

an extremely dangerous, but exhilarating, running game . . . that American and Soviet ships had been playing with each other for many years. Official [U.S.] Navy statements always have blamed the Russians for starting this game, but as any teen-aged boy knows, it takes two to make a drag race.[15]

Admiral Zumwalt has detailed a particular incident in which he was involved in 1962, and has stated that "[f]oolish episodes of this kind occurred all the time."[16] As Zumwalt has noted:

in addition to being juvenile, these incidents were terribly dangerous. Beyond the immediate damage to property and the loss of life any one of them might

---

11. Roy Varner and Wayne Collier, *A Matter of Risk* (Sevenoaks, Kent: Coronet Books, Hodder and Stoughton Limited, 1978), p. 38.
12. *The New York Times*, July 6, 1975, p. 26.
13. Historical Office, U.S. Department of State, *American Foreign Policy: Current Documents 1964* (Washington, D.C.: U.S. Government Printing Office, 1967), pp. 562–563.
14. See Fred Hiatt, "Soviet Sub Bumps Into U.S. Carrier," *The Washington Post*, March 22, 1984, p. 1; and "Soviet Sub and U.S. Carrier Collide in Sea of Japan," *The New York Times*, March 22, 1984, p. A7.
15. Elmo R. Zumwalt, Jr., *On Watch. A Memoir* (New York: The New York Times Book Co., 1976), p. 391.
16. Ibid., pp. 391–393.

cause, any one could lead people to shoot at each other with results that might be by that time impossible to control.[17]

HARASSMENT FOR MILITARY PURPOSES

Much harassment is deliberately designed for tactical military purposes. For example, Soviet interference with the flight operations of U.S. carriers and other obstruction of U.S. naval activities is often clearly intended to "prevent the launch of aircraft that might deliver an attack or, more probably, track a Soviet submarine."[18] Other obstructive activities clearly involve the practice of maneuvers that would need to be performed in actual naval conflict. Others enable the Soviet and U.S. navies to obtain information on likely responses to their attempts to disrupt combat operations.[19]

Some forms of harassment are difficult to distinguish from preparations for hostilities. This is particularly the case with simulated attacks involving the aiming of guns, missile launchers, torpedo tubes, other weapons, and sensor systems at adversary vessels. During the Jordanian crisis in 1970, for example, in an encounter between a U.S. ship and accompanying naval aircraft and several Soviet warships, the Soviet ships "went to battle stations . . . ran surface-to-air missiles out on their launchers, and appeared to track the departing U.S. aircraft with their fire control radars."[20] Countermeasures to these provocations, such as maneuvering away from the threatening vessels, jamming or deceiving the adversary's electronic equipment, or directly harassing the threatening forces, could increase apprehensions on the other side and thus "prompt the very preemptive attack that they were meant to avoid."[21]

Two agreements represent attempts to ameliorate this type of situation. The first is the "Agreement on Measures to Reduce the Risk of Outbreak of Nuclear War Between the USA and the USSR," which entered into force in 1971 and which commits each Party in situations involving unexplained nuclear incidents to act "in such a manner as to reduce the possibility of its actions being misinterpreted by the other Party."[22] This is particularly ger-

---

17. Ibid., p. 393.
18. Sean M. Lynn-Jones, "A Quiet Success for Arms Control: Preventing Incidents at Sea," *International Security*, Vol. 9, No. 4 (Spring 1985), p. 160.
19. Ibid., pp. 160–161.
20. Abram N. Shulsky, "The Jordanian Crisis of September 1970," in Bradford Dismukes and James M. McConnell, eds., *Soviet Naval Diplomacy* (New York: Pergamon Press, 1979), p. 176; quoted in ibid., p. 157.
21. Ibid., p. 166.
22. See Blair, "Arms Control Implications of ASW Programs," p. 117.

mane to such practices as Holystone-type activities and U.S. submarine ASW operations in so-called "forward areas."

The second is the "Agreement Between the Government of the United States of America and the Government of the Union of Soviet Socialist Republics on the Prevention of Incidents on and over the High Seas," signed on May 25, 1972, which establishes "rules of the road" for U.S. and Soviet ships and aircraft on and over the high seas. Although the Secretary of the Navy, John F. Lehman, has stated that this agreement has led to "a marked reduction in collisions and near collisions,"[23] it has clearly not stopped them entirely. In the late 1960s, the number of serious incidents at sea exceeded 100 per year, but according to Secretary Lehman, there were only about 40 potentially dangerous incidents between June 1982 and June 1983.[24] Moreover, there was an apparent decrease in the severity or degree of danger implicit in these incidents. On the other hand, serious collisions continue to occur at a rate of about half a dozen a year.

The Attractiveness of Ships as Nuclear Targets

Certain attributes of ships at sea suggest not only that they make lucrative targets but also that constraints on the use of nuclear weapons against them could well be weaker than those that pertain to land-based targets. Large capital vessels such as aircraft carriers represent an investment of more than two billion dollars; the destruction of naval vessels would involve few if any civilian casualties; and attacks against them could be clearly distinguished as a specific, limited operation, particularly if the strikes were launched from other ships rather than from land bases, thus profferring the possibility of containing a nuclear engagement to the sea. As Bernard Brodie once asked rhetorically, "what better targets are there for such [tactical nuclear] weapons than our nice, big aircraft carriers."[25]

On the other hand, U.S. authorities have explicitly sought to dismiss this possibility. As Richard Perle, Assistant Secretary of Defense for International Security Policy, testified in 1982, official U.S. policy is to "discourage the

---

23. See Hiatt, "Soviet Sub Bumps Into U.S. Carrier."
24. "Superpowers Maneuvering for Supremacy on High Seas," *The Washington Post*, April 4, 1984, p. A18.
25. Letter from Bernard Brodie to Admiral Stansfield Turner, January 16, 1976, in Collection 1223, Department of Special Collections, University Research Library, University of California, Los Angeles (UCLA), Box 9.

Soviets from believing that they could limit a nuclear war to forces at sea."[26] The destruction of large naval assets would disproportionately disadvantage the United States, both because of the enormous U.S. investment in its carrier forces and because of the greater U.S. dependence on sea lines of communication. The seven U.S. nuclear-powered carriers currently operational or under construction alone represent a total investment of some $15–$20 billion and have a total complement of about 50,000 officers and men (including air crews)—the loss of which would exceed the total number of U.S. fatalities suffered during the Second Indochina War! Moreover, attacks against U.S. vessels are likely to unleash emotional pressures for punitive strikes against higher-value targets in the attacker's homeland. As Chester Cooper has observed with respect to the Gulf of Tonkin incident in 1964, when the U.S. responded to alleged attacks on the destroyers *Maddox* and *Turner Joy* with the first bombing of North Vietnam:

There is something very magical about an attack on an American ship on the high seas. An attack on a military base or an Army convoy doesn't stir up that kind of emotion. An attack on an American ship on the high seas is bound to set off skyrockets and the "Star-Spangled Banner" and "Hail to the Chief" and everything else.[27]

More strategic considerations have been stressed by Richard Perle:

the Soviets retain a significant capability to attack ships at sea, and they may, as a consequence, be misled into believing that so long as civilian casualties are not involved in such attacks, as they presumably would not be, they could in fact limit a war to attacks on forces at sea. Given the vital importance we attach to maintaining the sea lines of communication, that would be a very dangerous development. . . .
The desire on our part [is] not to permit the Soviets to determine the scope of the battle to give whatever advantages would be inherent in their having the freedom to choose where the battle would be fought.[28]

And as the Department of Defense Guidance for FY 1984–88 reportedly states:

26. Committee on Armed Services, U.S. Senate, *Department of Defense Authorization for Appropriations for Fiscal Year 1983* (Washington, D.C.: U.S. Government Printing Office, 1982), Part 7, p. 4377.
27. Quoted in "The 'Phantom Battle' that Led to War," *U.S. News and World Report*, July 23, 1984, pp. 65–66.
28. Senate Committee on Armed Services, *Department of Defense Authorization for FY 1983*, Part 7, pp. 4377–4378.

It will be U.S. policy that a nuclear war beginning with Soviet nuclear attacks at sea will not necessarily remain limited to the sea.[29]

Hence, the U.S. response to nuclear attacks against its naval vessels would be likely to involve attacks against Soviet land-based forces, particularly if the attacks against its ships were being launched or supported from the land. Indeed, the threat to attack selected targets on Soviet territory is an explicit and integral component of the "horizontal escalation" deterrent strategy of the current Administration.[30]

The Launch Autonomy of Submarine Commanders

Unlike U.S. Air Force and Army nuclear weapons, which can only be fired following an electronic release authorization from the National Command Authority (NCA), the thousands of SLBMs, SLCMs, and other nuclear weapons deployed on submarines can be fired without any technical or other action by anyone outside the individual submarine.

The Navy has justified its resistance to any outside control with the following arguments:

1) Navy doctrine mandates that, at least in the case of SLBMs, no launch may proceed without a proper, authenticated order from the NCA—even in the event that communications are destroyed. According to Navy officials, "Our commanding officers have got to have positive direction. . . . They must have positive direction to launch, regardless of the scenario."[31] However, (a) the requirement for specific authorization does not seem to apply in the case of so-called "defensive" use of nuclear weapons, as evinced in the November 1969 incident involving the USS *Gato;* and (b) the necessity for positive direction is not physical/technical, but doctrinal and thus dependent on personnel discipline.

2) The Navy is confident that its careful personnel selection, training, and discipline are a strong guarantee against unauthorized launch.

3) Rigorous launch procedures exist. The Navy argues that although no technical release is required, the number of people that must be involved in

---

29. Cited in George C. Wilson, "Pentagon Guidance Document Seeks Tougher Sea Defenses," *The Washington Post*, May 25, 1982, p. 1.
30. Senate Committee on Armed Services, *Department of Defense Authorization for FY 1983*, Part 7, pp. 4377–4380.
31. See Lawrence Meyer, "AF Locks System Urged for Navy's Nuclear Missiles," *The Los Angeles Times*, October 14, 1984, p. 28.

launching a missile is sufficient to rule out any unauthorized launch. The launch procedures for SLBMs are reportedly as follows. The submarine's captain and the executive officer each have one of two keys necessary to open a safe containing specific launch instructions. In the event the submarine received an emergency war message, the two men would, in the presence of a third officer, remove these instructions from the safe. The captain would then open the lock on a red "fire" button, to which only he has the combination. This action would start a carefully coordinated launch sequence involving at least fifteen different individuals at various stations on the boat. To actually launch a missile, it takes four officers in different parts of the submarine to turn keys or throw switches. The navigation officer has a switch, the captain and launch control officer have keys, and the missile launch officer has a trigger. If one of these officers fails—or refuses—to do his part, the missile cannot be fired.[32]

4) The Navy has also argued that the ability of SSBN commanders to launch their missiles without their being subject to technical release means that the possibility of executing a successful "decapitation" strike is reduced and hence deterrence is enhanced.[33]

These arguments notwithstanding, however, the lack of any outside release mechanism allows at least the possibility of unauthorized launch of nuclear weapons.

Dual-Capable Systems and Platforms

Naval conflict is inherently escalative because of the common deployment of dual-capable weapons systems and platforms.

DUAL-CAPABLE WEAPONS SYSTEMS

The U.S. Navy has deployed several different types of dual-capable systems. The first of these was the ASROC RUR-5A anti-submarine rocket, which carries the 1 kt W44 warhead in its Mk-17 nuclear depth bomb and which became operational in 1961. More than 20,000 ASROC missiles were produced, but only about 850 of these were equipped with the Mk-17/W44 bomb.

---

32. Phil Stanford, "Who Pushes the Button?," *Parade Magazine*, March 28, 1976.
33. Meyer, "AF Locks System Urged," pp. 2, 28.

The dual-capable system is deployed aboard some 170 cruisers, destroyers, and frigates.[34]

Two dual-capable, surface-to-air and surface-to-surface weapons have also been produced. The Terrier-BTN RIM-2 is a short-range, surface-to-surface and surface-to-air missile carried by 31 cruisers and destroyers and three aircraft carriers. It is equipped with the 1 kt W50-0 warhead, of which about 310 have been deployed. Although numerous dual-capable Terrier launchers were deployed, the only missiles remaining in the inventory are nuclear-armed. These Terriers are to be phased out as the Standard SM-2 (N) RIM-67 becomes operational. Some 2044 Standard SM-2 missiles are planned for procurement, about 350 of which will be equipped with the W81 warhead with a low-kiloton yield. The SM-2 (N) is to be deployed on a variety of cruisers and destroyers. Because of its small size, it can be carried by vessels equipped with Tartar launchers (i.e., Virginia-class cruisers), giving them a nuclear air defense capability for the first time and increasing the number of ships carrying nuclear weapons.[35]

The most recent and most controversial dual-capable system is the Tomahawk sea-launched cruise missile (SLCM), the nuclear-armed version of which is the Tomahawk land-attack missile T-LAM(N) BGM-109A, which has a range of 2500 km and which is equipped with the 200–250 kt W80-0 warhead. Some 4068 SLCMs are currently planned, of which 750–1000 will be nuclear-armed. The T-LAM(N) is to be deployed aboard the Permit (SSN-594), Sturgeon (SSN-637), and Los Angeles (SSN-688) classes of attack submarines; the California (CGN-36), Virginia (CGN-38), and Ticonderoga (CG-47) classes of cruisers; the Spruance (DD-963) and Burke (DDG-51) destroyers; and the reactivated BB-61 Iowa-class battleships.[36]

DUAL-CAPABLE PLATFORMS

The range of weapons systems deployed aboard U.S. and Soviet major combatant vessels makes it impossible to delineate clearly between conventional and nuclear platforms or between tactical or theater nuclear as opposed to strategic nuclear platforms.

34. Thomas B. Cochran, William M. Arkin, and Milton M. Hoenig, *Nuclear Weapons Databook, Volume 1: U.S. Nuclear Forces and Capabilities* (Cambridge, Mass.: Ballinger, 1984), pp. 267–268.
35. Ibid., pp. 272–278.
36. Ibid., pp. 79–80, 184–187; and Tim Carrington, "Deadlier New U.S. Missiles About to be Deployed are More Urgent Soviet Concern Than Star Wars," *The Wall Street Journal*, January 25, 1985, p. 25 (which states that 758 SLCMs are to be nuclear-armed).

In the U.S. case, some 190 warships are certified to carry nuclear weapons of various types,[37] i.e., about one-third of the U.S. fleet. The distinction between conventional and nuclear platforms breaks down at the frigate level, since some 65 frigates are equipped with dual-capable ASROC launchers. All larger U.S. vessels are equipped to carry both conventional and nuclear armaments. (As noted below, the same is true in the Soviet case.) As George Quester has noted:

The navies of the nuclear powers have generally not yet adjusted to the idea of dividing and segregating themselves into a nuclear-weapons fleet and a conventional weapons fleet. Old maxims of "never divide the fleet" had a slightly different meaning, but admirals probably everywhere cherish a fundamental fungibility and interchangeability of the units in their navies.[38]

There are several important implications of this commonplace deployment of dual-capable systems and platforms.

INCREASING THE PREMIUM OF PREEMPTION. The possibility that adversary combatants might be equipped with nuclear weapons would generate strong pressures to destroy those combatants as soon as possible, thus increasing the tempo of conflict in the conventional phase. Further, the use of nuclear weapons against these combatants could be justified on the grounds that they needed to be totally destroyed rather than partially immobilized but still able to use their nuclear weapons.

EASING CONSTRAINTS ON NUCLEAR USE. Having nuclear weapons close at hand obviously makes it much easier to consider their employment—particularly in situations in which their use might provide the only means of achieving a particular objective or preventing one's own destruction. For example, it is difficult to imagine a commander on an ASW mission, having exhausted his supply of conventional depth charges and related anti-submarine munitions, not being seriously tempted to break open his cache of nuclear depth-charges.[39]

---

37. Senate Committee on Armed Services, *Department of Defense Authorization for FY 1983*, Part 6, p. 3601.

38. George H. Quester, "The Falklands and the Malvinas: Strategy and Arms Control," ACIS Working Paper No. 46 (Los Angeles: Center for International and Strategic Affairs, UCLA, May 1984), p. 21.

39. George Quester has posed for rumination the situation that might have pertained during the Falklands/Malvinas war had the Argentine submarine *San Luis* been able to sink or disable one or two major British ships while itself evading British conventional ASW efforts. He argues persuasively that a strong case would have been made for the use of a nuclear depth charge against the *San Luis* rather than risk the success of the Falklands expedition. Ibid., pp. 24–25.

BLURRING THE CONVENTIONAL/NUCLEAR AND TACTICAL/STRATEGIC DISTINC-
TIONS. The Tomahawk SLCM poses a particular problem for stability precisely
because of its versatility. The multi-purpose design of this weapon completely
obscures the distinctions between tactical and strategic weapons and between
nuclear and conventional weapons.

With regard to the firebreak between the tactical and strategic use of
nuclear weapons, it is possible to envision a number of uses of the Tomahawk
SLCM that could lead to miscalculations by the United States and misperceptions by the Soviet Union. For example, whereas the U.S. might regard
the SLCM as a tactical weapon because of the nature of its deployment, the
Soviet Union could well regard it as a strategic weapon because of its potential
to destroy Soviet strategic installations. And with respect to the conventional/
nuclear firebreak, the Soviet Union must consider any vessel equipped with
Tomahawks to be a nuclear threat even if in fact these missiles are carrying
only conventional payloads. The obfuscation of these distinctions is likely to
increase Soviet paranoia about U.S. naval deployments in the vicinity of the
Soviet homeland; it inevitably reduces the degree of certainty with which
Soviet responses can be predicted; it increases the likelihood of escalation
from actions that the U.S. might regard as being tactical; and it increases the
chances of miscalculation and misperception and hence of inadvertent escalation in general.[40]

USING OR LOSING THE "STRATEGIC RESERVE." Since National Security Decision Memorandum (NSDM)-242 was promulgated on January 17, 1974, there
has been a national requirement for the "maintenance of survivable strategic
forces in reserve for protection and coercion during and after major nuclear
conflict."[41] In 1980, it was decided that the Tomahawk T-LAM(N) BGM-109A
SLCMs would comprise a major component of this strategic reserve force.
As Admiral Kelso (the Director of the Strategic Submarine Division of the
Office of the Chief of Naval Operations) testified in 1981:

The sea launched nuclear land attack cruise missile (TLAM/N) is not planned
for commitment to the SIOP or the NATO general strike plan. TLAM/N will
be a theater nuclear weapon deployed on general purpose forces. It will be

---

40. Lieutenant Paul G. Johnson, "Tomahawk: The Implications of a Strategic/Tactical Mix," U.S.
Naval Institute *Proceedings*, Vol. 108, No. 4 (April 1982), p. 30.
41. See Jack Anderson, "Not-So-New Nuclear Strategy," *The Washington Post*, October 12, 1980,
p. C-7. See also Desmond Ball, *Targeting for Strategic Deterrence*, Adelphi Paper No. 185 (London:
International Institute for Strategic Studies, Summer 1983), pp. 35–36.

available for selective release in non-SIOP options and *in a post-SIOP environment it will contribute to the strategic reserve force.*[42]

Rear Admiral Stephen J. Hostettler, Director of the Joint Cruise Missiles Project, testified on March 14, 1984 on the Tomahawk weapon system as follows:

The increased strike range of a larger number of surface ships, operating under carrier air cover as well as independent covert forward-deployed submarines, presents the Soviets a formidable threat from 360 degree axis against which they have no reliable defense. Thus, *TLAM/N is ideally suited for a Nuclear Reserve Force role.*[43]

In other important respects, however, the T-LAM(N) SLCMs are far from ideal for the strategic reserve force role. For one thing, the submarine-based component of the force suffers severe C^3 problems. Communications from the NCA to U.S. submarines are problematical in wartime, and communications back from the submarines might well be impossible, particularly if the existence of the submarines was not to be revealed. For submarine-based systems to constitute a "secure reserve," the NCA would need to know the status of each submarine and its location in relation to particular targets at all times, but such information would be unlikely to be available without placing the security of the submarines in jeopardy.[44]

It is especially strange that there is no dedicated concern for the survivability of this "secure reserve." According to Admiral William A. Williams, the Director of the Strategic and Theater Nuclear Warfare Division of the Office of the Chief of Naval Operations:

[The T-LAM(N) is] to be a member of the theater commander's theater nuclear forces to be employed at his discretion with other theater weapons.
. . .
We cannot commit the general purpose forces, which will carry those weapons [i.e., Tomahawk land attack SLCMs], to a rigid SIOP role. So they tend to be a nuclear weapon carrier which are on call at the discretion of the

---

42. Committee on Armed Services, U.S. Senate, *Strategic Force Modernization Programs* (Washington, D.C.: U.S. Government Printing Office, 1981), p. 200. Emphasis added. See also the statement of Melvyn R. Paisley, Assistant Secretary of the Navy for Research, Engineering and Systems, cited in Michael R. Gordon, "Deployment of Tomahawk Cruise Missiles Stirs Arms Control Controversy," *National Journal*, May 26, 1984, p. 1030.
43. Statement of Rear Admiral Stephen J. Hostettler, Director, Joint Cruise Missiles Project, before the Procurement and Military Nuclear Systems Subcommittee of the Armed Services Committee, U.S. House of Representatives (Mimeo), March 14, 1984, p. 6. Emphasis added.
44. Ball, *Targeting for Strategic Deterrence*, p. 36.

theater commander who has operational control of them. He will make the tradeoff between whether they are best employed in launching the Tomahawk or doing a general purpose mission such as supporting the carrier battle group.[45]

And according to Admiral Kelso:

The sea launched nuclear land attack cruise missile (TLAM/N) will be deployed on general purpose forces. *This deployment will be on a not-to-interfere basis with the primary mission of these platforms.*[46]

It would not be unreasonable to expect that a large proportion of these platforms would be destroyed or incapacitated in pursuit of their "primary mission." Indeed, many would not survive the conventional phase of a conflict, let alone the initial counterforce rounds of a strategic nuclear exchange.

Losses would be especially heavy among those vessels operating in so-called "forward areas," since the Soviet forces would undoubtedly regard any T-LAM(N) platforms within a range of 2500 km of the Soviet homeland as high priority targets. Yet no special efforts would be made by the U.S. Navy to keep T-LAM(N) platforms away from the forward areas. As Admiral Kelso has stated:

Operations of general purpose naval forces in proximity of the Soviet coast is governed by operational requirements of the fleet commander and is not a function of whether they are carrying TLAM/N missiles.[47]

The expectation that the "secure reserve" might not survive into the post-SIOP period would thus generate strong pressures to employ these missiles before they were lost.

Anti-Submarine Warfare (ASW) Strategy

As Bruce G. Blair has argued:

ASW *strategy* creates a risk of Soviet–American confrontation, misinterpretation and escalation in peacetime, in crisis situations, and at various levels of actual conflict.[48]

---

45. Senate Committee on Armed Services, *Strategic Force Modernization Programs*, p. 189.
46. Ibid., p. 200. Emphasis added.
47. Ibid.
48. See Blair, "Arms Control Implications of ASW Programs," p. 116.

Three aspects of ASW strategy warrant consideration in this context:

FIRST, THE INABILITY OR RELUCTANCE OF THE U.S. NAVY TO FORMULATE A STRATEGY THAT DISTINGUISHES BETWEEN THE TYPES OF SOVIET SUBMARINES THAT ARE SUBJECT TO DESTRUCTION. The nature of the ocean medium makes it practically impossible to engage some forces while simultaneously indicating clearly and unambiguously the deliberate avoidance of others. Most ASW systems serve both tactical and counter-SSBN operations, and many cannot distinguish between attack (or "hunter-killer") submarines and fleet ballistic missile (FBM) submarines. As the Navy's former director of ASW and Ocean Surveillance programs testified in 1976, in a war-fighting situation the Navy "would not be in a position of differentiating their [i.e., Soviet] attack submarines from their SSBNs."[49] Moreover, Soviet nuclear-powered attack submarines and the SSBNs are equally subject to attack during a conventional as well as a nuclear conflict. As the Navy's Director of Command, Control and Communications testified in 1977, "in a conventional war all submarines are submarines. They are all fair game."[50]

One reason for this, besides the technical problems of differentiating between the various types of submarines detected, is that SSBNs themselves have a significant ASW capability in the numerous torpedoes with which they are equipped. As Admiral Kelln testified in 1976, although SSBNs have not been designed for ASW missions, "the fact remains that it is not inconceivable that the SSBN, if the situation became necessary, could be used as an offensive tactical weapon, that is to seek out other submarines."[51] Another reason is that Soviet SSBNs may be targeted against surface ships rather than against land targets in the United States.[52] The pressure to attack these submarines (including the SSBNs) as soon as possible after the outbreak of a conflict, before the Soviet Union could destroy U.S. ASW sensor systems and thus blind U.S. ASW forces, could be irresistible. Certainly, the notion that the Navy should refrain from attacks on SSBNs for fear of sending a false signal of impending escalation "has an aura of unreality for many

49. Committee on Armed Services, U.S. Senate, *Fiscal Year 1977 Authorization for Military Procurement* (Washington, D.C.: U.S. Government Printing Office, 1976), Part 4, p. 1972.
50. Committee on Armed Services, U.S. Senate, *Fiscal Year 1978 Authorization for Procurement* (Washington, D.C.: U.S. Government Printing Office, 1977), Part 10, p. 6699.
51. Committee on Armed Services, U.S. Senate, *Fiscal Year 1977 Authorization for Military Procurement, Research and Development, and Active Duty, Selected Reserve and Civilian Personnel Strengths* (Washington, D.C.: U.S. Government Printing Office, 1976), Part 12, p. 6609.
52. Blair, "Arms Control Implications of ASW Programs," p. 115.

professional officers."[53] Indeed, Secretary of the Navy John Lehman has stated, though probably with some hyperbole, that U.S. SSNs would attack Soviet SSBNs "in the first five minutes of the war."[54]

SECOND, THE U.S. NAVY'S STRATEGY OF FORWARD OPERATIONS. A primary assignment of the U.S. nuclear-powered attack submarine force is the conduct of offensive operations in forward areas, including enemy-controlled waters—particularly in the early stages of a conflict, even if that conflict is in those stages still nonnuclear. As Navy officers have testified: the nuclear-powered attack submarine is the only platform "that can operate where the enemy controls the air and the surface"[55] such as waters near the U.S.S.R.; and the SSN "often puts itself in and operates in areas which are very contiguous with the home bases of an adversary."[56]

The high risks of this strategy have been recognized by both senior Pentagon officials and Navy officers. As Admiral Worth Bagley has written, "the apparent American practice of employing SSN covertly for such operations [i.e., ASW operations in forward areas] risks an incident in time of tension."[57] It obviously also creates risks of escalation once conflict has been initiated.

THIRD, INADEQUATE CONTROL OF SSN OPERATIONS BY THE NCA. The inadequate control of SSN operations derives both from technical difficulties and from U.S. Navy doctrine and operational practice.

The principal technical difficulty is that of timely, reliable communications. In order to remain secure, SSNs normally operate at depths that preclude both transmission and reception of communications. Even in the case of VLF radio, which is the primary shore-to-ship means of communication and the only means that can be received by a submarine without protruding an antenna above the surface, an SSN cannot be contacted at normal operational depths:

[SSNs] must be within tens of feet of the surface of the water [for VLF reception], and that means you cannot talk to the submarine when you want

53. Captain Linton F. Brooks, "Pricing Ourselves Out of the Market: The Attack Submarine Program," *Naval War College Review,* September–October 1979, p. 5.
54. Cited in Melissa Healy, "Lehman: We'll Sink Their Subs," *Defense Week,* May 13, 1985, p. 18.
55. Committee on Armed Services, U.S. Senate, *Department of Defense Authorization for Appropriations for Fiscal Year 1980* (Washington, D.C.: U.S. Government Printing Office, 1979), Part 6, p. 2927.
56. Senate Committee on Armed Services, *Fiscal Year 1977 Authorization for Military Procurement,* Part 12, p. 6609.
57. Worth H. Bagley, *Sea Power and Western Security: The Next Decade,* Adelphi Paper No. 139 (London: International Institute for Strategic Studies, Winter 1977), p. 12.

to. You have to have an appointment for the submarine to come up and listen.[58]

Moreover, as Blair has noted:

[I]n an emergency situation the amount of time required to get a submarine to communicate with higher authority depends on its location. But both in wartime and peacetime operations, US SSNs have complete freedom in depth and speed while carrying out ASW, trail, escort, and other sensitive missions. Consequently, communications between higher authority and an SSN may be as infrequent as once every 12 hours; essential instructions cannot be reliably received by the SSN in a timely fashion. . . .

Even in advance of the initiation of hostilities, both attack and ballistic missile submarines run the risk of engaging Soviet SSN's and SSBN's in tactical combat about which higher authorities could not be quickly informed. While it is precisely under such circumstances that communications are most desirable from the standpoint of national-level force management, submarine commanders are supposed to deal with the tactical situations before reporting their predicament to a higher authority. The status of a submarine, the state of mind of its crew, and the location of a submarine within its large patrol area is not known at all times, even by the National Military Command Center at the Pentagon. (U.S. SSBN's are assigned to large patrol areas and only when the submarine leaves port does the commanding officer draw up his patrol plans.)[59]

Hence, Blair has concluded:

The prompt termination of ASW activities appears especially problematic. The inability to exercise precise, national-level control over submarines—for instance during a crisis when the likelihood of tactical engagements between forces on high alert increases, and during a conventional conflict when Soviet SSBN's become subject to attack—is a potentially grave problem that publicly receives scant official attention.[60]

Two examples illustrate the difficulty of exercising centralized control over ASW operations and, in particular, of promptly terminating those operations—as well as indicating potential consequences of this difficulty.

First, U.S. decision-making during the Cuban missile crisis of October 1962 is commonly regarded as being the model of application of the principles of centralized control and measured response, but it was effectively negated by

---

58. Senate Committee on Armed Services, *Fiscal Year 1978 Authorization for Military Procurement,* Part 10, p. 6735.
59. Blair, "Arms Control Implications of ASW Programs," pp. 116–117.
60. Ibid., p. 117.

U.S. Navy ASW operations. As John Steinbruner has described it, "there was an extraordinary effort to co-ordinate the actions of the government and to subject those actions to exhaustive deliberation."[61] This extraordinary effort notwithstanding, however, the U.S. military response to the crisis developed further in the direction of global strategic operations than the President or the Executive Committee either intended or imagined in advance. The strongest instance of this concerned ASW operations in the North Atlantic, which the Navy pursued as a normal operational measure but which were much more provocative than anything the President and his advisers had either approved or wanted. One set of the targets of these ASW operations was Soviet submarines carrying cruise missiles, which at that time were the principal element of the Soviet strategic nuclear deterrent forces. These operations constituted extremely strong strategic coercion and violated the spirit of the Executive Committee policy, but they probably represented the strongest signal received in Moscow in the course of the crisis. Hence, as Steinbruner has concluded, "the efforts to bring American policy under central direction must be said to have failed."[62]

In the second example, the *Belgrano*, an Argentine cruiser, was sunk by a British SSN, HMS *Conqueror*, on the afternoon of May 2, 1982 during the Falklands/Malvinas war. This represented a significant escalation in the scale and bounds of the conflict.

Communications with the *Conqueror* were conducted every two hours by means of a surfaced UHF antenna. This meant that orders issued from Whitehall and/or the command center at Northwood, Middlesex, were inevitably two hours behind the appreciation of the local tactical situation as comprehended by the *Conqueror*, while the clarification of any orders that might be desired by the *Conqueror* necessarily required a further two-hour wait. The instruction to attack the *Belgrano* was transmitted at 2 p.m. GMT (10 a.m. in the South Atlantic) on May 2. The signal was either garbled, or the submarine commander thought that the instruction might be affected by the information he transmitted back at the same time that the *Belgrano* had been sailing west—i.e., away from the British Task Force and the Exclusion Zone—for some five hours. The attack instructions were repeated at 4 p.m. GMT (12 noon South Atlantic) and at 6 p.m. GMT (2 p.m. South Atlantic);

---

61. John Steinbruner, "An Assessment of Nuclear Crises," in Franklyn Griffiths and John C. Polanyi, eds., *The Dangers of Nuclear War* (Toronto: University of Toronto Press, 1979), pp. 37–38.
62. Ibid., pp. 38–39.

the *Conqueror* went to action stations at 7 p.m. GMT and attacked at 8 p.m. GMT (4 p.m. South Atlantic).[63]

It is now clear that the British War Cabinet had determined to attack and sink the *Belgrano*, despite the fact that it knew by at least 2 p.m. GMT that it posed no threat to the British task force at the time and despite the active U.S. and Peruvian diplomatic efforts to effect a peaceful solution to the conflict, and that the *Conqueror*'s actions were not due to the lack of adequate or timely means of communications. Nevertheless, the incident suggests some disturbing possibilities. Even if the U.S. and Peruvian diplomatic initiatives had been more seriously received in London and Buenos Aires, there was a real possibility that any attempt to rescind the *Conqueror*'s orders would not have been received by the submarine, and that this would have nullified those diplomatic efforts.

The Vulnerability of ASW and C³I Support Systems

The U.S. and its NATO allies have deployed an extensive network of ASW and C³I support systems designed to serve both tactical and strategic ASW operations. Essential elements of these systems are highly vulnerable to nonnuclear as well as nuclear attacks.

The principal systems include:[64]

1) The SOSUS network of large fixed-bottom sonar arrays. Some three dozen such installations are distributed at key locations around the world (although strictly speaking only about two-thirds of these are part of the SOSUS network proper), each of which is connected by cable to shore-based processing facilities.

2) The network of ocean surveillance signal intelligence (SIGINT) stations, which are designed to monitor shore-to-ship and ship-to-shore communications and locate the positions of ships by direction-finding (DF) techniques. There are about 100 ocean surveillance SIGINT stations deployed around the world, although these are coordinated by only about half-a-dozen net control stations and feed into about a dozen processing facilities.

3) The ground control network for the U.S. Navy's Ocean Surveillance Satellite System (Classic Wizard). There are five ground stations in this network,

---

63. See "Mrs. Thatcher's Watergate," *New Statesman*, September 14, 1984, p. 2.
64. See Jeffrey T. Richelson and Desmond Ball, *Ties that Bind: Intelligence Cooperation between the UKUSA Countries—the United Kingdom, the United States of America, Canada, Australia and New Zealand* (Sydney, London and Boston: George Allen and Unwin, 1985), chapter 9.

located at Guam, Diego Garcia, Adak (Alaska), Winter Harbor (Maine), and Edzell (Scotland).

4) The major bases for U.S. and allied long-range maritime patrol (LRMP) aircraft, such as the P-3C Orion and the British Nimrod aircraft.

5) The ground stations for the Defense Satellite Communications System (DSCS) and Fleet Satellite Communications (FLTSATCOM) satellites used to relay ocean surveillance intelligence from these facilities and bases to central processing stations and Fleet Ocean Surveillance Information Centers (FOS-ICs).

6) The central processing stations, FOSICs, and naval intelligence centers.

All of these installations are quite soft, and most of them could be destroyed or incapacitated by sabotage or conventional ordnance as well as by nuclear means. The whole system could be rendered useless by attacks on fewer than 50 installations.

It must be expected that elements of this system would be attacked in the early stages of any conflict. For example, any attempt by the Soviet navy to move its SSBNs and/or surface fleet through the Greenland-Iceland-United Kingdom (GIUK) gap would doubtless be accompanied by attacks on SOSUS, SIGINT, and related facilities in Iceland, Norway, and Scotland. The expectation that these systems would soon be unavailable would cause many commanders to strike against any Soviet ships that were in their "sights," be they SSBNs or whatever, while they still had the opportunity.

U.S. Navy Doctrine for the Conduct of Offensive Operations in Forward Areas

The strategy of the U.S. Navy is to engage any enemy "as far forward as possible."[65] A recent Chief of Naval Operations (CNO) has testified to his preference to "seek out and destroy" Soviet naval forces "wherever they may be, even in Soviet coastal waters."[66] The United States must "contain and attrite the Soviet Navy as close to their home waters in a conflict as possible."[67]

---

65. Senate Committee on Armed Services, *Fiscal Year 1977 Authorization for Military Procurement*, pp. 1944, 1956.

66. Committee on Armed Services, U.S. Senate, *Fiscal Year 1980 Authorization for Military Procurement* (Washington, D.C.: U.S. Government Printing Office, 1979), p. 1292.

67. Committee on Armed Services, U.S. Senate, *Fiscal Year 1981 Authorization for Military Procurement* (Washington, D.C.: U.S. Government Printing Office, 1980), Part 2, p. 867.

As described above, "offensive operations in forward areas" is a primary assignment of the SSN force.[68] Under the Reagan/Lehman naval program, it has also been explicated as a primary mission of the Navy's carrier task forces. These forward offensive missions are not limited to ASW operations. As a former CNO has stated:

Our plan would be as the first line of defense to strike the airbases from which the Backfire bombers fly and the submarine bases from which the nuclear-powered submarines operate.[69]

And the current CNO, Admiral J.D. Watkins, has testified that the Navy plans "to catch the Backfires on the ground. . . . [For example], in the Northwest Pacific our feeling is that at the very front end of [a] conflict, if we are swift enough on our feet, we would move rapidly into an attack on [the Backfire base at] Alekseyevka."[70]

This strategy contains the seeds of extremely rapid escalation. It is not just that it puts Soviet SSBNs at risk and hence could cause inadvertent escalation. Such a strategy also makes it difficult for National Command Authorities to forgo preemption. Being quite familiar with the U.S. Navy's strategic predilections, the Soviet NCA would have to move to disperse the Backfire force and to "surge" the SSBNs at the outset of any conflict. This would, in turn, put the U.S. NCA under strong pressure to preempt.

U.S. Navy Doctrine for the Employment of Tactical Nuclear Weapons

The U.S. Navy currently has approximately 3500 tactical nuclear weapons available for employment.[71] (The details of these are given in Table 1.) However, the production and deployment of these weapons has never been accompanied by any clear or coherent doctrine for their use.

---

68. Committee on Armed Services, U.S. House of Representatives, *Department of Defense Appropriations for Fiscal Year 1970* (Washington, D.C.: U.S. Government Printing Office, 1969), Part 4, p. 277.
69. Committee on Armed Services, U.S. Senate, *Fiscal Year 1979 Department of Defense Authorization* (Washington, D.C.: U.S. Government Printing Office, 1978), Part 5, p. 4321.
70. Committee on Armed Services, U.S. Senate, *Department of Defense Authorization for Appropriations for Fiscal Year 1985* (Washington, D.C.: U.S. Government Printing Office, 1984), pp. 3875, 3887.
71. Cochran, Arkin, and Hoenig, *Nuclear Weapons Databook, Volume 1*, various entries; and Norman Polmar, "Tactical Nuclear Weapons," U.S. Naval Institute *Proceedings*, Vol. 109, No. 7 (July 1983), pp. 125–126.

Table 1. U.S. Navy Tactical Nuclear Weapons

Weapon	Weapon Designation	Warhead	Number Deployed	Warhead Yield	Launch Platform	Targets	Date Operational	Comments
1. Terrier-BTN	RIM-2	W45	310	1 Kt	Cruisers, destroyers	Aircraft, land targets, limited anti-cruise missile and limited anti-ship capability.	1962	Dual capable launcher system. Scheduled for replacement by Standard SM-2 (N).
2. Bomb		B43	500(?)	1 Mt	A-6E Intruder, A-7E Corsair	High-value urban-industrial targets and moderately hard military targets.	1961	Being replaced by B61
3. ASROC	RUR-5A	W44	850	1 Kt	65 frigates, 78 destroyers, and 27 cruisers	Primarily submarines but also capable against surface ships and land targets.	1961	Dual capable.
4. SUBROC	UUM-44A	W55	400	1–5 Kt	73 attack submarines (SSN-688, SSN-685, SSN-671, SSN-637, & SSN-594 classes).	Primarily submarines with surface ships secondary. Can also be used for surface-to-surface missions against land targets.	1964	4–6 SUBROCs normally deployed aboard each submarine on patrol.
5. Depth bomb	Mk-105	B-57	1000	5–10 Kt	P-3 Orion, S-3 Viking, & SH-3 Sea King	Primarily submarines, but also surface targets.	1963	Being replaced by B61.
6. Bomb		B-61	500(?)	100–500 Kt	Usable on any U.S. aircraft that can deliver a nuclear bomb.		1968	Replacing B43 and B57 bombs.
7. Tomahawk T-LAM (N)	BGM-109A	W80-0	750–1000	200–250 Kt	Battleships, cruisers, destroyers, attack submarines.	Land targets, primarily naval related (such as ports and bases). Also surface ships.	1984	Major component of Strategic Reserve Force.
8. Standard	RIM-67	W81	350	Low Kt	Cruisers, destroyers.	Aircraft, anti-ship cruise missiles, and surface ships.	1988	Designed to replace Terrier-BTN RIM-2

In the case of U.S. Army and Air Force tactical nuclear weapons, there are detailed, comprehensive, and officially endorsed employment doctrines. There are certainly some worrisome aspects of these doctrines, but they are at least subject to public scrutiny and criticism, and there is no doubt that this has improved both their internal consistence and their compatibility with national strategic objectives. Navy doctrine is, in comparison, quite inchoate and incoherent.

As Douglass and Hoeber have written:

The Navy has developed very little in the way of doctrine, tactics, strategy, or policy for the use of [tactical nuclear weapons] capabilities. Further, there is almost no training for warfighting either in support of a land theater nuclear war, in a nuclear war limited to the ocean theater, or in a general nuclear war. There are very few U.S. studies of such warfighting and the problems it presents.[72]

And as a Navy Captain who has served with the submarine force has written:

[U.S. Navy planners] have consistently failed to consider a factor which might radically alter the character of a future war—the existence of tactical nuclear weapons. Despite repeated declarations in the Chief of Naval Operations' annual posture statements that it ". . . is essential that the U.S. Navy maintain a capability to use tactical nuclear weapons if the United States is to be able to fight and win at sea," we have given little serious thought to the naval implications of tactical nuclear war. There are many illustrations of our neglect. The senior course at the U.S. Naval War College—an institution whose stated mission is to "enhance the professional capabilities" of future military leaders and to develop "advanced strategic and tactical concepts for the future employment of naval forces"—contains essentially nothing on any aspect of tactical nuclear weapons employment. Sea Plan 2000, now enshrined as the principal articulation of the Navy's vision of a future war, pictures a global contest between the United States and the Soviet Union in which nuclear weapons play no part. . . . [Navy officers] speak and act as though war could have three possible outcomes: we win, we lose, or the war "goes nuclear." But what happens once that nuclear boundary is crossed is a possibility we have set aside in our thoughts and in our planning.

The Navy is not alone in ignoring the impact of nuclear weapons on war at sea. Secretary of Defense Harold Brown, for example, spent nine pages of his Fiscal Year 1980 Report to Congress in a discussion of tactical and theater nuclear weapons. Exactly 33 words are devoted to fleet systems. A recent

---

72. Joseph D. Douglass, Jr., and Amoretta M. Hoeber, "The Role of the U.S. Surface Navy in Nuclear War," U.S. Naval Institute *Proceedings*, Vol. 108, No. 1 (January 1982), p. 58.

study by the Congressional Budget Office, *Planning U.S. General Purpose Forces: The Theater Nuclear Forces*, totally ignores nuclear war at sea. The silence extends to unofficial writings as well. The *Air University Library Index to Military Periodicals* publishes an annual index of articles appearing in 77 different journals specializing in military, naval, or general defense matters. An examination of this index for the past five years fails to reveal a single article remotely related to tactical nuclear war at sea.[73]

A number of studies of naval tactical nuclear weapons requirements and the utility of nuclear weapons in a war at sea have been conducted by various Pentagon and Navy agencies since the late 1970s. In 1978, President Carter issued a presidential directive (PD) that ordered an examination of the utility and arms control impact of new naval nuclear warheads under development. Subsequent studies were conducted by the Under Secretary of Defense for Policy, the Navy Research Advisory Committee, and the Center for Naval Analysis. These studies generally concluded, *inter alia*, that developments in naval nuclear weapons should proceed; that naval officers should increase their awareness of the nuclear capabilities and doctrine of the Soviet navy; that the U.S. navy should conduct more realistic exercises that place greater emphasis on tactical nuclear weapons employment concepts; and that more realistic career programs should be developed for naval officers specializing in nuclear weapon fields. However, there was no clarification of the strategic and tactical concepts and doctrines for the employment of these weapons.[74]

Soviet Doctrine for War at Sea

Any assessment of the prospects for limiting a war at sea to conventional operations obviously requires some discussion of Soviet naval capabilities and doctrine concerning the employment of nuclear weapons. However, it is not possible to be authoritative on this subject, at least on the basis of publicly available information. Soviet writings tend, on the one hand, to be very general or philosophical (such as Admiral Gorshkov's discussion of the impact of nuclear weapons on "naval art," which seems to suggest that sea denial is a more appropriate objective than sea control under conditions

---

73. Captain Linton F. Brooks, "Tactical Nuclear Weapons: The Forgotten Facet of Naval Warfare," U.S. Naval Institute *Proceedings*, Vol. 106, No. 1 (January 1980), p. 29.
74. Polmar, "Tactical Nuclear Weapons," p. 126; Cochran, Arkin, and Hoenig, *Nuclear Weapons Databook, Volume 1*, p. 246; and William M. Arkin, "Nuclear Weapons at Sea," *Bulletin of the Atomic Scientists*, October 1983, p. 6.

where nuclear warfare is possible[75]); on the other hand, they tend to focus on naval tactics, such as the importance of radio-electronic combat (REC). Western writings are invariably concerned with Soviet naval capabilities, force structure development, and deployment patterns, with little attempt at extrapolation of operational strategy and doctrine.

Most large Soviet naval combatants are equipped to carry nuclear weapons—including ship-to-ship weapons such as the Shaddock SS-N-3 (the standard warhead for which has a yield of about 250 kt), deployed on the Kresta-1 and Kynda class guided missile cruisers; and anti-submarine weapons such as the FRAS-1 anti-submarine rocket (with a 5 kt nuclear depth charge) deployed on the Moskva class helicopter carriers and the Kiev class aircraft carriers, and the nuclear-capable SS-N-14 Silex anti-submarine torpedo deployed on the Kirov, Kresta-II, and Kara class guided missile cruisers and the Krivak class missile frigates. A version of the standard Soviet 533 mm torpedo has also been fitted with a nuclear warhead with an estimated yield of about 15 kt and deployed in some submarines—including the Whiskey class submarine (No. 137) that ran aground in Swedish waters in October 1981.[76]

The major Soviet surface combatants are generally quite inferior to their U.S. counterparts. In particular, they are less capable of engaging in any extended conventional operations. As a study prepared for the Atlantic Council has argued:

> The dominant characteristics of many Soviet surface combatant ships—high speed; great striking power; and relatively limited cruising ranges and reload and resupply capabilities—all suggest that their employment in a long drawn-out conventional war was not foreseen as a major mission when they were built.[77]

Moreover Soviet naval forces are less designed for and generally not deployed in Western-style task forces, so that the possibility of destroying U.S. carriers in matched-fleet battles is not a real option for them.[78]

---

75. S.G. Gorshkov, *The Sea Power of the State* (Oxford: Pergamon Press, 1979), pp. 232–233.
76. Ronald T. Pretty, ed., *Jane's Weapon Systems 1983–84*, 14th ed. (London: Jane's Publishing Company Limited, 1983), pp. 84, 166, 167.
77. Paul H. Nitze, Leonard Sullivan, Jr., and the Atlantic Council Working Group on Securing the Seas, *Securing the Seas: The Soviet Naval Challenge and Western Alliance Options* (Boulder, Colo.: Westview Press, 1979), p. 74.
78. Ibid., pp. 20, 74.

In these circumstances, the Soviet navy must be expected to resort to the use of nuclear weapons at a fairly early stage in any major engagement at sea, particularly when it is called upon to destroy U.S. carrier task forces— and particularly if it is believed that the use of nuclear weapons could be confined to the sea.

Resistance of the U.S. Navy to Contingency Planning Concerning the Control of Escalation

George Quester has recently recalled an old American aphorism: "There are three ways of doing anything: the right way, the wrong way, and the Navy way."[79] Despite the fact that escalation control has been the predominant concept in U.S. national strategic policy for more than a decade now, the U.S. Navy has still to regard the concept seriously.

There are several reasons for the Navy's resistance. First, and most generally, the U.S. Navy is much more self-contained than the other services, and its autonomy is cherished as a primary value. Naval planners do not give much thought to the operational concepts and plans of the other services and do not see why outside influences should be directed at their concepts and plans. Second, naval officers generally accept a traditional, uncompromising view of their mission in the event of conflict: to "seek out and destroy" the opposing naval forces, by the most effective means possible. Third, naval officers generally do not believe that escalation control is a serious proposition anyway, and hence to sacrifice operational effectiveness in its pursuit is extremely foolish.

Conclusions

The possibility of nuclear war at sea must be regarded as at least as likely as the occurrence of nuclear war in other theaters. Indeed, there is probably a greater likelihood of accidental or unauthorized launch of sea-based nuclear weapons, and the constraints on the authorized release of nuclear weapons are possibly more relaxed than those that pertain to land-based systems. Further, there are several important factors that make it likely that any major conflict at sea would escalate to a strategic nuclear exchange relatively

---

79. Quester, *The Falklands and the Malvinas*, p. 21.

quickly. However, neither the possibility of nuclear war at sea nor the factors that would impel the escalation process have ever received adequate attention from either policymakers or strategic analysts.

Given that there is a high risk of nuclear weapons being used fairly quickly in any major East–West conflict at sea, and that once nuclear weapons are used the risk of escalation to all-out nuclear war is also quite high, it follows that greater attention must also be accorded the avoidance of any conflict at sea. Even a conventional conflict should not be joined without national decision-makers having clearly and consciously determined that the purpose served by such action justifies the real risk of an all-out nuclear exchange. Much more consideration should also be given to the means by which a war at sea might be kept limited, since such a war has a high propensity to escalate unless great care is taken to keep it under control. Moreover, it must be recognized that, even if great care is exercised, a war at sea may still escalate—if for no other reason than that the Soviets might escalate despite the best efforts of Western planners to control the conflict.

There are several particular issues that require the most serious consideration. These include:

A REASSESSMENT OF THE BASIC NATIONAL STRATEGIC POLICY OF ESCALATION CONTROL.

There are several technical and doctrinal aspects of naval strategy and operational practice that are likely to cause inadvertent escalation in the event of any major conflict at sea. Some of these may be amenable to change by policy direction, but others are essentially immutable. Since it would be realistic to expect that any war would have a significant naval component, the reliance on the prospect of controlling escalation as a basis for national strategic policy would seem to be unreasonable. Prudence alone should dictate that the focus of thinking and analysis with respect to strategic policy should be directed much more toward avoiding incidents and managing crises in the first place, as well as toward the means for rapidly disengaging from conflicts should these nevertheless eventuate.

STRENGTHENING MEASURES TO LIMIT INCIDENTS AT SEA.

Although some attention has been accorded this issue, it has clearly been inadequate. Consideration should be given to strengthening the 1971 and 1972 Agreements between the U.S. and the Soviet Union as well as to unilateral measures designed to avoid or at least reduce the likelihood of such incidents. U.S. submarine deployments and ASW practices in forward areas warrant review in this context.

THE PREVENTION OF UNAUTHORIZED LAUNCH.

The arguments advanced by the U.S. Navy to justify its refusal to equip naval nuclear weapons with "permissive action" devices are unpersuasive. The rigorous arming and firing procedures established by the Navy are salutary, but they are no substitute for physical/technical systems.

THE SEGREGATION OF NUCLEAR/CONVENTIONAL AND TACTICAL/STRATEGIC WEAPONS SYSTEMS AND PLATFORMS.

The removal of dual-capable weapons systems and the segregation of ships into conventional, tactical nuclear, and strategic nuclear are obviously practically impossible. The costs would be prohibitive and the operational problems for the Navy would be innumerable. However, it should be possible to effect some improvements to the current situation. In particular, a special case can be made against the deployment of the Tomahawk T-LAM(N) SLCM on the general purpose ships.

EXEMPTION OF SSBNS FROM ASW OPERATIONS.

The U.S. Navy should be induced to make serious efforts to distinguish between Soviet SSNs and SSBNs, and to avoid actions that could be seen as threatening the latter submarines. Making this distinction under operational conditions is obviously a technically difficult exercise, particularly where sensor systems have been destroyed or degraded, but the risks of inadvertent escalation make the attempt imperative.

RECONSIDERING THE POLICY OF OFFENSIVE FORWARD OPERATIONS.

The military/strategic advantages of offensive operations by the U.S. Navy in forward areas are probably irrefutable. However, because of the danger of triggering escalation, the Navy should be directed to exempt certain targets from these operations. In addition to SSBNs at sea, Soviet SSBN bases and land-based forces such as Backfire bombers should be exempted from attack for as long as possible.

IMPROVING NCA CONTROL OVER NAVAL OPERATIONS.

There are clearly grave deficiencies in the ability of the NCA to exercise precise and timely control over significant aspects of naval operations. Some of these deficiencies are technical and while some of these are, in turn, intrinsic to the nature of the ocean medium, there are others that could be susceptible to technical solution. Moreover, improving the physical survivability of the ASW C^3I support infrastructure would also reduce the escalatory pressures that are produced by the current vulnerability of this infrastructure. In addition, the control of the NCA can be strengthened by improving the means by which the NCA monitors naval deployments and practices.

CONCEPTS AND DOCTRINE FOR THE EMPLOYMENT OF NUCLEAR WEAPONS AT SEA.

It is clear that much more attention needs to be accorded general questions concerning the utility of nuclear weapons at sea, Soviet naval nuclear weapons capabilities and doctrine, and the implications of nuclear war at sea for escalation control. The resolution of these questions would enable the development of concepts and the promulgation of doctrine that should serve to enhance the strategic maritime position of the West while contributing to the fundamental objective of maintaining deterrence with respect to all possible levels of conflict.

Inadvertent Nuclear War?

Escalation and NATO's Northern Flank

Barry R. Posen

Could a major East–West conventional war be kept conventional? American policymakers increasingly seem to think so. Recent discussions of such a clash reflect the belief that protracted conventional conflict is possible, if only the West fields sufficient conventional forces and acquires an adequate industrial mobilization base. Indeed, the Reagan Administration has embraced the idea of preparing for a long conventional war, as evidenced by its concern with the mobilization potential of the American defense industry.[1] Underlying this policy is the belief that the United States should be prepared to fight a war that, in duration and character, resembles World War II. American decision-makers seem confident of their ability to avoid nuclear escalation in such a war if they so desire.

That confidence is dangerous and unwarranted. It fails to take into account that intense conventional operations may *cause* nuclear escalation by threatening or destroying strategic nuclear forces. The operational requirements (or preferences) for conducting a conventional war may thus unleash enormous, and possibly uncontrollable, escalatory pressures despite the desires of American or Soviet policymakers. Moreover, the potential sources of such escalation are deeply rooted in the nature of the force structures and military strategies of the superpowers, as well as in the technological and geographical circumstances of large-scale East–West conflict. If the escalatory pressures

While the author bears sole responsibility for all views expressed here, he is grateful to Bruce G. Blair, Joshua Epstein, William W. Kaufmann, John Mearsheimer, Stephen Van Evera, and Kenneth Waltz for their suggestions.

Barry Posen is a Council on Foreign Relations Fellow. This article was written while he was a Postdoctoral Fellow at the Center for International Affairs, Harvard University.

1. See, for example, the accounts of Secretary Weinberger's views in George Wilson, "Weinberger Order: Plan for Wider War," *The Boston Globe*, July 17, 1981; and Richard Halloran, "Weinberger Tells of New Conventional-Force Strategy," *The New York Times*, May 7, 1981. For further indications of the Administration's views on this subject, see also Richard Halloran, "Needed: A Leader for the Joint Chiefs," *The New York Times*, February 1, 1982; Richard Halloran, "Reagan Selling Navy Budget as Heart of Military Mission," *The New York Times*, April 11, 1982; and Caspar W. Weinberger, Secretary of Defense, *Annual Report to Congress, FY 1983*, pp. I-13, 16–17, 28–29.

International Security, Fall 1982 (Vol. 7, No 2) 0162-2889/82/010028-27 $02 50/0
© 1982 by the President and Fellows of Harvard College and of the Massachusetts Institute of Technology

that could attend a major conventional war are to be prevented from driving decision-makers toward decisions they neither intend nor wish to make, those pressures must be recognized and guarded against by the leaders of both superpowers.[2]

Moreover, underestimating the escalatory risks that would accompany conventional war has several significant negative consequences, even in peacetime. First, American decision-makers pay insufficient attention to the details of conventional posture and operations essential to the limitation of war. Too many agree with the observation that "both sides understand conventional warfare, they know that it can be controlled in the present age."[3] Second, leaders who fail to appreciate fully the dangers of nuclear escalation may not be cautious enough about both the initiation and the conduct of direct confrontations between Soviet and American military power. Third, nuclear deterrence may be undermined by excessive public confidence in the limitability of superpower conventional war. The "threat to lose control" is an important element of NATO's flexible response strategy, and must be preserved. It would be unfortunate if the public pronouncements of Western strategists encouraged the Soviets to believe that they could easily avoid nuclear punishment for "conventional" aggression. Fourth, emphasis on protracted conventional conflict weakens Western Europe's confidence in America's nuclear guarantee. Emphasizing instead the difficulty of keeping conventional war conventional might ameliorate Alliance fears that the U.S. nuclear umbrella no longer shields them.

Unfortunately, surprisingly little attention has been devoted to analyzing and understanding this path to nuclear war. Escalation has generally been conceived of as either a rational policy choice, in which a leadership decides to preempt or to escalate in the face of a conventional defeat, or as an accident, the result of mechanical failure, unauthorized use, or insanity. But escalation arising out of the normal conduct of intense conventional conflict falls between these two categories: it is neither a purposeful act of policy nor an accident. What might be called "inadvertent escalation" is rather the *unintended* consequence of a decision to fight a *conventional* war. Defense

---

2. Because the American defense posture and war plans are more easily studied and influenced than those of the Soviet Union, this essay will focus mainly on American choices of plans and deployments that affect escalation to the use of nuclear weapons. It should be emphasized that unless Soviet leaders are also sensitive to this problem, their decisions about the conduct of conventional war could as easily cause escalation as those of their American counterparts.
3. Morton Halperin, *Limited War in the Nuclear Age* (New York: John Wiley, 1963), p. 64.

analysts have long believed that nuclear war is most likely to emerge out of a conventional conflict, but the idea of *inadvertent* escalation has somehow escaped their attention. It nevertheless is as plausible a route to nuclear conflict as many of the scenarios that have received extensive scrutiny, and probably more likely than many. Because nuclear escalation could destroy everything that we value, the West must be careful that the way it intends to defend itself conventionally does not bring about the very destruction it hopes to avoid.

Since inadvertent escalation is a consequence of conventional military practices, its likelihood is affected by choices regarding force postures and operational plans. There is little evidence, however, that the problem of escalation control has *influenced* such choices. As a result, the huge investments that the United States has made in conventional forces for the purpose of buffering against precipitate nuclear escalation may be negated by imprudent operational planning and reckless execution of those plans.

Under the prevailing conceptions of escalation, most NATO strategists have focused their attention on improving the conventional balance in Central Europe as the most important barrier to the early use of nuclear weapons (and, indeed, the collapse of NATO's conventional defenses could lead to nuclear escalation under the strategy of flexible response). Concern about inadvertent escalation, however, causes NATO's Northern Flank to loom much larger in importance, for it illustrates vividly the escalatory pressures inherent in many conventional operations. It is from a thorough examination of this area that I shall principally draw in this essay.

The danger of inadvertent escalation arises with particular acuteness in the Northern Flank, especially in the region comprised of the Norwegian and Barents Seas, Northern Norway, and the Kola Peninsula in the Soviet Union. In Northern Norway, NATO territory and NATO conventional capabilities are situated near a critical element of Soviet strategic nuclear power—the ballistic missile submarine force based at Murmansk in Kola. Conventional military operations in the region could put this force at risk. The U.S. Navy frequently argues in favor of attacks with conventionally armed, carrier-based aircraft against Soviet naval bases and naval forces in this area, a strategy endorsed by the Reagan Administration.[4] If such attacks should threaten the

---

4. Congressional Budget Office, *Building a 600-Ship Navy: Costs, Timing, and Alternative Approaches* (Congress of the United States, March 1982), pp. xi–xii, xxiii, 17; Halloran, "Reagan Selling Navy Budget"; John Lehman, Secretary of the Navy, "America's Growing Need for Seaborne

survivability of Soviet ballistic missile submarines, as they might, how would the Soviet Union react? It could decide that a nearly-certain-to-succeed nuclear strike against those threatening carriers was both lucrative and necessary.

The Causes of Inadvertent Escalation

Such scenarios raise the possibility of inadvertent escalation because the two superpowers have closely identified their security with the maintenance of survivable strategic nuclear forces. It thus seems probable that both would be willing to run great risks in order to preserve the safety of those forces, even including the use of nuclear weapons themselves. Conventional operations that might threaten strategic nuclear forces are therefore a plausible path to nuclear escalation. Both before and during a conventional conflict, political leaders may have difficulty exerting enough control over military operations to avoid posing such threats. If they occur, Soviet or American leaders could easily see them as a prelude to nuclear escalation. In general, conventional operations can bring nuclear forces into jeopardy in three ways:

1. THE OFFENSIVE INCLINATION. Offensive military actions can cause, or require, hostile contact between strategic and conventional forces, as, for example, the offensive operations under consideration by the U.S. Navy would take Western military forces close to the bases of Soviet strategic nuclear power on the Northern Flank—with unpredictable consequences. This problem is hard to avoid because military organizations have a proclivity for offensive operations and because they generally resist civilian intervention in any operational planning.[5]

Air Bases" (letter), *The Wall Street Journal*, March 30, 1982; and Weinberger, *Annual Report 1983*, pp. I-15–16, 30, III-19, 21. The Secretary remarks on how to deal with Soviet aircraft dedicated to anti-shipping missions "Our preferred approach is to destroy enemy bombers before they can reach ASCM [air-to-surface cruise missile] launch range by *striking their bases* or destroying them in transit," and that "We are also studying the use of *long-range strategic* bombers to attack Soviet surface ships and naval targets ashore" (p. III-21, my emphasis). Soviet bombers are normally based in the Soviet Union.

5. The offensive proclivities of military organizations can be deduced from organization theory, the civil-military relations literature, and from the instrumental problems of combat. Many historical examples of militaries striking out on offensive action unbeknownst to their civilian superiors can be found. The U.S. Navy, acting on its own authority, forced Soviet submarines to surface during the Cuban Missile Crisis. General John Lavelle conducted twenty unauthorized bombings of North Vietnam in 1971–1972. British Generals sent to protect the Abadan oil facilities at the outbreak of World War I decided to attempt the capture of Baghdad. See Barry R. Posen, *The Systemic, Organizational, and Technological Origins of Strategic Doctrine: France, Britain,*

Military organizations, like all large organizations, tend to seek autonomy from outside influences. Thus, in peacetime, civilians are seldom exposed to the intricacies of military planning, and, in wartime, when civilian intervention in the details of military policy is much more likely, soldiers often interpret policymakers' injunctions in ways that allow them maximum operational discretion.

There are many historical examples which demonstrate military evasion of civilian control over military operations. During the Cuban Missile Crisis, the U.S. Navy ran its blockade according to its traditional methods, disregarding President Kennedy's instructions.[6] Prior to World War I, German civilians knew little about the details of the Schlieffen Plan. When the General Staff denied that the plan could be changed, during the German mobilization in the crisis immediately preceding that war, civilians had no way of knowing that the soldiers had dissembled. As a consequence, Germany's leaders were compelled to plunge ahead into a war on two fronts.[7] In an even earlier instance, prior to and during the Prussian War with France in 1870, the Prussian General Staff tried to prevent Bismarck from having any contact "with the operational aspects of the war."[8] These few examples are suggestive of the difficulty that civilian leaders may have in keeping war limited.

Historically, offensive action, which requires complicated, detailed, expert planning, has been a way for militaries to evade civilian control. Under current conditions, this pattern suggests that American civilian policymakers may have the least influence over the most escalatory operations.

2. THE FINE LINE BETWEEN OFFENSIVE AND DEFENSIVE ACTS. Inadvertent escalation may occur because offensive and defensive actions are frequently indistinguishable.[9] The defensive needs of attack submarines, for example, may cause them to destroy Soviet strategic submarines—acts likely to be interpreted by the Soviet Union as very offensive. In more general terms, measures that one state takes to defend itself may seem offensive to the

---

and Germany Between the World Wars (Ph.D. dissertation, University of California, Berkeley, 1981); Graham Allison, *Essence of Decision* (Boston: Little, Brown, 1971), p. 138; Richard K. Betts, *Soldiers, Statesmen, and Cold War Crises* (Cambridge: Harvard University Press, 1977), p. 49; and Norman Dixon, *On the Psychology of Military Incompetence* (London: Jonathan Cape, 1976), p. 96.

6. Allison, pp. 129–130.

7. Gordon Craig, *The Politics of the Prussian Army 1640–1945* (London: Oxford University Press, 1955), pp. 294–295.

8. Ibid., p. 204.

9. Robert Jervis, "Cooperation Under the Security Dilemma," *World Politics*, Volume 30 (January 1978), pp. 167–214.

adversary against whom they are directed. The defender may have no choice but to take such actions, even if he understands that they threaten assets that the adversary values highly. Even more dangerous, however, is that the defender frequently does not understand how threatening his behavior, though defensively motivated, may seem to the other side. Thus, when the adversary reacts in a violent or escalatory fashion, the defender is surprised, and may respond even more extremely. This is one way that escalation spirals start.

Geography often contributes to this identification problem: territory necessary for defense may also facilitate offense. One geographic problem, for example, that would plague efforts to limit an East–West war is the proximity to Soviet borders of much of what the United States seeks to defend. A conventional conflict in Europe would involve large-scale military engagements near or over the Soviet Union which could be (or be perceived to be) threatening to Soviet strategic nuclear forces. Commanders of Soviet *strategic* forces may fear that surprise nuclear attacks could be camouflaged by the confusion and tumult of intense conventional combat. In an air battle over Central Europe, for example, thousands of planes would be in the air in circumstances that could easily make Soviet leaders nervous: Soviet air defenses would probably be degraded, NATO would almost surely have nuclear-capable aircraft in the air, and the Soviets might well feel that important strategic assets such as command, control, communications, and intelligence facilities were threatened.

In short, what the West does conventionally to *defend* itself can produce an *offensive* threat against Soviet strategic nuclear forces. Because the United States does not have an analogous geographical problem, it is difficult for American planners to recognize the stress that conventional war might put on Soviet leaders concerned about the survivability of their strategic forces.

Technology, like geography, can blur the line between offense and defense. Weapons useful for defense are often equally useful for attack. The United States, for example, maintains substantial anti-submarine warfare (ASW) forces to protect the sea lanes to Europe; those forces, however, could also attack Soviet ballistic missile submarines. The Soviets would put those submarines to sea in a serious crisis, in an effort to protect their sea-based deterrent and to deter attack on the Soviet Union. Once submerged, however, they are largely indistinguishable from attack submarines to Western ASW sensors. And because they are armed with torpedoes, they are viewed as an offensive threat to the sea lanes and to American attack submarines.

Under these circumstances, Soviet ballistic missile submarines could easily become targets in a conventional naval battle.

3. THE FOG OF WAR. Inadvertent escalation may also result from the extreme difficulty of gathering and understanding the most relevant information about a war in progress and using it to control and orchestrate the war. Not only might this difficulty help to cause inadvertent escalation, but it may exacerbate potentially escalatory situations created by offensive operations or by the indistinguishability of offensive and defensive acts. The disarray of command, control, communications, and intelligence, often called the "fog of war," would assume global proportions in an East–West war. Although modern technology may provide reams of intelligence data, it will not always be timely or accurate. Analysis is difficult under the pressure of intense conventional conflict. Communications, to and from the theater of operations, are likely to be uncertain and intermittent, as critical links are jammed or destroyed. Forces may end up in the wrong place, and events may be misreported. Civilians retaining the image of direct communication and control in the Cuban Missile Crisis or the Iran rescue mission may be shocked at how hard it is to follow, much less manage, a global war.

There are numerous examples of inaccurate or incomplete understanding by policymakers of ongoing military operations. General Lavelle's bombing of North Vietnam was apparently unknown to American leaders, and damaged peace negotiations with North Vietnam.[10] During the Cuban Missile Crisis, orders to cease U2 flights near the Soviet border were either not received or were ignored; Soviet detection of these flights hindered the negotiations to end the crisis.[11] During World War II, the British and Germans both apparently misappraised the accuracy and reliability of each other's bombing systems, so that accidental bombings of population centers were assumed to be deliberate; what were thus perceived to be deliberate attacks on cities justified deliberate attacks on cities in response.[12] In these examples (and there are many others), the fog of war resulted in uncontrolled operations and incorrect interpretations of unfolding wartime actions. In the nuclear age, the likelihood of inadvertent escalation might be increased because misperceptions, misunderstandings, poor communications, and unauthorized or unrestrained offensive operations could reduce civilian control of the war and may precipitate unexpected but powerful escalatory pressures.

---

10. Betts, pp. 49–50.
11. Allison, p. 141.
12. George Quester, *Deterrence Before Hiroshima* (New York: John Wiley, 1955), pp. 115–122.

In sum, these three factors—the offensive operations preferred by the American military, the ambiguities of offense and defense that geography and technology present, and the difficulties raised by the fog of war—make it likely that Soviet strategic forces will be placed in some jeopardy in an East–West conventional war. There is little reason for confidence that a large conventional war will be managed in such a way as to avoid threats to the strategic forces of the Soviet Union. It may not be easy to avoid making such threats, to control the situations they may create, or to make clear our limited intentions to the Soviet Union, especially with hostilities underway. Rather, inadequate political control of operations, accidents, misperception, and spirals of action and reaction seem just as probable. These difficulties may emerge with special force on NATO's Northern Flank.

The Northern Flank: Interests and Capabilities

The possibility of inadvertent escalation in the Northern Flank derives primarily from the combination of the geography of the region and the character of the military forces deployed there. The key problem from the Soviet perspective is that Soviet naval forces must pass through geographic "choke points" to reach open water from virtually all of their principal ports. This is a particular liability for ballistic missile submarines (SSBNs) trying to reach open water undetected. Western anti-submarine warfare (ASW) sensors can monitor these relatively narrow channels to detect the passage of the somewhat noisy Soviet submarines.

Consequently, the Soviet Union prefers to deploy its strategic submarines in the Barents Sea, adjacent to the Kola Peninsula and the major Soviet Navy base at Murmansk. Roughly two-thirds of Soviet SSBNs, protected by roughly two-thirds of their nuclear attack submarines (SSNs), and probably the best quarter of their remaining naval surface combatants, diesel submarines, and naval aviation capabilities are based in Kola.[13] The Soviet Navy has apparently decided to establish a defended wartime "sanctuary" in the Barents Sea for SSBNs carrying long-range ballistic missiles.[14] However, its older Yankee-class SSBNs (with shorter range missiles), as well as Soviet attack submarines that might try to attack the Atlantic sea-lines of commu-

---

13. Johan J. Holst, "Norway's Search for a Nordpolitik," *Foreign Affairs*, Vol. 60, No. 1 (Fall 1981), p. 66.
14. Michael MccGwire, "The Rationale for the Development of Soviet Seapower," *U.S. Naval Institute Proceedings*, May 1980, pp. 155–183.

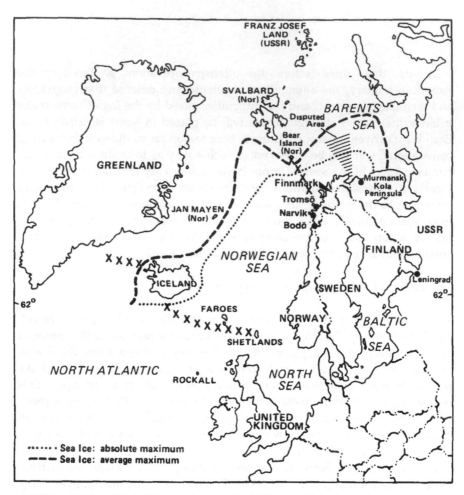

SOURCE: Geoffrey Kemp, "The New Strategic Map," *Survival,* March/April 1977 (London. The International Institute for Strategic Studies).

XXXX: Presumed location of SOSUS arrays Much of the Barents, Greenland, and Norwegian Seas, as well as the North Atlantic, is said to be within the detection ranges of these arrays

There are said to be four military airfields in Northern Norway The main fighter base is at Bodo The main base for P3 ASW aircraft is at Andoya, on an island just off the coast, near Narvik Bardufoss, near Tromsö, is the main airfield for the receipt of NATO reinforcements, but also may be used as a fighter base Banak, northwest of the Altafjord, is an air/sea rescue base in peacetime Its wartime function is unclear

Bardufoss is the headquarters of Brigade North, the ready formation charged with defending Norway against surprise attacks

The Norwegian Navy has a forward operating base in caves, at Olavsvern, near Tromso The coastline from Tromsö to Narvik is said to be well protected by shore batteries of artillery, torpedos, and mines.

nication (SLOC), would have to transit the narrow passages of the North Cape–Bear Island Gap and the better-known Greenland–Iceland–United Kingdom Gap (GIUK), both of which will be heavily defended by NATO forces.

NATO deploys extensive naval capabilities in this area in order to protect its Atlantic SLOC (which run south of the GIUK Gap). To do so, it seeks to bottle up and destroy Soviet naval forces (particularly submarines) in the Barents and Norwegian Seas. Failing that, NATO wishes to harry their southward passage to the Atlantic, exacting as high a toll as possible. NATO places four defensive barriers between the Soviet Navy and the SLOC, including forces in the North Cape Gap and the GIUK Gap, ASW hunter-killer groups, and finally each convoy's own escorts. NATO would also like to monitor Soviet SSBNs in the area, to prepare for strategic ASW should it be so ordered. Moreover, any SSBNs trying to run the Western barriers are likely to be considered fair game.[15]

In Norway, NATO's day-to-day assets in the north do not appear formidable at first glance, but they are not insignificant and could be perceived as a threat to the Soviet Northern Fleet.[16] Norwegian air bases, for example, are well situated for conducting air strikes against the Kola Peninsula. In addition, intelligence-gathering facilities located in Northern Norway allow NATO to plot the movements of Soviet aircraft, surface vessels, and submarines.

---

15. The actual effectiveness of Western ASW forces is said to be high, but I am aware of no unclassified detailed estimates of the rate at which NATO can destroy Soviet submarines either in the Barents Sea, or in the barriers to the south. It would be misleading to suggest that finding and destroying Soviet submarines will be an easy task, even for the very capable ASW forces of the United States and its allies. The process is largely one of attrition, with each barrier taking a modest toll of Soviet submarine forces passing through. It is the cumulative effect of the barriers on the Soviet force, going and coming, that eventually grinds it down to a state of comparative weakness. If NATO forces go looking for Soviet submarines in the Barents Sea, finding and killing them will probably be somewhat more difficult than waiting for the Soviet submarines to try running the barriers. This is because the sensors and C^3 in the barriers should be more effective than anything the West could take into the Barents, either on or under the surface. See also footnote 17.

16. Details on the military situation in the Northern Flank are drawn from "Soviets' Buildup in North Exceeds Protection Level," *Aviation Week and Space Technology*, June 15, 1981, pp. 101–107; "Norway Formulating Long-Range Defense Plans," *Aviation Week and Space Technology*, July 6, 1981, pp. 42–48; R.D.M. Furlong, "The Threat to Northern Europe," *International Defense Review*, April 1979, pp. 517–525; R.D.M. Furlong, "The Strategic Situation in Northern Europe," *International Defense Review*, June 1979, pp. 899–910; General Sir John Sharp, "The Northern Flank," *RUSI Journal*, December 1976, pp. 10–16; and *Military Balance 1981–1982* (London: International Institute of Strategic Studies, 1982). No non-Norwegian NATO forces are permanently based in the country in time of peace as a matter of national policy.

An underwater moored sonar array (SOSUS) in the Bear Island Gap allows the Norwegians to monitor the movements of Soviet submarines in the Barents Sea region.[17] Data from this network is said to be collected near Tromso in Northern Norway. Land-based radar help watch the activity of the Soviet Northern Fleet's air arm, forcing it, in time of war, to fly time- and fuel-consuming evasive legs to avoid Norwegian or other NATO interceptors. Electronic intelligence-gathering facilities near the Soviet border and elsewhere in Norway monitor the communications of Northern Fleet headquarters.[18]

These intelligence assets provide the West with substantial early warning of Soviet preparations for general war—conventional or nuclear. If war should break out, they would provide valuable tactical warning and tactical intelligence. This information will likely be provided to NATO naval forces present in the Norwegian Sea and in the GIUK, presumably increasing their effectiveness and thereby increasing the threat to the Soviet Navy.

Thus, the airbase structure and intelligence-gathering facilities in Northern Norway could be construed by the Soviets to be (and could well be) an offensive threat to Soviet control of the Barents Sea. Northern Norway's four military airfields are within tactical fighter range of the Kola Peninsula and Barents Sea. The intelligence facilities would reveal much about Soviet operations in the area. These facilities would aid any NATO air, surface, subsurface, or combined offensive against the Northern Fleet and its bases.

Consequently, whether the Soviet Union wishes to attack NATO's SLOC in the Atlantic or simply to protect its SSBN sanctuary in the Barents Sea, elimination of NATO capabilities in Northern Norway would appear attractive. To do so, the Soviet armed forces will have to deal with Norwegian forces and substantial NATO land, sea, and air reinforcements. The Soviets would probably like to eliminate the North Cape barrier through aerial bombardment and naval attack, but it is unlikely that this would give high

17. Discussions of U.S. ASW capabilities in this essay are drawn largely from Norman Friedman, "SOSUS and U.S. ASW Tactics," *Proceedings*, March 1980, pp. 120–123; U.S., Congress, Foreign Affairs and National Defense Division, Congressional Research Service, *Evaluation of Fiscal Year 1979 Arms Control Impact Statements* (Washington, D.C.: U.S. Government Printing Office, 1978), pp. 103–119; Joel S. Wit, "Advances in Antisubmarine Warfare," *Scientific American*, Number 244 (February 1981), pp. 31–41.

18. Owen Wilkes and Nils Petter Gleditsch, "Intelligence Installations in Norway: Their Number, Location, Function, and Legality," photocopy (Oslo: International Peace Research Institute, February 1979, revised July 1979).

confidence of destroying the bulk of NATO capabilities in the North. Invasion and occupation of Northern Norway would then be necessary to assure the permanent removal of NATO forces. Moreover, Soviet planners themselves might find Norwegian bases useful both to defend the Barents sanctuary and to attack the Atlantic SLOC. This motivation alone could prompt a Soviet attack on Norway, even if no Norwegian or other NATO forces were present in the northern part of the country.

It is apparent, then, even from a general examination of the interests of the two superpowers in the Northern Flank and of the array of forces that each has deployed there that a conventional clash between the two sides in this region would mix together strategic and conventional forces in a potentially escalatory way.[19] The risk of inadvertent escalation in the Northern Flank is revealed even more clearly, however, in the operational inclinations of the U.S. Navy, for it contemplates performing its missions in ways that are bound to raise this problem.

Northern Flank Scenarios

The U.S. Navy publicly identifies three possible Northern Flank military operations: 1) forward anti-submarine warfare operations by Western nuclear attack submarines (SSNs) in the Norwegian and possibly the Barents Seas; 2) offensive-carrier battle-group operations in the same region to destroy the Soviet fleet and its bases in the Kola Peninsula; and 3) carrier battle-group operations to help defend Northern Norway from a Soviet invasion. In varying degrees, these plans possess inherent and dangerous escalatory pressures.

FORWARD SSN OPERATIONS
The U.S. Navy views the Soviet Northern Fleet, particularly its nuclear and conventional attack submarines, as the principal threat to the Atlantic SLOC. It intends to complement NATO's barrier strategy for interdicting any at-

---

19. In general, Soviet military writings on conventional war include more explicit expectations of nuclear escalation than do analogous Western writings. Indeed, Soviet soldiers are likely to strive with their conventional operations early in any East–West war to eliminate or reduce perceived threats to Soviet strategic forces. See Joseph D. Douglas and Amoretta Hoeber, *Conventional War and Escalation: The Soviet View* (New York: Crane, Russak, 1981).

tempted southward passage of these forces with offensive operations against the Northern Fleet and argues that this can further contribute to the security of the SLOC by forcing the Soviets onto the defensive. The Soviet Navy is to be so occupied defending itself as to preclude attacks on the SLOC.

In particular, the U.S. Navy sees several benefits to attacking Soviet submarines as far north as possible. For one thing, many potentially threatening Soviet attack submarines could be destroyed before they can even attempt the journey south into the Atlantic. In addition, the mere presence of Western SSNs in far northern waters would force Soviet submarines to slow down (and thereby run more quietly) to avoid detection—thus limiting the speed with which they could reach the SLOC, and the rate at which the total Soviet force could be cycled back and forth to the mid-Atlantic. Moreover, forward deployment of Western attack submarines may be favored because it makes possible strategic ASW against Soviet strategic submarines in the Barents Sea. In the event of nuclear war, it seems probable that Soviet SSBNs in the Barents will be attacked at some point. To prepare in advance for *strategic* ASW, Western ASW assets might wish to pre-deploy to the Soviet Barents Sea SSBN sanctuary during the conventional phase of a war.

These operational preferences are frequently evident in the statements of Navy spokesmen. The Chief of Naval Operations, for example, wishes to "seek out and destroy" Soviet naval forces, "wherever they may be, even in Soviet coastal waters."[20] The United States must "contain and attrite the Soviet Navy as close to their home waters in a conflict as possible. . . . Accordingly, *early* concepts of attack submarine employment in *far forward* area offensive operations, barriers, and vectored intercept roles has grown in scope and asset requirements."[21] Vice Admiral Doyle, Deputy Chief of Naval Operations, has put "Offensive Operations in Foward Areas" at the top of his list of SSN requirements—above "Barrier Operations."[22] Because

20. U.S., Congress, Senate, Committee on Armed Services, *Hearings on Department of Defense Appropriations for Fiscal Year 1980* (Washington, D.C.: U.S. Government Printing Office, 1979), p. 1292.

21. U.S., Congress, Senate, Committee on Armed Services, *Hearings on Department of Defense Appropriations for Fiscal Year 1981* (Washington, D.C.: U.S. Government Printing Office, 1980), Part 2, p. 867.

22. U.S., Congress, House, Committee on Armed Services, *Hearings on Department of Defense Appropriations for Fiscal Year 1970* (Washington, D.C.: U.S. Government Printing Office, 1969), Part 4, p. 277.

the U.S. SSN is believed to be highly survivable, he indicates, it "will be assigned to conduct offensive operations in forward areas in the *early stages* of conflict. . . ."[23] Forward operations would be conducted in "areas which are *very contiguous* with the *home bases* of an adversary."[24]

Such operations could cause inadvertent escalation by destroying Soviet strategic submarines. The "forward operations" of American attack submarines seem to include the Barents Sea—the Soviet SSBN sanctuary. It will be difficult for Western SSNs to distinguish between enemy attack submarines and SSBNs. Navy spokesmen have said that Western ASW platforms "would not be in a position of differentiating their [Soviet] attack submarines from their SSBNs."[25] Vice Admiral Kaufman, the Navy's Director of Command, Control, and Communications, has declared that "in a conventional war all submarines are submarines. They are all fair game."[26] Of Soviet SSBNs, it is said, "Of course the ballistic missiles are their main reason for being, but they do carry torpedoes and mines."[27] The Navy thus blurs the distinctions between Soviet submarine types. There is a risk, therefore, that Soviet ballistic missile submarines will inadvertently be sunk in the conventional phase of an East–West war. The Soviet Union could see such sinkings as a deliberate attempt to degrade the Soviet Union's *nuclear* retaliatory capability rather than as "accidents" to be accepted with equanimity.

This suggests how Western operations to *defend* the SLOC may appear *strategically offensive* to the Soviets, even as Soviet deterrent forces (SSBNs) appear offensive (i.e., just another torpedo-carrying submarine) to Western forces. Whenever Soviet and American sub-surface conventional and nuclear forces are mixed together in wartime, the potential for nuclear escalation exists.[28] This problem is exacerbated to the extent that the West pushes into

---

23. Ibid.
24. U.S., Congress, Senate, Committee on Armed Services, *Hearings on Department of Defense Appropriations for Fiscal Year 1977* (Washington, D.C.: U.S. Government Printing Office, 1976), p. 6609.
25. Ibid., Part 4, p. 1972.
26. U.S., Congress, Senate, Committee on Armed Services, *Hearings on Department of Defense Appropriations for Fiscal Year 1978* (Washington, D.C : U.S. Government Printing Office, 1977), p. 6699.
27. U.S., Congress, House, Committee on Armed Services, *Hearings on Department of Defense Appropriations for Fiscal Year 1980* (Washington, D.C : U.S. Government Printing Office, 1979), pp. 663, 679.
28. This discussion has ignored any predeployment of Soviet or American SSN or SSBN across these barriers in the period leading up to war. Such predeployments might actually cause the

the Barents or the Soviets try to break out of the Barents into the Norwegian Sea. Moreover, since Soviet SSNs seem to protect Soviet SSBNs as a primary mission, an ASW campaign confined to Soviet attack submarines might still be perceived by the Soviets as putting their SSBNs in jeopardy.

In addition to inadvertent attacks, there is the possibility that American subs may deliberately stalk and/or kill Soviet SSBNs during the conventional phase of an East–West war. Since U.S. strategic nuclear war plans have long included plans for attacks on Soviet strategic nuclear capabilities, it seems plausible if not probable that steps will be taken to keep track of Soviet SSBNs, and possibly to sink them, during the conventional phase of a war. Although this mission is seldom discussed publicly, there have been official allusions to it.[29]

One possible motive for the deliberate sinking of Soviet SSBNs even during the conventional phase of the war is that Western ASW assets appear to work better in a non-nuclear than a nuclear environment, a fact which bolsters already existing escalatory tendencies. Rear Admiral Metzel, Director of ASW for the Chief of Naval Operations, has noted the extreme importance of Command, Control, and Communications (C³) for effective ASW; how-

---

outbreak of hostilities by mixing together Soviet and Western submarine commanders under conditions of poor communications with National Command Authorities. Events similar to the Libyan–American fighter clash of Summer 1981 are not impossible. Moreover, predeployed Soviet Yankee SSBNs would probably be considered fair game in wartime, since they constitute a conventional threat due to their substantial torpedo load, and a nuclear first-strike threat due to their proximity to Western strategic bomber bases. (Additionally, because some Yankees are having their missiles removed and may be converted to pure SSNs due to continued Soviet adherence to SALT I, Western submariners cannot afford to avoid attacks on "strategic nuclear" Yankees. They may treat anything with a Yankee signature as an SSN—a threat to themselves and to the SLOC. This point may be moot, however, since as noted in the text, Western ASW sensors are reported to have difficulty distinguishing between different Soviet submarine types. The option of avoiding attacks on Soviet SSBNs may not exist.)

29. *Evaluation of Fiscal Year 1979 Arms Control Impact Statements*, p. 113. The possibility of deliberate strategic anti-submarine warfare raises an important related point. In general, the greater the counterforce capabilities in Soviet and American strategic nuclear forces and the greater their commitment to counterforce strategies for nuclear war fighting, the greater the likelihood that the factors discussed in this essay will lead to nuclear escalation. What might ordinarily seem an accidental or ambiguous conventional threat to one's strategic forces is more likely to be seen as deliberate and direct if one's adversary is believed to have a counterforce nuclear doctrine. What might seem a minor loss if one had a large, invulnerable second-strike capability could appear as a major loss if one's adversary were known to have many counterforce options. In this sense, large counterforce capabilities, which are often presented as a tool to control and limit damage in a superpower conflict, may become a cause of escalation from conventional to nuclear war. Insofar as the best way to limit damage in a nuclear war is not to have one, one might wish to re-examine the case for large counterforce capabilities.

ever, C³ facilities are unlikely to survive even limited Soviet nuclear strikes.[30] As Richard Garwin has observed, "Nuclear war confined to the sea appears to degrade ASW capability far more than it degrades the survivability of the SLBM [sub-launched ballistic missile]."[31] A relatively small number of nuclear warheads (perhaps well under fifty) directed at SOSUS data collection points, military communications facilities, and P3 ASW aircraft bases might significantly degrade the West's ability to conduct ASW operations. The electromagnetic pulse generated by nuclear detonations would certainly interfere with radio communications—at least for a short time. Underwater detonations could reduce the detection capability of passive sonars, and might destroy moored sonar arrays (SOSUS). In short, if the West wishes to reduce Soviet sea-based retaliatory capability, there is an incentive to do so during the conventional phase of an East–West war. At the same time, with the detonation of relatively few nuclear weapons, on naval targets where there would be comparatively low collateral damage to civilians, the Soviets might quickly reverse any Western success at sea.[32]

A deliberate conventional campaign against Soviet SSBNs could be understood by the Soviets as the beginning of a damage-limiting strategic first-strike. Given the importance of nuclear weapons and nuclear war in Soviet doctrine, even the appearance of such a campaign could trigger dire consequences. American leaders may be surprised by the Soviet response, since they seem to believe that so long as nuclear weapons have not been used in destroying Soviet strategic forces, the prospect of Soviet escalation is not raised.

Regardless of whether Western SSNs are in the Barents Sea or nibbling along its edges, whether they are sinking Soviet SSBNs accidentally or deliberately, and whether they are doing so quickly or slowly, the activity is likely to cause considerable disquiet in the Soviet Union. In wartime, such actions

---

30. U.S., Congress, House, Committee on Armed Services, *Hearings on Department of Defense Appropriations for Fiscal Year 1980* (Washington, D.C.: U.S. Government Printing Office, 1979), pp. 710–711; Desmond Ball, "Can Nuclear War Be Controlled?" *Adelphi Paper*, Number 161 (London: International Institute for Strategic Studies, 1981), pp. 14–25.

31. Richard Garwin, "The Interaction of Anti-Submarine Warfare with the Submarine-Based Deterrent," in Kosta Tsipis, Anne H. Kahn, and Bernard T. Feld, eds., *The Future of the Sea-Based Deterrent* (Cambridge: M.I.T. Press, 1973), p. 89.

32. Such a strike could severely reduce the West's ability to control the SLOC and move conventional reinforcements to Europe. NATO might then face defeat due to growing Soviet conventional superiority in Central Europe. NATO's use of nuclear weapons in the theater, or against Soviet submarine bases on the Kola Peninsula, could then be its only alternative to surrender.

are likely to be interpreted in the worst possible light.[33] Accidents or confusion that appear to be consistent with a deliberate Western strategic ASW campaign may be interpreted as such. If so, a Soviet nuclear response cannot be ruled out.

Even if American leaders are sensitive to these problems, it could prove impossible to prevent forward SSN operations from having some of these effects. Submarine missions are difficult to control because communications with SSNs under wartime conditions are likely to be tenuous. These boats tend to operate as lone wolves; radio silence and concealment are critical to their survival—particularly in areas as heavily defended as the Barents. SSN commanders in Soviet waters are unlikely to come near the surface frequently either to send or receive messages. Once war begins, Western SSNs would presumably try to sink any Soviet submarine in their vicinity. Since submarines have no armor or "defensive" weapons, they survive in combat by stealth or by shooting first. The SSN therefore may have an incentive to sink enemy submarines before they have a chance to detect and sink it. The SSN commander cannot afford to be particularly discriminating in his decisions to shoot or not to shoot. Extricating American SSNs from this situation before they have taken potentially escalatory actions is likely to be difficult. Carrying twenty or more torpedoes, such subs could do a lot of damage if *each* is not quickly restrained. If communication problems prevent the imposition of such restraint, the continued destruction of Soviet subs could harm any attempted wartime negotiations. In short, putting SSNs in far-forward areas is likely to establish a situation that will be difficult to control—whether the U.S. wishes to fine-tune the destruction of Soviet SSBNs for its own reasons, or to stop their destruction because a grim warning has been received from the Soviet Union.

THE BATTLE-GROUP ASSAULT ON THE NORTHERN FLEET
The risk of inadvertent escalation is also raised by a second type of operation advocated by the U.S. Navy: offensive-carrier battle-group operations which would launch air attacks against the Northern Fleet and its bases in the Kola Peninsula. Such a procedure would include four to six carriers, as well as a substantial number of American and British surface and submarine escorts.[34]

---

33. Robert Jervis, *Perception and Misperception in International Politics* (Princeton: Princeton University Press, 1975), pp. 319–329.
34. "NATO's Sinking Feeling," *The Economist*, June 6, 1981, p. 51.

The Navy is vague as to when in a war such attacks would occur, but its spokesmen talk about them with great enthusiasm.

The Navy argues that this operation would tie down Soviet forces that would otherwise attack the SLOC. With luck, the Northern Fleet might be decisively defeated, securing command of the sea for the West, and allowing easy resupply of Western Europe. In the words of the Chief of Naval Operations:

We must fight on the terms which are most advantageous to us. This would require taking the war to the enemy's naval forces with the objective of achieving the *earliest possible destruction of his capability* to interfere with our use of the sea areas essential for support of our overseas forces and allies. In this sense *sea control is an offensive rather than a defensive function.* The *prompt destruction* of opposing naval forces is the most economical and effective means to assure control of the sea areas required for successful prosecution of the war and support of the U.S. and allied war economies. Our current offensive naval capabilities, centered on the *carrier battle forces* and their supporting units, are well-suited for the execution of this strategy.[35]

In addition to destroying Soviet naval forces at sea, according to Senator Gary Hart, the Admiral advocates "the use of carrier attack aircraft against Soviet land targets," presumably Soviet naval bases and airfields on the Kola Peninsula.[36] If this operation were ordered, it would probably unfold in close cooperation with the SSN operation outlined above. Carrier strike aircraft would seek out Soviet naval assets afloat and ashore; SSNs would defend the carrier from Soviet submarines and hunt them down, as would surface escorts.

If successful, this operation could increase the risks of nuclear escalation in several ways. By destroying the bulk of Soviet conventional naval capabilities, the attack could reduce, if not eliminate, the Soviet Navy's ability to protect its strategic submarines in their Barents Sea sanctuary from a concentrated attack by Western ASW forces. This would be equivalent to the United States suddenly discovering in wartime a great new vulnerability in its SSBN forces, and one that could be expected to grow at an unpredictable rate.

---

35. U.S., Congress, House, Committee on Armed Services, *Hearings on Department of Defense Appropriations for Fiscal Year 1980* (Washington, D.C.: U.S. Government Printing Office, 1979), p. 841. Emphasis mine.
36. U.S., Congress, Senate, Committee on Armed Services, *Hearings on Department of Defense Appropriations for Fiscal Year 1980* (Washington, D.C.: U.S. Government Printing Office, 1979), p. 557.

This operation might also directly destroy Soviet SSBNs—either accidentally or on purpose. Carrier-based ASW aircraft, long-range shore-based ASW aircraft, and surface escorts would add to the threat posed to Soviet SSBNs by Western SSNs. The synergistic effect of these forces might cause some quite rapid degradations of the Soviet strategic submarine force.

In addition, such air strikes, even with conventional ordnance, might hit critical vulnerabilities in the Soviet military system. The Soviet Navy, like the rest of the Soviet armed forces, is highly centralized and extremely dependent on battle management from higher authorities.[37] Thus, destruction of naval tactical Command, Control, Communications, and Intelligence (C^3I) facilities, SSBN C^3 facilities, and Kola-based strategic early warning systems might all have a disproportionate effect on Soviet capabilities and confidence. Destruction of any of these assets could cause a sudden unravelling of Soviet forces, or a sudden increase in Soviet perceptions of strategic vulnerability.

Moreover, if such air attacks were to succeed for any length of time, Soviet air defense commanders might begin to lose confidence in their ability to detect and stop nuclear surprise attacks by tactical fighters, strategic bombers, or cruise missiles. Repeated, undetected penetration of Soviet airspace by U.S. carrier-based fighters would be stark evidence of the inadequacy of Soviet air defenses. Indeed the conventional, carrier-aircraft campaign might be perceived by the Soviets as a deception operation designed to accustom them to a certain number of Western air strikes every day. How then could a nuclear surprise attack be avoided?[38]

Finally, U.S. air strikes on the Soviet homeland—the Motherland—could have unpredictable consequences. They could trigger an emotional response from Soviet leaders, even if they did not do much real military damage.

---

37. Ball, p. 45; "The Soviet Navy Declaratory Doctrine for Theatre Nuclear Warfare," a report prepared by the BDM Corporation, September 30, 1977, for the Defense Nuclear Agency.

38. At least one high-level Soviet publication has noted the possibility that the West might use a "local war" as camouflage for the preparation of a nuclear strike:

We cannot exclude attempts to achieve surprise by means of unleashing a local war. . . . The local war can be used by the aggressor for the additional mobilization of forces. In the guise of moving troops to the regions of the military conflict, a strike grouping of forces and means can be created for an attack. Such a war gives rise to an increase in the combat readiness of all armed forces of the aggressor, an intensification of strategic reconnaissance, the deployment of control points and communications centers in the territory of the dependent countries, and the carrying out of an entire series of other measures.

—Major General N. Vasendin and Colonel N. Kuznetsov, "Modern Warfare and Surprise Attack," *Voyennaya Mysl*, Number 6 (1968), p. 45.

In short, it would be easy for the offensive military operations that the U.S. Navy justifies on *defensive* grounds to threaten Soviet strategic capabilities. Since the Soviets value their nuclear capabilities, any of these eventualities could elicit a nuclear response. Moreover, Western naval assets, particularly the aircraft carriers, are terribly attractive targets for a Soviet nuclear attack, both because of their relative vulnerability and because attacking them would not require use of nuclear weapons against the American homeland. Soviet use of nuclear weapons at sea therefore seems quite plausible, insofar as the purpose would be to protect *their* strategic nuclear capabilities from immediate and future attack. The Soviet nuclear response could be limited to a small number of naval and land targets in the Northern Flank region and still achieve its critical objectives.

Thus, the combination of military preferences for offensive operations and the inability of both superpowers to distinguish between conventional and strategic, offensive and defensive forces may lead to a potentially escalatory clash in the north Norwegian and Barents Seas.[39] This clash will not be easy for policymakers to control. Communications with the forces engaged may be quite tenuous. Intelligence is likely to be faulty and slow; communications links will be jammed or destroyed; commanders will frequently observe radio silence to avoid detection. Moreover, the great firepower and range of modern *conventional* weaponry makes possible sudden changes in the fortunes of each side—changes that may cause panic and over-reaction.[40] In such circumstances, attempts at negotiation will be complicated and difficult.

---

39. The Navy's enthusiasm for the carrier assault notwithstanding, such an operation would be very risky for Western naval forces. Some analysts privately doubt that the U.S. Navy is serious about it. Soviet forces are rather strong in the Barents Sea and Kola Peninsula, and could be reinforced with a great deal of land-based defensive and offensive airpower. Of course, this makes for a demanding mission, one that gives the Navy a strong argument for more forces in the competition for resources among the services. There is, however, a fine line between arguments employed to win more forces and beliefs about how a war should actually be fought. The Navy's frequent discussion of offensive carrier operations should at minimum be taken as evidence of a strong tendency or preference. Moreover, the operation might be attempted several weeks into a war, after additional naval power could be mustered from other theaters, and after forward SSN operations and land-based air operations from Northern Norway degraded Soviet defenses. The operation could also be supported by other NATO aircraft based in Northern Norway and farther south. Thus, the operation must be taken as a serious possibility.

40. At Midway, a World War II naval battle fought with substantially less-capable conventional weaponry than that deployed today, the Japanese lost three large aircraft carriers in less than ten minutes and a fourth less than eight hours later—one-half of their entire carrier force. Samuel Eliot Morison, *The Two Ocean War* (Boston: Little, Brown, 1963), p. 157.

NATO'S NAVAL COUNTERATTACK

NATO's efforts to defend Norway against Soviet attack constitute yet a third way that inadvertent escalation could occur on the Northern Flank. Thus, Soviet actions, too, may create serious escalatory pressures, for a Soviet assault on Northern Norway could (and probably would) precipitate a powerful NATO naval counterattack which, unless carefully controlled, could become a cause of escalation.

Norway's military capabilities, as well as NATO reinforcements likely to arrive early in a mobilization period or a war, may be perceived by the Soviet Union as an offensive threat. Some of these NATO assets are essential to the defense of Norway and NATO, but this is little comfort to the Soviets. Consequently, a Soviet attempt to take Northern Norway and to destroy NATO's northern ASW barriers is likely.[41]

Contrary to the conventional wisdom, however, a quick and easy victory for Soviet forces is by no means assured, which means that NATO could well have the time to mount its naval counterattack. Although Soviet ground forces *would* have an easy time crossing the Soviet–Norwegian border, it would be difficult for them to capture NATO's major base areas, which lie over 600 road miles away through rough country. To have any hope of rapid success, the Soviets would have to attempt a combined arms, airborne-amphibious assault *early* in the war. Even this attack could still fail, or at least fail to produce a rapid victory, because such operations are notoriously tricky, Soviet airlift and sealift capabilities are limited, and the Norwegians appear well prepared to thwart just such an attack. Free passage through Northern Finland, *by no means assured*, would ease the problem of providing a ground component for a Soviet invasion, but even this would not guarantee a quick win, because Norwegian forces are stationed close to the Finnish wedge, the likely Soviet invasion route. The road from the wedge to the main Norwegian bases runs through very mountainous terrain, and is vulnerable to air interdiction, ground demolition, ambush, and other delaying tactics. Thus, a Soviet attack on Northern Norway seems possible, but a quick victory does not. It is entirely plausible that Norwegian defenders would hold on long enough to permit the arrival of NATO reinforcements.

The West is likely to attempt a vigorous defense of Norway, particularly because a Soviet attack there may be seen as part of a Soviet campaign against NATO's SLOC, aiming to destroy NATO's ability to wage conven-

---

41. McccGwire, pp. 155–183.

tional war in Europe. (This might indeed be one of the Soviet goals.) U.S. Navy carrier battle groups would probably play an important role in any Western counterattack designed to thwart Soviet designs in Norway.

This counterattack is likely to be somewhat less adventurous than the offensive operations designed to attack the Kola Peninsula and the Barents Sea. However, it runs many of the same risks, because any great naval battle in the Norwegian Sea will, if NATO is successful, have the same implications for the Soviets as a full-fledged attack on the Northern Fleet. U.S. Navy commanders would probably want to exploit any initial success with a follow-up attack against Soviet naval units fleeing back to Kola. In short, there are other ways for the U.S. Navy to end up in a "decisive battle" with the Soviet Northern Fleet than a simple, early, all-out offensive against the Kola Peninsula. While Norway *must* be defended, any naval engagement of this kind needs to be carefully planned and controlled if outcomes that could prompt the Soviet Union to use nuclear weapons are to be avoided.

Conclusion: Avoiding Inadvertent Escalation

These three Northern Flank scenarios—the forward operation of American attack submarines in the Norwegian and Barents Seas, offensive aircraft-carrier operations against the Kola Peninsula, and a NATO naval counter-attack in defense of Norway—illustrate the potential problems that might arise in trying to prevent escalation in the context of large-scale conventional conflict. While not inevitable, these operations do seem to have widespread support in the Navy and among some civilians. If undertaken, they may not be successful. If successful, they may not precipitate a nuclear response. However, insofar as such things can be thought about in advance, these Northern Flank operations appear to be plausible paths to nuclear escalation.

This is true because survivable strategic nuclear forces are critical to the security of both superpowers. States fight hardest—and may be willing to risk the most dire consequences—when assets essential to the preservation of their sovereignty are at stake. Thus, conventional operations that threaten strategic nuclear forces can be extremely provocative. Both sides are frightened of nuclear weapons, and seem in no hurry to use them. However, although it may seem paradoxical, the use of some nuclear weapons to protect the remainder of one's strategic deterrent is probably among the most plausible scenarios for nuclear escalation. Such escalation is unlikely to be easy to manage or control.

It should be stressed that this problem is not specific to the Northern Flank. Though it is the best example of the general escalation tendencies of an East–West conventional war, there are others. For instance, the U.S. Navy frequently argues for attacks directed against the Eastern Soviet Union, especially Vladivostok and Petropavlovsk, that are similar to those aimed at the Kola Peninsula. Since the other one-third of Soviet SSBN assets are based in the East, such attacks could be almost as destabilizing as attacks against the Kola. Of course, if adequate forces were available, simultaneous attacks at both ends of the Soviet Union would be possible, creating still greater escalation pressures because the entire Soviet strategic submarine force would then be threatened.

The great air battle that would attend any NATO–Warsaw Pact war is another campaign fraught with escalation potential. Including upwards of six or eight thousand aircraft, such a battle may create huge problems for the strategic air defenses (PVO) of the Soviet Union. The electronic "noise" of such a battle may make small surprise nuclear attacks from Western Europe difficult to detect. Soviet leaders might fear that such an attack against its strategic command and control capabilities was a prelude to a large-scale American strategic nuclear assault against its forces. This fear would be reinforced by deliberate Western attempts to destroy portions of the Warsaw Pact's air defense in Eastern Europe, such as radars and command posts, particularly since these defenses are integrated with and provide early warning for air defenses of the Soviet Union proper.

Additional problems would be created by the "electronic battle," because various capabilities may quickly be discovered and countered in an intense conventional air battle. Since many of the same air defense systems protect the Soviet Union and Eastern Europe, electronic and other countermeasures devised to degrade air defenses over Eastern Europe may be useful for air attacks against the Soviet Union. If Western air forces mounted deliberate conventional strikes against the air defense network in the Western military districts of the Soviet Union, Soviet fears of nuclear attack would be even more intense, and the pressure to escalate could be great.

Not only are conventional operations and force posture poorly designed to avoid nuclear escalation, but elements of the nuclear forces are poorly designed to last out an extended conventional war. Perhaps the most serious difficulties for the West would result from the limited endurance of the U.S. bomber force and airborne C^3 systems. In an East–West conventional war, bomber forces would be brought up to high levels of alert—from 25 percent to over 50 percent. It would be difficult, however, to sustain such a surge for very long. Nuclear-armed "Quick Reaction Aircraft" based in Europe might face similar problems, as might Soviet strategic forces, especially SSBNs, which normally sustain much lower peacetime alert rates than their Western counterparts. Because it is difficult to sustain nuclear forces at high levels of alert for protracted periods of time, concerns about the survivability of these forces could arise, and for both sides "windows of vulnerability"—and of opportunity—could open and close at an uneven rate as a conventional war unfolds. This is still another way that nuclear escalation could result from conventional war.

MINIMIZING ESCALATORY RISKS

How can the risk of escalation inherent in these situations be minimized? There are several measures which would help to ameliorate the *general* tendencies towards escalation in an East–West conventional war. First, there should be a constant, ongoing project to audit all conventional war operational plans for their escalatory potential. A group of civilian and military officials should be constituted for this purpose. The group would have the responsibility of bringing the most critical dangers to the attention of responsible authorities and of suggesting less escalatory alternatives.

This suggestion will not be easy to implement. As it stands today, few civilians have access to military war plans on a regular basis, and the American military may resist attempts by civilian authorities to see the war plans. Those civilians with access to the plans, such as the Secretary of Defense, have too many other responsibilities to give the escalation problem the attention it deserves. Nevertheless, an effort should be made to institutionalize a process of peacetime civil-military cooperation aimed at reducing the risk of escalation.

Second, the West's force posture should be reviewed on an ongoing basis to ensure that civilian decision-makers are not compelled by the capabilities on hand to act in dangerous ways. For example, NATO should never have so many aircraft carriers and so few convoy escort vessels and land-based ASW patrol aircraft that, once war breaks out, offensive operations against the Kola Peninsula appear to be the only feasible way to effectively protect the SLOC. The current U.S. Navy procurement program already seems somewhat weighted in favor of costly, heavy, offensive vessels such as carriers, cruisers, and nuclear attack submarines, over lighter, less costly, convoy escorts and land-based ASW patrol aircraft. If financial limits overtake the Administration's naval buildup, escorts and ASW aircraft are likely to be the first items to suffer cuts. From the perspective of escalation control, this would be an unfortunate outcome.

Third, an appreciation of the escalatory potential of East–West conventional war should be consciously introduced into Western declaratory policy. Even if the West takes steps to *reduce* the tendencies and difficulties that contribute to nuclear escalation, important dangers will remain. The West should never publicly display excessive confidence in the ease with which any conventional war involving the superpowers can be controlled. Instead, conventional war should be portrayed as Schelling's "threat that leaves something to chance"; conventional war rolls the nuclear dice. If the Soviets can be made to believe this, then deterrence is enhanced.

These three steps would diminish the likelihood that conventional war would cause nuclear escalation. There are also more specific measures that could be taken to reduce the escalation potential of the Northern Flank.

For one thing, if escalation control is our goal, then the necessity of offensive operations of any kind in the Barents Sea and against the Kola Peninsula should be re-examined. Even if giving up these offensive operations somehow increases the Soviet ability to threaten the SLOC, it may be wise to permit that to happen and to compensate for it in some other way. Carrier task forces that might be used offensively against Soviet base areas could be reconfigured to reinforce the defensive barriers that NATO intends to maintain in any case. For instance, some of the strike aircraft (bombers) in carrier air wings could be replaced by anti-submarine warfare aircraft and air-defense aircraft. Carriers so configured could be deployed to strengthen the air defenses and anti-submarine defenses in the area of the GIUK Gap.

Different forces, more suitable to defensive missions, could be acquired and employed to strengthen barrier defenses. Recent proposals that the United States Navy adopt a "sea-control ship," a small aircraft carrier that would operate a mix of VSTOL (vertical/short take-off and landing) aircraft and ASW helicopters would be consistent with this suggestion.[42] Such a ship

---

42. Most "sea-control ship" proposals have suggested an airwing made up of VSTOL fighters and ASW helicopters. To cope with the growing threat of Backfire bombers and cruise missiles, however, it would probably be wise to include some austere AWACS-type aircraft in the wing to do the kind of job the E2C Hawkeye does for current large carriers. Perhaps a commercial STOL (short take-off and landing) design could serve as the basis for such an AWACS.

could either form the core of an ASW hunter/killer group of the kind deployed so successfully in World War II, or directly beef up convoy defenses. Several such ships could be bought for the price of a nuclear carrier, and although collectively they would probably lack the carrier's *offensive* punch, it is not implausible that they would make a superior overall defensive contribution to the SLOC battle. Sea control ships would not, of course, replace the large carrier in "power projection" missions. But because the U.S. already has a dozen large carriers with attached air wings, the Congress should consider substituting sea-control ships for the two new nuclear carriers proposed by the Reagan Administration.

Similarly, NATO might exercise great care in its reinforcement of Norway during either a mobilization or a war. Initial reinforcements could be confined to "tactically defensive" forces such as surface-to-air missiles, anti-aircraft guns, air defense fighters, close support aircraft, light naval units, ground troops, and the like. The movement of heavy naval units, such as carrier task forces, and long-range, high-payload, land-based strike aircraft, such as U.S. Marine A-6Es, to the north of Norway could be deferred in order to reduce the Soviet incentive to mount a major attack on Northern Norway. The Norwegians have more or less adopted such a policy for their day-to-day peacetime deployments. Extending it into the mobilization phase of any East–West clash could slow the pace of the crisis.[43] If the Soviet forces attack, NATO's naval counteroffensive should be executed with a combination of vigor and restraint. Soviet forces must be driven off, but it may not be wise to pursue them into the Barents Sea, or to attack their bases.

Exercising restraint in conventional operations on the Northern Flank may not be the ideal way to conduct a *conventional* war. But it is preferable to the nuclear escalation that more offensive operations might provoke.

THE DANGER OF MISCALCULATION

We now live in the worst of all possible worlds. We plan potentially escalatory conventional operations, but we do not seem to understand their implications. Rather, our utterances reveal an implicit confidence that superpower conventional war will not escalate to the use of nuclear weapons. This may fool both our adversaries and ourselves about the great inherent dangers of an East–West clash. If the nuclear specter loses its fearsome immediacy, each

---

43. Holst, pp. 69–74.

side may be too willing to risk direct confrontation with the other. Mutual deterrence is eroded. The same misunderstanding of superpower conventional war that has spawned our current overconfidence in its limitability ensures that we will operate our conventional forces imprudently should war come. NATO must defend its territory, but the risks of nuclear escalation present us with difficult choices as to how a conventional war should be fought. Nobody really knows if an East–West conventional war can be controlled, but poor analysis of the special problems of *conventional war between nuclear superpowers* increases the probability that deterrence will fail, and that potentially catastrophic nuclear escalation will occur. A rethinking of the special problems of conventional war in the nuclear age is essential.

A Quiet Success for Arms Control

Preventing Incidents at Sea

Sean M. Lynn-Jones

The 1972 Agreement on the Prevention of Incidents at Sea between the United States and the Soviet Union[1] is a virtually forgotten remnant of an era that produced dozens of U.S.–Soviet accords. Although it was almost ignored in both the American and Soviet announcements of the various agreements that emerged from the May 1972 Moscow summit, the agreement has helped to avert potentially dangerous incidents between the U.S. and Soviet navies. Most of the achievements of détente have lost their luster with the passage of time, but the agreement's effectiveness appears to have survived the deterioration of U.S.–Soviet relations. The agreement deserves to be considered more closely, not only because it has reduced the oft-overlooked dangers of naval incidents, but also because of its potential utility as a model for other agreements to govern incidents in the air or in space. More generally, the success of the agreement demonstrates that confidence-building measures—constraints on military activities and improvements in communications that are intended to reduce the risk of inadvertent war or surprise attack—represent a workable alternative to traditional arms control proposals that impose quantitative or qualitative limits on weapons.

The 1972 agreement has reduced the number of dangerous incidents and cases of harassment at sea. Before the agreement, encounters between U.S. and Soviet warships on the high seas frequently led to tense situations as opposing vessels maneuvered to disrupt one another's formations or ha-

The author would like to thank Joseph S. Nye, Eliot A. Cohen, and members of the Avoiding Nuclear War working group for their helpful comments on earlier drafts of this article. For support during the research and writing of this article, he thanks the Avoiding Nuclear War Project and the Josephine de Karman Fellowship Trust.

Sean M. Lynn-Jones is a doctoral candidate in the Department of Government at Harvard University and a member of the Kennedy School of Government's Project on Avoiding Nuclear War.

1. The full title is: Agreement Between the Government of the United States of America and the Government of the Union of Soviet Socialist Republics on the Prevention of Incidents on and over the High Seas, May 25, 1972. The complete text of the English and Russian versions appears in U.S. Department of State, *United States Treaties and Other International Agreements*, Vol. 23, Pt. 1, 1972 (Washington, D.C.: U.S. Government Printing Office, 1973), pp. 1168–1180.

International Security, Spring 1985 (Vol. 9, No 4) 0162-2889/85/040154-31 $02.50/0

rassed each other with searchlights and low-flying aircraft. These incidents became particularly intense in the late 1960s when the Soviet navy asserted itself on the world's oceans. Since the agreement was signed, fewer serious naval confrontations have occurred, and those that take place have not generated dangers of escalation or political crises.

In the wake of the failure to ratify the SALT II treaty, the breakdown of other arms control negotiations, and the slow progress of arms control during the Reagan Administration, several observers have suggested that the Incidents at Sea Agreement exemplifies an alternative approach. In their examination of possible U.S.–Soviet procedures for nuclear crisis control, William Ury and Richard Smoke note the success of the agreement and argue that it could serve as a model for further steps to control superpower crises.[2] An assistant secretary of state in the Reagan Administration has also identified the agreement as an example of the manner in which U.S.–Soviet relations could be made safer.[3] Although much of the literature on confidence-building measures has concentrated on their potential applications in Europe, the Incidents at Sea Agreement provides an important example of how such measures can work in other areas. Neither the effectiveness of the agreement nor the general utility of confidence-building measures should be overstated; improved communications and constraints on threatening military activities can rarely prevent the deliberate initiation of war. But they can reduce the dangers of inadvertent war and, in some cases, contribute to the prevention of crises.

As interest in confidence-building measures increases, the Incidents at Sea Agreement clearly merits a reexamination to assess its effectiveness and value to both superpowers. It provides a useful case study of a measure that has been in operation for almost thirteen years during a period of increasing superpower rivalry on the oceans. This article will review the history of the agreement, including the nature, causes, and dangers of the naval incidents that led to its negotiation and signing. The provisions of the agreement will

---

2. See William Langer Ury and Richard Smoke, *Beyond the Hotline: Controlling a Nuclear Crisis* (Cambridge: Nuclear Negotiation Project, Harvard Law School, 1984). Senators Sam Nunn (D.-Georgia) and John Warner (R.-Virginia) also see the agreement as "an excellent example of what can be accomplished" in reducing the risks of war and suggest that it might be a precedent for a joint U.S.–Soviet Crisis Control Center. See "A Nuclear Risk Reduction System," excerpts from the *Report of the Nunn/Warner Working Group on Nuclear Risk Reduction*, *Survival*, Vol. 26, No. 3 (May/June 1984), p. 135.
3. Paul D. Wolfowitz, "Preserving Nuclear Peace," *Naval War College Review*, Vol. 36, No. 2 (March-April 1983), p. 78.

be examined to determine their utility, and the implications of recent naval incidents will also be considered. After a discussion of some of the factors that may have contributed to the successful negotiation and implementation of the agreement, some possible extensions and other applications of the agreement will be offered.

The Nature of Incidents at Sea

The term "incident at sea" can be applied to a variety of situations resulting from different maritime activities. At the most general level, it means an action on the high seas by a ship or plane that endangers, or is alleged to endanger, another vessel or aircraft. Some actions of U.S. and Soviet naval units that fit into this category include violations of the International Regulations for Preventing Collisions at Sea ("Rules of the Road"),[4] close, high-speed reconnaissance by aircraft ("buzzing"), simulated attacks on ships or planes, accidental firing upon vessels during naval exercises, and other actions that interfere with the safe navigation of ships, such as shining searchlights on the bridge of vessels.

DANGEROUS MANEUVERS. Naval units in close proximity can engage in a variety of maneuvers that force other vessels to take evasive action to avoid imminent collisions. Many incidents of this sort have occurred between U.S. and Soviet ships, including some that have led to collisions. Although the Rules of the Road are intended to prevent such maneuvers, Soviet vessels have often exploited ambiguities in the rules to play "chicken" or to "shoulder" U.S. vessels off course, particularly during carrier operations or refueling at sea. Prior to the 1972 agreement, the United States and the Soviet Union often issued diplomatic protests that accused the other side of violations of the Rules of the Road.

CLOSE AIR SURVEILLANCE ("BUZZING"). U.S. and Soviet aircraft have frequently flown close passes near opposing vessels for reconnaissance purposes. The Soviets have protested that such actions violate the Rules of the Road and threaten the safe navigation of their vessels; but their own actions are not significantly different.[5] They have been particularly concerned over

---

4. The rules of the road govern nautical lighting, maneuvering, and signaling procedures to ensure safe navigation. For a discussion of recent revisions see T.J. Cutler, "More Changes to the Rules of the Road," *U.S. Naval Institute Proceedings*, Vol. 109, No. 6 (June 1983), pp. 89–93.
5. D.P. O'Connell, *The Influence of Law on Sea Power* (Annapolis: Naval Institute Press, 1975), p. 165.

the U.S. practice of dropping sonar buoys from aircraft to track Soviet submarines, as such actions, they claim, endanger the safety of their submarines.[6]

SIMULATED ATTACKS. Soviet and U.S. naval vessels have often simulated attacks by aiming guns, missile-launchers, torpedo tubes, other weapons, and sensor systems at each other's ships and planes. For example, in an encounter between a U.S. ship and accompanying naval aircraft and several Soviet warships during the 1970 Jordanian crisis, the Soviet ships "went to battle stations . . . ran surface-to-air missiles out on their launchers, and appeared to track the departing U.S. aircraft with their fire control radars."[7] Such actions are at least unnerving to the naval officers and men who experience them. In times of acute international tension, they could increase the incentives for preemptive strikes.

ACCIDENTAL FIRING DURING EXERCISES. Naval exercises involving the use of live ammunition obviously can endanger warships or merchant vessels that stray into the area. Such an incident apparently occurred on March 8, 1963, when a Soviet trawler allegedly came under fire from U.S. naval vessels seventy miles east of Norfolk, Virginia. The United States dismissed the possibility that there was any danger to the Soviet vessel, noting that the incident took place in a U.S. Navy operations area.[8] The absence of prior notification of exercises and the interest of both countries in observing the other's maneuvers as closely as possible increase the probability of such incidents. In this case, the Soviet trawler was almost certainly engaged in surveillance of U.S. naval activities.

OTHER HARASSMENT. There are a variety of ways in which U.S. and Soviet vessels can harass one another at sea. The warships of both countries have illuminated the bridges of opposing vessels with powerful searchlights. The Soviets, for example, alleged that a U.S. patrol vessel off Florida passed

---

6. This issue arose in the negotiations leading to the 1972 agreement. See below pp. 22, 25. The 1983 incident in the Atlantic in which a Soviet submarine was apparently disabled and forced to surface after U.S. sonar equipment became entangled in its propellor demonstrates the validity of Soviet concerns, although that sonar array was towed by a surface ship, not dropped from the air. Even if buoys and cables do not become entangled in Soviet submarines, they may be extremely irritating. According to a U.S. official, when such buoys are dropped near submarines, "the pinging really drives them crazy." See *The Washington Post*, June 8, 1984, p. A15.
7. Abram N. Shulsky, "The Jordanian Crisis of September 1970," in Bradford Dismukes and James M. McConnell, eds., *Soviet Naval Diplomacy* (New York: Pergamon Press, 1979), p. 176.
8. See the exchange of U.S. and Soviet notes on this incident in Historical Office, U.S. Department of State, *American Foreign Policy: Current Documents 1964* (Washington, D.C.: U.S. Government Printing Office, 1967), pp. 562–563.

within 100 meters of the Soviet passenger steamer *Turkeniya* on May 28, 1964, "and repeatedly lit up the hull and the captain's bridge of the vessel with a powerful searchlight, blinding the navigating personnel and creating a danger of collision."[9] Soviet vessels have engaged in similar behavior on a number of occasions. U.S. and Soviet ships can also engage in harassment by firing flares at one another.

There have been hundreds of incidents involving U.S. and Soviet surface ships and aircraft, although the details of these encounters have not always been made public. Many of these incidents occurred in the late 1960s and early 1970s. Additional incidents have involved collisions between Soviet and U.S. submarines, some of which were carrying nuclear weapons.[10] The 1972 Agreement on Incidents at Sea, however, covers only surface and aerial maneuvering. Both the United States and the Soviet Union deliberately excluded submerged submarines from the agreement.[11] This article will therefore concentrate on the causes and dangers of incidents involving surface ships and aircraft.

The Motivations and Causes of Incidents at Sea

The most difficult question to be addressed in considering the causes of incidents at sea is whether the United States or the Soviet Union has deliberately authorized its naval commanders to harass the vessels of the other. Some incidents doubtless are the result of the excessive zeal or incompetence of local naval commanders. Given the intense rivalry between the two countries, it would be surprising if naval personnel did not occasionally play chicken with their opponents at sea. The increase of incidents that seems to have occurred in the late 1960s and early 1970s, by this account, was the inevitable result of the larger Soviet presence at sea and more frequent interaction with U.S. forces.

But incidents may also serve some political purpose, possibly justifying their deliberate use as instruments of policy. Seen in this light, naval harassment and dangerous maneuvering are variations on the venerable practice

---

9. Note from the Soviet Embassy in Washington to the U.S. Department of State, in ibid, p. 672.
10. See "Operation Holystone," *Nation,* July 19, 1975, pp. 35–36, and Dan Caldwell, *American–Soviet Relations: From 1947 to the Nixon–Kissinger Grand Design* (Westport, Conn.: Greenwood Press, 1981), p. 128.
11. See below, p. 19.

of gunboat diplomacy.[12] These actions constitute limited applications of naval force to achieve some political objective. The Soviet Union and the United States may have purposefully engaged in some instances of harassment or dangerous maneuvering at sea, or at least given such actions de facto approval by not disciplining the commanders involved.

Although any discussion of Soviet motivations is necessarily speculative, Soviet harassment of U.S. naval vessels appears to have served several purposes. First, and most generally, Soviet harassment of U.S. ships, particularly simulated attacks on aircraft carriers, demonstrates to the U.S. Navy that Soviet warships can deny the United States the freedom of action that it has traditionally enjoyed on the high seas. The awareness of the Soviet presence and capabilities may constrain U.S. actions in a crisis, as former U.S. Chief of Naval Operations Elmo Zumwalt has suggested.[13] The Soviets appear to have at times had this objective in mind. Admiral Sergei Gorshkov, Commander-in-Chief of the Soviet navy, has written:

In a series of instances, our ships and naval aviation have demonstrated operational and active actions as a result of which some foreign governments became convinced that they could not consider their aircraft carriers and submarines "invisible," "untouchable," and in the event of war "invulnerable" in whatever areas they may be located.[14]

Second, simulated attacks on U.S. warships, particularly carrier forces, in periods of relative international calm have increased the ability of the Soviet Union to communicate with the United States during crises. Exercises that demonstrate Soviet tactics for attacking U.S. vessels enable U.S. commanders to recognize these actions during crises, thereby allowing the Soviet navy to send strong signals to their U.S. counterparts. In effect, simulated attacks give the Soviet navy an "action language" for signaling their U.S. adversaries and American political leaders.[15]

Third, the Soviets have clearly used harassment as a means of conveying their resentment over U.S. naval operations in the Black Sea and the Sea of Japan—areas the Soviet Union apparently regards as home waters. The

---

12. For a seminal discussion of this subject, see James Cable, *Gunboat Diplomacy, 1919–1979: Political Applications of Limited Naval Force* (New York: St. Martin's Press, 1981).
13. Elmo R. Zumwalt, Jr., "Gorshkov and his Navy," *Orbis*, Vol. 24, No. 3 (Fall 1980), pp. 491–510.
14. Quoted in David R. Cox, "Sea Power and Soviet Foreign Policy," *U S. Naval Institute Proceedings*, Vo. 95, No. 6 (June 1969), p. 41.
15. Charles C. Petersen, "Showing the Flag," in Dismukes and McConnell, eds., *Soviet Naval Diplomacy*, p. 105.

United States has sent two Sixth Fleet destroyers into the Black Sea every six months to assert its rights under the Montreux Convention. Regarding these actions as provocative, the Soviets have dispatched naval vessels to shadow and harass U.S. ships throughout their voyage through the Black Sea.[16] Similarly, in the Sea of Japan, Soviet land-based attack aircraft routinely simulate missile attacks on U.S. carriers.[17] Many incidents of harassment of U.S. naval vessels have occurred in this region, including the well-known collisions of the U.S. destroyer *Walker* with Soviet vessels in 1967 and the March 1984 collision of the U.S. carrier *Kitty Hawk* and a Soviet submarine. Although these actions could be attributed to the aggressiveness or incompetence of local Soviet commanders, it seems more likely that at least some of them are the result of deliberate policy. Soviet diplomatic protests following incidents in the Sea of Japan claim that "U.S. ships show no regard for existing international norms, grossly violate international norms for the prevention of collisions of ships at sea, and take a number of illegal actions against the Soviet ships in this area, coming dangerously close to them." But the real motivation for Soviet actions seems to be their belief that "the very fact of U.S.–Japanese exercises close to Soviet shores cannot be regarded as anything but a premeditated, organized provocative military demonstration."[18]

Fourth, the Soviet Union may use harassment for tactical military purposes. Interfering with the flight operations of U.S. carriers and otherwise obstructing U.S. naval activities can prevent the launch of aircraft that might deliver an attack or, more probably, track a Soviet submarine. Soviet vessels may attempt to disrupt U.S. naval operations even when there is no immediate U.S. threat, as such maneuvers would need to be practiced in order to be performed successfully in actual naval combat. Moreover, harassment may

---

16. Richard T. Ackley, "The Soviet Navy's Role in Foreign Policy," *Naval War College Review*, Vol. 24, No. 9 (May 1972), p. 55. This practice apparently continued after the agreement was signed in 1972. In 1979, Soviet planes, including Backfire bombers, conducted more than 30 mock attacks against the U.S. destroyers *Caron* and *Farragut*. See *The New York Times*, August 11, 1979, p. 4. More recently, on February 18, 1984, the destroyer *David R. Ray* was harassed in the Black Sea near Novorossiysk when a Soviet plane fired cannon rounds into its wake and a Soviet helicopter came within 30 feet of its deck. See *The Washington Post*, June 8, 1984, p. A15.
17. Abram N. Shulsky, "Coercive Diplomacy," in Dismukes and McConnell, eds., *Soviet Naval Diplomacy*, p. 123.
18. See statement issued by the official Soviet News Agency TASS, May 13, 1967, in Historical Office, U.S. Department of State, *American Foreign Policy: Current Documents 1967* (Washington, D.C.: U.S. Government Printing Office, 1969), pp. 457–458.

enable the Soviets to obtain information on the likely U.S. responses to attempts to interfere with combat operations.

Finally, Soviet harassment, especially simulated attacks, may demonstrate naval capabilities to internal Soviet audiences. Overflights of U.S. aircraft carriers, for example, demonstrate the value of naval aviation and the need for increased spending in this area. Many Soviet naval activities and writings may be intended for domestic consumption by political decision-makers, as naval expenditures have not always been given priority in the Soviet Union.[19]

The U.S. reasons for harassment of Soviet warships probably do not differ significantly from Soviet motivations. Like the Soviets, U.S. units have harassed Soviet vessels to impede their operational effectiveness. U.S. warships may also have harassed Soviet merchant ships or trawlers on the grounds that such vessels were probably engaged in surveillance of U.S. naval forces and operations. Some naval officers have even suggested that harassment be employed deliberately to reduce the ability of Soviet vessels to launch a surprise attack.[20]

There are, however, some differences in the possible U.S. motivations for provoking incidents at sea. The U.S. Navy has long been established as a preeminent force on the world's oceans; it need not engage in deliberate harassment to make its presence felt. In addition, U.S. forces have different missions and capabilities than their Soviet counterparts. Air surveillance of Soviet warships may be considered vital due to the paucity of naval information released by the Soviet Union. The United States also has a much greater capability to engage in aerial reconnaissance, given its monopoly on large aircraft carriers. U.S. naval units are able to track Soviet submarines and force them to the surface. The Soviets lack comparable antisubmarine warfare (ASW) capabilities. These differences in capabilities and objectives tend to produce asymmetries in motivations.

Incidents are also produced by an action-reaction process. In the Sea of Japan, for example, the Commander-in-Chief of the U.S. Seventh Fleet, exasperated by continued Soviet harassment, issued instructions that his ships were to maintain course and speed even if a collision resulted, and he

19. Steven E. Miller, "Assessing the Soviet Navy," *Naval War College Review*, Vol. 32, No. 5 (September–October 1979), p. 65.
20. Frank Andrews, "The Prevention of Preemptive Attack," *U.S. Naval Institute Proceedings*, Vol. 106, No. 5 (May 1980), p. 139.

recommended claiming damages from the Soviet Union in the event of a collision.[21] This sort of reaction to incidents at sea can create a vicious cycle in which incidents begin to take on a life of their own and provoke continued reprisals. Regardless of the potential utility of some forms of harassment, there are clearly inherent dangers in incidents at sea.

The Dangers of Incidents at Sea

Incidents at sea strike most observers as clear examples of risky, confrontational behavior. The political benefits may not outweigh the potential risks of unregulated superpower confrontation at sea. Both the United States and the Soviet Union eventually recognized that the risks of naval harassment undermine any justification for its continued unconstrained practice. The 1972 agreement was the eventual result of this mutual recognition of the dangers of naval incidents. These dangers can be divided into three categories: (1) the physical danger to lives and vessels posed by a collision; (2) the possibility that an incident, even if relatively minor in itself, will provoke a crisis or even war; and (3) the risk of direct and immediate combat and escalation as a result of misperception or misinterpretation of an incident by local commanders.

THE PHYSICAL DANGERS OF COLLISIONS
The most obvious danger of incidents at sea is the threat to sailors and vessels posed by practices that interfere with safe navigation. Although collisions in times of relative international calm may not have grave political repercussions, they still threaten injury to men and damage to ships involved. With the exception of several incidents in the 1950s and early 1960s in which the Soviets shot down U.S. military aircraft over international waters, few naval incidents appear to have caused significant damage or loss of life. A Soviet TU-16 bomber attempting to buzz U.S. vessels apparently crashed into the Norwegian Sea in May 1968,[22] but most actual collisions between ships appear to have involved the scraping of hulls, not broadside or head-on collisions. Nevertheless, both governments have stressed the

---

21. O'Connell, *The Influence of Law on Sea Power*, p. 178.
22. Thomas W. Wolfe, "Soviet Naval Interaction with the United States and its Influence on Soviet Naval Developments," in Michael MccGwire, ed., *Soviet Naval Developments Capability and Context* (New York: Praeger Publishers, 1973), p. 266.

threat to human life posed by acts that impede safe navigation.[23] The importance of avoiding collisions to minimize risk to ships and men should not be underestimated as a factor leading to the 1972 agreement, as both navies presumably regard their vessels as significant military assets that should not be jeopardized.

INCIDENTS INCREASING TENSIONS AND RAISING THE RISK OF WAR

A second threat posed by incidents involving U.S. and Soviet warships is the possibility that collisions or other confrontational naval encounters might increase tensions or even lead to war. Although an incident itself might not directly escalate to major armed conflict, it could create grounds for political demands or reprisals.

Historically, naval incidents have often increased tensions and provided the catalyst for the outbreak of war. The bizarre incident that triggered the War of Jenkins' Ear between Great Britain and Spain (1739–1741) is perhaps the most unusual example of such an encounter on the high seas.[24] Incidents at sea precipitated the War of 1812 and brought the United States and Germany into conflict in both world wars. More recently, the Gulf of Tonkin incident and the seizure of the *Mayaguez* provoked significant U.S. military responses. Although these two incidents did not involve Soviet forces and, strictly speaking, entailed different types of actions than those defined as incidents above, they do indicate that even low-level naval clashes could lead to escalatory U.S. military actions.

The Dogger Bank affair between Great Britain and Russia during the Russo–Japanese War is a classic case of a naval incident that brought the countries involved to the brink of war. After Japanese torpedo boats had attacked the Russian fleet in the Far East in 1904, the Russian Baltic Fleet began the long voyage that eventually ended on the bottom of the Tsushima Straits. Fleet commanders feared further attacks by Japanese torpedo boats— possibly disguised as trawlers—in the North Sea. A jittery Russian captain, confused in a fog at night, bombarded British trawlers, sinking one, damaging five, and leaving two fishermen dead and six wounded.

---

23. See diplomatic notes in Historical Office, U.S. Department of State, *American Foreign Policy: Current Documents 1964*, pp. 669–673.
24. The unfortunate Captain Robert Jenkins, a British smuggler, allegedly lost his ear in 1731 when it was cut off during a fracas after his ship had been boarded by a Spanish vessel. The incident became an issue in Britain in 1738, when Jenkins exhibited his severed ear to Parliament and the uproar was exploited by proponents of war with Spain.

The British public was incensed. King Edward VII, Admiral John Fisher, and others urged a military response. Even Foreign Secretary Lansdowne, an advocate of entente with Russia, was outraged, and felt that Britain's reputation as a great power was at stake. Arthur Balfour estimated the probability of war to be about 50 percent. The British fleet was poised to intercept the Russians as they steamed southward. Eventually, however, the Russians complied with the British demand that those responsible for the incident be put ashore to face a tribunal. They also pledged to avoid any repetition of such errors in judgment.[25]

Although the Dogger Bank affair did not end in war, it illustrates many of the dangers inherent in U.S.–Soviet naval incidents. Under conditions of international tension and superpower rivalry, public opinion in a liberal democracy is likely to demand retaliation after a provocation by a major rival. Naval incidents seem to elicit particularly emotional responses in the United States. Reflecting on the Gulf of Tonkin incident, former White House aide Chester Cooper observed:

There is something very magical about an attack on an American ship on the high seas. An attack on a military base or an Army convoy doesn't stir up that kind of emotion. An attack on an American ship on the high seas is bound to set off skyrockets and the "Star-Spangled Banner" and "Hail to the Chief" and everything else.[26]

The reaction to an incident, even if it is clearly accidental, will depend on the subjective interpretation of events, which may be independent of the level of damage or loss of life. American reactions to the sinking of the *Maine* indicate that a probable accident can be given the most negative possible interpretation by a nation predisposed to war.

It is, of course, relatively unlikely that a naval incident could provoke a nuclear exchange between the United States and the Soviet Union. Political leaders would have the opportunity for reflection—provided that the incident did not immediately escalate—and it is difficult to imagine them acting with the bellicosity of the British in the Dogger Bank affair. An incident could, however, increase tensions and needlessly disrupt negotiations or other political discourse, much as the U-2 incident of 1960 forced the cancellation of

---

25. For a discussion of this incident, see Richard Ned Lebow, "Accidents and Crises: The Dogger Bank Affair," *Naval War College Review*, Vol. 31, No. 1 (Summer 1978), pp. 66–75.
26. Quoted in "The 'Phantom Battle' that Led to War," *U.S. News and World Report*, July 23, 1984, pp. 65 66.

the Khrushchev–Eisenhower summit. The downing of Korean Air Lines Flight 007 has again demonstrated the sensitivity of superpower relations to military incidents. A naval incident could easily have similar effects. The cumulative impact of a succession of incidents could undermine the stability of superpower relations even further. As Thomas Schelling has written, "nothing is more threatening to the nuclear fate of the world than the loss of confidence on each side in the other's restraint, patience, and security."[27] Public reaction to incidents at sea could create such a loss of confidence.

INCIDENTS AND ESCALATION DURING CRISES

The most alarming, although not necessarily the most likely, danger posed by U.S.–Soviet incidents at sea is the possibility that harassment of warships or aircraft will accidentally escalate to actual combat. As Ury and Smoke have suggested: "Perhaps the most likely path to nuclear war today is through a crisis that escalates out of control because of miscalculation, miscommunication, or accident."[28] Initial hostilities involving only local naval units could escalate to a more general conventional or nuclear conflict, particularly if the incident occurred during a period of acute international tension. Although the probability of such escalation is very low, "the pressures for pre-emptive nuclear strikes would likely be enhanced after the line between superpower peace and war was crossed."[29] This type of escalation is distinct from the possibility that an incident might increase international tensions and raise the risk of war. In that case, the incident serves only as a catalyst that triggers or increases hostility between the parties involved; it does not immediately lead to sustained fighting. Political leaders would have the opportunity to assess the incident and then act. In a crisis, however, the incident itself might lead to significant hostilities and direct escalation. The political leadership of both countries might eventually have the opportunity to reflect and decide upon further military measures, but major combat already would have taken place at the local or theater level.

An incident between U.S. and Soviet naval vessels could lead to accidental and unintended hostilities if some form of harassment was interpreted as a

---

27. Thomas C. Schelling, "Confidence in Crisis," *International Security*, Vol. 8, No. 4 (Spring 1984), p. 57.
28. Ury and Smoke, *Beyond the Hotline*, p. iii.
29. Albert Carnesale, Paul Doty, Stanley Hoffmann, Samuel P. Huntington, Joseph S. Nye, Jr., and Scott D. Sagan, *Living with Nuclear Weapons: Report of the Harvard Nuclear Study Group* (New York: Bantam Books, 1983), p. 58.

sign of imminent attack. An American commander might, for example, view Soviet actions such as the training of weapons and fire-control radars on U.S. vessels as the prelude to a large-scale attack. Under such circumstances, he might react by launching a preemptive attack against the threatening Soviet warships or aircraft, or, more likely, he might engage in countermeasures to reduce the likelihood of a successful Soviet first strike.[30] Options include maneuvering away from the threatening Soviet vessels, jamming Soviet equipment or deceiving it with feints and decoys, and harassing the Soviet forces. This last category of acts includes shouldering the Soviet ships onto a nonthreatening course, forcing submarines to surface, and escorting Soviet aircraft away from the U.S. vessel. Although these actions might reduce the probability of a successful Soviet attack, they could also increase Soviet apprehensions over the possibility of imminent hostile actions by U.S. forces. Indeed, some of these acts might even end with the destruction of Soviet units.[31] Thus U.S. countermeasures intended to reduce the chances of a successful attack could increase Soviet fears of a U.S. attack and prompt the very preemptive attack that they were meant to avoid. The nature of the interaction between hostile naval units creates an inherent instability at the tactical level. This instability is likely to be most pronounced during crises when tension is heightened and actual fighting appears more likely. Under such circumstances, the temptation to strike first could increase.

Several factors contribute to the inescapable instability of contemporary naval interaction. First, current naval technology gives an overwhelming advantage to the side that strikes first.[32] This condition increases the military temptation to launch a preemptive attack when threatened. Moreover, most observers believe that Soviet naval doctrine emphasizes the importance of striking first in any naval engagement. Soviet forces, lacking aircraft carriers and their attendant tactical air support, are configured for a first strike against U.S. aircraft carriers. The writings of Soviet naval commanders, including Admiral Gorshkov, stress "decisive, offensive actions" and "the struggle for the first salvo."[33] Soviet naval exercises usually involve attacks on passive

---

30. Present NATO rules of engagement prohibit preemptive strikes and preparations for attacks during crises, but the United States is apparently attempting to change this doctrine. See "NATO Issue: When to Let Its Ships Fire," *The New York Times*, April 2, 1984, p A5.

31. Andrews, "The Prevention of Preemptive Attack," p. 133.

32. See George H. Quester, "Naval Armaments: The Past as Prologue," in Quester, ed., *Navies and Arms Control* (New York: Praeger Publishers, 1980), pp. 1–11.

33. Sergei Gorshkov, quoted in Raymond G. O'Connor and Vladimir P. Prokofieff, "The Soviet Navy in the Mediterranean and Indian Ocean," *Virginia Quarterly Review*, Vol. 49, No. 4 (Autumn 1973), pp. 491–492.

targets, indicating a desire to achieve surprise in war at sea. During Okean 75, a major Soviet exercise, the Soviet navy simulated a surprise attack by over 200 warships and 300 submarines within 90 seconds of each other.

Some analysts question whether the Soviets actually plan to launch preemptive strikes to initiate naval warfare with the United States. George Quester suggests that the Soviet naval literature emphasizes preemption to impress domestic audiences and to increase the Soviet navy's share of military expenditures. Michael Klare argues that the Soviet navy lacks the capability to engage the United States in conflict at sea. And apparent changes in Soviet tactics may indicate a reduction in Soviet confidence in their ability to preempt U.S. naval forces.[34]

Despite the uncertainty over the possibility of Soviet preemptive strike on U.S. vessels, U.S. commanders probably will not become complacent. The proximity of U.S. and Soviet forces during a crisis would enable the Soviets to launch an attack from point-blank range without advance warning.[35] Under such conditions, dangerous maneuvering or simulated attacks by Soviet vessels could provoke a U.S. response.

Second, incidents at sea remain unstable because information can easily be misinterpreted in the confusion caused by the proximity of large numbers of vessels from various countries, including third parties. The Israeli attack on the American ship *Liberty* in the June 1967 war could, for example, have been misinterpreted as a Soviet attack.[36] Similarly, in several of the wars in the Middle East, Soviet vessels might have mistakenly reacted to Israeli naval actions, assuming that Israeli units were hostile U.S. forces. Attempts by national command authorities to control local units might be disrupted or otherwise prove unsuccessful, as occurred in the case of the *Liberty* and the Cuban missile crisis.[37] The "fog of crisis" could also lead to a misinterpretation of a nonhostile act. Harassment meant as a political signal might be misread

---

34. See Quester, "Naval Armaments"; Michael T. Klare, "Superpower Rivalry At Sea," *Foreign Policy*, No. 21 (Winter 1975-76), pp. 86–89, 161–167; and Charles C. Petersen, "About-Face in Soviet Tactics," *U.S. Naval Institute Proceedings*, Vol. 109, No. 8 (August 1983), pp. 57–63.
35. Stansfield Turner, "The Naval Balance: Not Just a Numbers Game," *Foreign Affairs*, Vol. 55, No. 2 (January 1977), p. 350.
36. The first reaction of then Secretary of Defense Robert McNamara was to consider the possibility that the Soviet Union was responsible for that attack and to plan retaliation against Soviet forces in the area. Fortunately, however, the United States soon concluded that a Soviet attack was unlikely and the Hot Line was used to assure the Soviets that U.S. aircraft dispatched to assist the *Liberty* were not going to threaten Soviet forces. See Phil G. Goulding, *Confirm or Deny* (New York: Harper and Row, 1970), pp. 97–98.
37. Repeated orders instructing the *Liberty* to move away from the battle zone were misrouted or delayed. During the Cuban missile crisis, a U.S. plane strayed over Soviet territory and naval commanders initially established the blockade further from Cuba than Kennedy had wanted.

by a local commander. An attempt by U.S. or Soviet vessels to shake enemy "tattletales" (surveillance ships) might be viewed as a prelude to offensive action, not as a legitimate defensive maneuver. The use of decoys to deceive trailing vessels could be misinterpreted, as such decoys can sound like anti-submarine torpedoes.[38]

Finally, the danger of naval conflict and escalation is heightened by the tendency of naval confrontations to assume a life of their own, prolonging competitive deployments after the crisis has abated. Although the October 1973 Middle East war ended in late October, the U.S. Sixth Fleet continued to operate at DEFCON III readiness until mid-November.[39] The 1971 Indo–Pakistani war ended on December 17, but intense U.S.–Soviet naval inter-action did not begin until December 22 and ended on January 8, 1972.[40] These extended confrontations multiply the risks inherent in shorter crises. More-over, naval units may enjoy greater scope for autonomous action after polit-ical authorities are no longer preoccupied with the crisis.

Neither the United States nor the Soviet Union has deliberately and con-sistently attempted to raise the risks of naval confrontations. Indeed, many American observers argue that Soviet naval activity, like Soviet crisis behavior in general, has been remarkably circumspect.[41] Nevertheless, dangerous in-cidents have taken place during international crises. Although the Soviets maneuvered carefully in the Mediterranean in June 1967 to avoid any incident with the U.S. Sixth Fleet that might be misinterpreted or escalate out of control, on June 8 a Soviet escort and destroyer interfered with the operations of the carrier *America*'s task group.[42] (This incident may have been the result of indiscretion or an attempt to stop U.S. ships from tracking a Soviet sub-marine.) In the aftermath of the seizure of the *Pueblo* by North Korea in January 1968, Soviet vessels engaged in harassment of U.S. warships in the Sea of Japan. U.S. Navy records show a dozen violations of the nautical rules of the road by Soviet vessels during this period, as well as a collision between

---

38. O'Connell, *The Influence of Law on Sea Power*, p. 180.
39. F.C. Miller, "Those Storm-beaten Ships, Upon which Arab Armies Never Looked," *U.S. Naval Institute Proceedings*, Vol. 101, No. 3 (March 1975), p. 24.
40. James M. McConnell and Anne Kelly Calhoun, "The December 1971 Indo–Pakistani Crisis," in Dismukes and McConnell, eds., *Soviet Naval Diplomacy*, p. 191.
41. See, for example, Adam Yarmolinsky, "Department of Defense Operations During the Cuban Crisis," *Naval War College Review*, Vol. 32, No. 4 (July-August 1979), p. 88; and Dismukes and McConnell, "Conclusions," in *Soviet Naval Diplomacy*, p. 289.
42. Anthony R. Wells, "The June 1967 Arab–Israeli War," in Dismukes and McConnell, eds., *Soviet Naval Diplomacy*, p. 165.

the Soviet merchant ship *Kapitan Vislobokov* and the U.S. destroyer *Rowan*.[43] Following the global alert of U.S. forces during the October 1973 Arab–Israeli war, Soviet units began anticarrier exercises using U.S. units as targets—the most intense signal they have ever transmitted with naval forces during a crisis.[44] U.S. forces have also engaged in provocative naval acts during crises, including following, harassing, and forcing Soviet submarines to surface during the Cuban missile crisis. According to Robert Kennedy, President John F. Kennedy was extremely concerned over the possible dangers of this harassment and sought to control the actions of local naval commanders as much as possible.[45]

The risks of superpower naval confrontation should not be overestimated or exaggerated. The overall stability of the strategic nuclear balance adds to the incentives for caution and reduces the possibility that either side will see any advantage in initiating war at sea. Moreover, naval officers and political leaders are aware that the stakes in any crisis probably do not justify the launching of a surprise attack at sea. In most cases, even the most extreme forms of harassment are likely to be regarded as a bluff. Nevertheless, naval commanders on the scene will not become complacent. Even if the probability of war at sea is low, it may be higher than the chance of U.S.–Soviet hostilities in Europe or other regions. It certainly appears greater than the odds of a "bolt from the blue" nuclear strike by either side. Although most scenarios envisage the start of U.S.–Soviet hostilities on land, the risks of naval incidents were apparently great enough to induce both sides to negotiate the 1972 Agreement on the Prevention of Incidents at Sea. In the nuclear age, even a relatively low risk of superpower conflict can justify significant precautions.

Negotiating the Agreement

The increasing frequency and severity of U.S.–Soviet naval incidents led the United States to propose negotiations on the subject in 1967.[46] This overture

---

43. Donald S. Zagoria and Janet D. Zagoria, "Crises on the Korean Peninsula," in Stephen S. Kaplan, ed., *Mailed Fist, Velvet Glove: Soviet Armed Forces as a Political Instrument* (Washington, D.C.: U.S. Department of Commerce, National Technical Information Service, 1979), p. 9-6.
44. Stephen S. Roberts, "The October 1973 Arab–Israeli War," in Dismukes and McConnell, eds., *Soviet Naval Diplomacy*, p. 210.
45. See Robert F. Kennedy, *Thirteen Days: A Memoir of the Cuban Missile Crisis* (New York: W.W. Norton, 1971).
46. The following chronology is based on a personal interview with Ambassador Herbert Okun,

was ignored for over two years until the Soviets surprised U.S. officials in 1970 by proposing that negotiations be opened in the spring of 1971. The United States did not respond immediately, but initiated an interagency review of the problem to formulate a position. The interagency process was chaired by Ambassador Herbert Okun, State Department Soviet expert, and involved representatives of the Navy, Department of Defense, and the National Security Council (NSC). National Security Adviser Henry Kissinger and senior NSC staff member Helmut Sonnenfeldt were directly involved. The interagency review team compiled and analyzed information on all previous U.S.–Soviet incidents at sea and the subsequent protests by either party. It sought to develop a negotiating position that would not constrain U.S. or allied naval missions or activities while preventing dangerous Soviet maneuvers. The United States could not accept the inclusion of limitations on submarine activities, since such a provision, it was felt, might lead the Soviet Union to propose the establishment of submarine operating zones in which ASW would be prohibited.[47] Given the U.S. lead in ASW technology, this step would benefit the Soviet Union. Moreover, the United States was reluctant to discuss submarine incidents, as any negotiated provisions might force the disclosure of submarine locations and compromise strategic and reconnaissance missions.

Navy representatives were also concerned over the possible negotiation of a distance formula that would govern how closely U.S. vessels and aircraft would be able to approach their Soviet counterparts. Any form of distance limitation could interfere with naval operations and complicate aerial surveillance of Soviet warships. Soviet protests following previous incidents indicated that they were particularly irritated by U.S. close air surveillance. The question of a distance formula thus became a critical issue in the subsequent negotiations.

In contrast to other negotiations, where the U.S. representatives came to listen to the Soviets before offering their own position, the U.S. delegation formulated detailed proposals prior to the start of any talks and considered the likely Soviet responses. The U.S. negotiating position essentially called for clarifying and expanding the rules of the road. The United States was

a principal negotiator of the 1972 agreement, Washington, D.C., November 29, 1983, and Anthony F. Wolf, "Agreement at Sea: The United States–USSR Agreement on Incidents at Sea," *Korean Journal of International Studies*, Vol. 9, No. 3 (1978), pp. 57–80.

47. For an example of such a proposal, see Ken Booth, "Law and Strategy in Northern Waters," *Naval War College Review*, Vol. 34, No. 4 (July–August 1981), pp. 3–21.

partially interested in preventing Soviet vessels from shouldering U.S. aircraft carriers and thus disrupting their flight operations. Having formulated its position, the United States accepted the Soviet offer to negotiate in June 1971, and discussions were scheduled to begin in Moscow in October of that year.

The U.S. delegation to Moscow consisted primarily of participants in the interagency review process. It was headed by John Warner, then undersecretary of the Navy. Okun was the vice-chairman. Vice-Admiral Harry Harty, senior military advisor to the delegation, represented the Joint Chiefs of Staff. Other members of the U.S. delegation included Charles Pittman of the State Department's legal office, Commander William Lynch, special assistant to Warner for law of the sea, Captain Edward Day, an expert in naval aviation assigned to the State Department's Bureau of Politico–Military Affairs, Captain Robert Rawlins of the Navy's policy planning branch, Op-61, and Captain Robert Congdon of the Defense Department's Office of International Security Affairs.

The U.S. negotiating team was the highest-ranking U.S. military delegation to visit the Soviet Union since 1945. But the Soviet delegation consisted of even higher-ranking officials, conveying a clear interest in the talks. Headed by Admiral Vladimir Kasatonov, deputy commander of the Soviet navy, the team also included Admiral V.A. Alexeyev of the Naval Staff, Rear-Admiral Motrokhov, chief navigator of the Soviet navy, and Colonel-General Vishinsky of Naval Air. The delegation thus included the second, third, fourth, and fifth highest-ranking officers of the Soviet navy. In addition to demonstrating their interest, the high rank of the Soviet delegation probably reduced the constraints that central authorities generally impose on Soviet negotiators.

Although many naval officers and some U.S. experts on Soviet foreign policy had been skeptical about the prospects,[48] the Soviet delegation warmly welcomed their American counterparts. In the negotiations, the Soviets accepted the U.S. agenda and agreed that submarines would not be discussed. The announcement of Nixon's planned visit to Moscow on October 12 created an even more propitious political atmosphere.

The actual negotiations were conducted in surface and air working groups. The surface working groups discussed the rules of the road, signaling, disruption of flight operations, the definition of naval platforms, and the training

---

48. Some experts on the Soviet Union were optimistic, however, including former U.S. Ambassador Llewelyn Thompson.

of weapons and sensor systems on enemy vessels. Among these issues, the most contentious was the matter of a distance formula that would modify the rules of the road by, for example, prohibiting maneuvers within a certain distance of opposing warships. In contrast to the general U.S. pattern of seeking highly specific agreements, the American negotiators refused to accept Soviet proposals for a distance formula, preferring to stress good judgment and general principles. Similar disagreements arose in the talks between U.S. and Soviet negotiators in the air working group.

Despite the lack of agreement on any distance limitations, the negotiations made considerable progress. In talks that often lasted up to 16 hours per day, agreement was reached on most issues. The U.S. delegation was not surprised by the Soviet proposals, having anticipated them in its preparation. This advance preparation, as well as Soviet concessions on critical issues such as the dropping of sonar buoys from aircraft, enabled the two delegations to initial a memorandum of understanding covering points on which agreement had been reached and listing outstanding issues, including the distance formula, which were to be discussed in Washington in May 1972.

The negotiations and the memorandum of understanding were subjected to a second, more intensive, interagency review before the talks resumed. As the initially skeptical military departments now confronted the possibility that an agreement would actually be reached, they increased their level of representation and vigorously objected to any suggestion that a distance formula be accepted. The Navy argued that such a provision would interfere with surveillance activities. As the Navy was concerned that the State Department might be overly conciliatory, Lawrence Eagleburger of the Pentagon's Office of International Security Affairs chaired the second interagency review.

The seemingly intransigent attitude of the Navy at this stage of the review might be traced to naval opposition to the idea of any agreement with the Soviets. Naval officers accustomed to thinking of their Soviet counterparts as rivals might understandably be reluctant to enter into any form of cooperation with the "enemy." Indeed, some naval officers might even favor the continuation of incidents at sea that demonstrate the intensity of U.S.–Soviet hostility there and thus provide a basis for increased naval appropriations. They might fear that an agreement would exert a "lulling effect" on Congressmen who would interpret any evidence of reduced tension and rivalry at sea as grounds for reductions in U.S. naval expenditures.

The distance formula was the principal issue in the second round of talks. Although the Soviets initially seemed adamant on the distance issue, they eventually dropped their objections in return for an agreement to discuss the issue in the future. They apparently continued to feel strongly about the issue, as indicated by a Soviet captain in the September 1972 issue of *Morsky Sbornik*:

It is quite evident that the Agreement would more fully serve its purpose if it contained fixed maximal permissible distances for the approach of ships and aircraft . . . . Therefore the Commission appointed in accordance with Article X will have to develop practical recommendations relative to concrete fixed distances which must be observed when approaching warships and aircraft.[49]

Despite their misgivings, the Soviets were apparently satisfied with the rest of the agreement. Their interest in negotiating some form of constraints on dangerous maneuvers and harassment at sea overrode their desire for a distance limitation. Soviet negotiators wanted an agreement and were reconciled to serious concessions to obtain it. Soviet interest was demonstrated not only by the warm welcome accorded to the U.S. delegation in Moscow, but also by the complacent Soviet naval reaction to the U.S. mining of Haiphong harbor, which occurred in the middle of the Washington negotiations. Admiral Kasatonov actually watched Nixon's speech announcing the mining at John Warner's Georgetown home. After a pause, he remarked: "This is a very serious matter. Let us leave it to the politicians to settle this one." This comment implicitly acknowledged that the naval talks were too important to be disrupted even by U.S. actions that endangered Soviet merchant ships. As Okun later remarked of the incident: "We were highball to highball, and they were the first to clink."[50]

The agreement was initialed by Warner and Kasatonov in Washington and formally signed by Warner, then secretary of the Navy, and Admiral Gorshkov on May 25, 1972, during the Moscow summit meeting. It represented the first important military agreement between the two superpowers since World War II. In announcing the results of the summit to Congress, Nixon

49. Captain First Rank V. Serkov, quoted in Anne Kelly Calhoun and Charles Peterson, "Changes in Soviet Naval Policy: Prospects for Arms Limitations in the Mediterranean and Indian Ocean," in Paul J. Murphy, ed., *Naval Power in Soviet Policy* (Washington, D.C.: U.S. Government Printing Office, 1978), pp. 244–245.
50. Marvin Kalb and Bernard Kalb, *Kissinger* (Boston: Little, Brown, 1974), p. 306.

contended that the agreement would have a "direct bearing on the search for peace and security in the world" and was "aimed at significantly reducing the chances of dangerous incidents between our ships and aircraft at sea."[51]

The Provisions of the Agreement

The 1972 Agreement on the Prevention of Incidents at Sea serves four basic purposes: (1) regulation of dangerous maneuvers; (2) restriction of other forms of harassment; (3) increased communication at sea; and (4) convening regular naval consultations and exchanges of information. Its provisions address many of the possible causes of U.S.–Soviet naval incidents, particularly those arising from misunderstanding or misinterpretation of the other party's action.

REGULATION OF DANGEROUS MANEUVERS. The agreement reaffirms the rules of the road, and Article II specifically requires ships to remain well clear of one another to avoid hindering the evolution of formations, and to show particular care in approaching ships engaged in launching or landing aircraft, as well as ships engaged in replenishment underway.

RESTRICTION OF OTHER HARASSMENT. Articles III and IV of the agreement also prohibit simulated attacks or the launching of objects in the direction of passing ships of the other party, the use of searchlights to illuminate navigation bridges, the performance of "various aerobatics" over ships, and the dropping of various objects that would be hazardous to ships or constitute a hazard to navigation. This last provision apparently reflects Soviet concern over the U.S. practice of dropping sonar buoys from aircraft, but the ambiguous wording probably allows the United States to continue to act as it did before the agreement. It is always possible to claim that the sonar buoys did not actually endanger safe navigation.

INCREASED COMMUNICATION AT SEA. Article III of the agreement requires the use of internationally recognized signals to convey information about operations and intentions and to warn ships of the presence of submarines in an area. Article V requires signals to announce the commencement of flight operations and also mandates that aircraft flying over the high seas display navigation lights "whenever feasible." Finally, Article VI requires three to five days' advance notification of actions (naval exercises or missile

---

51. Richard M. Nixon, "The Moscow Summit: New Opportunities in U.S.–Soviet Relations," *Department of State Bulletin*, June 26, 1972, p. 856.

test launches) on the high seas that represent a danger to navigation or to aircraft, as well as requiring increased use of informative signals to signify intentions of vessels maneuvering in close proximity to one another. These provisions reduce the danger of accidental attacks during exercises and limit the possibility of collisions arising from misunderstanding or misinterpretation. Significantly, this article represents the first time that the Soviets have agreed in principle to give advance notification of some naval exercises.[52]

REGULAR CONSULTATIONS AND INFORMATION EXCHANGES. Article VII stipulates that the Soviet and U.S. naval attachés in each other's capitals shall serve as the channel for the exchange of information concerning incidents and collisions. This provision may help to minimize the diplomatic consequences of incidents at sea by ensuring that such matters are handled primarily by the two navies. The Soviets during the negotiations were apparently pleased that incidents could be discussed between "brothers at sea"; this type of attitude may provide a basis for fruitful exchanges on the details of any incidents. Article IX provides that the United States and the Soviet Union shall conduct annual reviews of the agreement. Article X specifically establishes a committee to meet within six months to "consider the practical workability of concrete fixed distances to be observed in encounters between ships, aircraft, and ships and aircraft."

U.S. and Soviet negotiators did not reach any agreement on a distance formula, but they did produce a protocol to the original agreement on May 22, 1973.[53] This protocol extends some provisions of the 1972 agreement to nonmilitary ships. Article I states that measures shall be taken to notify nonmilitary ships of each party of the provisions of the agreement directed at securing mutual safety. Article II prohibits simulated attacks on nonmilitary ships and the dropping of objects near them in a hazardous manner. The protocol seems to be a logical extension of the original agreement, as several incidents at sea have involved merchant or fishing vessels. Although this protocol obviously is meant to safeguard the nonmilitary ships of both parties, it may confer somewhat greater benefits on the Soviet Union. Its provisions could serve to protect Soviet trawlers or other nonmilitary vessels

---

52. Okun interview.
53. Protocol to the Agreement Between the Government of the United States of America and the Government of the Union of Soviet Socialist Republics on the Prevention of Incidents on and over the High Seas, Signed May 25, 1972. English and Russian texts can be found in U.S. Department of State, *United States Treaties and Other International Agreements*, Vol. 24, Pt. 1, 1973 (Washington, D.C.: U.S. Government Printing Office, 1974), pp. 1063–1066.

that actually serve some military purpose, particularly surveillance. More-over, the large and growing Soviet merchant fleet is more exposed to Western harassment. The bulk of Soviet merchant and fishing vessels operate in waters that can be controlled by the West, whereas only an insignificant portion of U.S. shipping is usually found in Soviet-controlled waters.[54]

Assessing the Agreement

The 1972 Agreement on the Prevention of Incidents at Sea is generally re-garded as a success. Incidents have continued since it was signed, but they have become less frequent and less severe. The number of serious incidents exceeded 100 per year in the late 1960s, but Secretary of the Navy John Lehman, Jr., reported that there were only about 40 potentially dangerous incidents between June 1982 and June 1983.[55] Lehman has attributed this substantial reduction in collisions and near-collisions to the 1972 agreement.[56] The most dangerous maneuvers and attempts to disrupt formations are no longer commonplace. When incidents do occur, they often are resolved by the U.S. and Soviet navies without becoming diplomatic controversies. Leh-man feels that the annual meetings to review the accord have produced a "stable pattern" of dealing with incidents and that they provide "pretty good resolution" of any disputes in a "rather businesslike" manner. The annual meetings continued to take place even after the Soviet invasion of Afghani-stan, when most U.S.–Soviet contacts were suspended.

U.S.–Soviet naval interaction in the October 1973 Arab–Israeli war appar-ently exemplifies the positive impact of the agreement. The Soviets deployed a peak of 96 vessels during the war, confronting a slightly smaller number of U.S. ships. Despite the heightened political tensions and the increased probability of incidents due to the proximity of so many hostile vessels, incidents were relatively rare. Some Soviet warships trained guns or search-lights on U.S. vessels, fired flares near U.S. aircraft, or engaged in close maneuvering. But, as Stephen Roberts notes, these actions were probably "clumsy efforts at reconnaissance" or were performed for "operational rea-sons." The "gun movements (which tended to occur around 8:00 AM)" may

54. Michael MccGwire, "Soviet Naval Policy for the Seventies," in MccGwire, ed., *Soviet Naval Developments*, p. 509.
55. "Superpowers Maneuvering for Supremacy on High Seas," *The Washington Post*, April 4, 1984, p. A18.
56. "Soviet Sub Bumps into U.S. Carrier," *The Washington Post*, March 22, 1984, p. A28.

have been "routine checks of equipment." On the whole, Soviet ships observed the agreement and avoided harassment of their U.S. counterparts.[57] Admiral Worth Bagley, Commander-in-Chief of U.S. Naval Forces in Europe at the time, remarked that the "Soviets weren't overly aggressive. It looked as though they were taking some care not to cause an incident."[58]

More recent occurrences may suggest that the agreement is not faring so well. Soviet vessels reportedly interfered with salvage operations by U.S. and allied vessels in the Sea of Japan in the wake of the downing of the Korean airliner in 1983. A Soviet guided-missile frigate apparently attempted to disrupt flight operations of the U.S. carrier *Ranger* in the Arabian Sea before colliding with the U.S. frigate *Fife* in November 1983. The U.S. Navy protested the incident through its attaché in Moscow, claiming that the Soviets had clearly violated the agreement. Although the Soviet ship signaled that it had steering difficulties, it appeared to be under full control at all times.[59]

Additional incidents were reported in early 1984. In March, a Soviet Victor-I class submarine running without lights collided with the U.S. carrier *Kitty Hawk* in the Sea of Japan. Although the carrier was not seriously damaged, the submarine was apparently disabled. The U.S. Navy began an inquiry to determine whether the Soviet submarine was at fault, although Lehman indicated that the collision appeared to be "inadvertent."[60] Several days later, the Soviet carrier *Minsk* fired eight flares at the U.S. frigate *Harold E. Holt*. Three hit the U.S. vessel, including one that passed within three feet of the captain. The *Holt* was within thirty yards of the *Minsk*, which had stopped for unexplained reasons. The *Holt* had apparently signaled that it planned to pass the *Minsk* on the starboard side and did so despite several warnings from the Soviet vessel. The U.S. Navy decided to raise the incident at the annual meeting in May 1984.[61]

These incidents do not necessarily signal the demise of the agreement. Soviet vessels in the Sea of Japan may have been reacting to the extraordinary tension that followed the downing of the Korean airliner, or they may have

---

57. Roberts, "The October 1973 Arab–Israeli War," p. 196. The Soviets did, however, simulate attacks after the U.S. alert. See above, p. 18.
58. Quoted in Caldwell, *American–Soviet Relations*, p. 228.
59. "Soviet Warship, US Navy Vessel Collide in Mideast," *The Boston Globe*, November 18, 1983, p. 6.
60. See "Soviet Sub Bumps into U.S. Carrier," p. 1; and "Soviet Sub and U.S. Carrier Collide in Sea of Japan," *The New York Times*, March 22, 1984, p. A7.
61. William E. Smith, "Moscow's Muscle Flexing," *Time*, April 16, 1984, pp. 28–30.

been attempting to prevent the United States from recovering the flight recorder. U.S. naval officials believe that the harassment was politically motivated, at least in part because of the proximity to Soviet borders and bases. Naval commanders privately acknowledge that incidents tend to become more frequent as U.S.–Soviet political tensions increase. The collisions with the *Fife* and *Kitty Hawk* and the firing of flares at the *Holt* are all potentially dangerous incidents, but Lehman believes that relations between the U.S. and Soviet fleets are still "very professional and workmanlike." Commenting on the *Minsk* incident, he said: "I don't see anything sinister in the incident with the *Minsk*. Let's say there are two plausible sides to that story. The *Minsk* skipper may not have been all on the wrong side."[62] The agreement may still be functioning successfully by reducing the number of incidents and facilitating their diplomatic resolution, even if some continue to take place. Senior U.S. officials have reaffirmed that the incidents of 1983 and 1984 have not changed their interpretation of Soviet behavior, arguing that the "Soviets have made it very clear that they believe in the Incidents at Sea agreement. They want it to continue. They want it to work. They want to live up to it." U.S. officials also point out that each year we've seen basically a decrease" in the number of incidents.[63] In addition, the agreement may not fully cover all the actions that might create incidents. More explicit rules for submarines and a distance formula might have prevented the *Kitty Hawk* and *Holt* incidents, but such provisions were ruled out by U.S. negotiators in 1971 and 1972.

The May 1984 meetings in Moscow of U.S. and Soviet naval representatives to review the agreement provided further evidence of its success. The talks were reportedly conducted in an open, frank, and professional manner. Both sides acknowledged the concerns of the other and avoided political rhetoric and unreasonable demands. U.S. admirals, who said the sessions were the best such meetings in memory, and their Soviet counterparts agreed to renew the agreement for three years. Announcing this renewal to a conference on U.S.–Soviet exchanges, President Reagan described the agreement as "useful." In addition, the Soviet delegation reportedly proposed extending the principles of the agreement to cover additional activities of military aircraft.[64]

---

62. Ibid., p. 30.
63. "High Seas Diplomacy Continuing," *The Washington Post*, June 8, 1984, p. A15.
64. William Beecher, "Election Clouds Weapons Talks," *The Boston Globe*, July 17, 1984, p. 4. The full text of Reagan's speech appears in *Weekly Compilation of Presidential Documents*, Vol. 20, No. 26 (July 2, 1984), pp. 944–946.

The only significant criticism that seems to have been directed against the agreement is that it has allegedly accorded the Soviet navy symbolic parity with the U.S. fleet.[65] This argument suggests that the Soviets see the agreement as a U.S. acceptance of Soviet equality on the oceans, and it is supported to some extent by the writings of Admiral Gorshkov. In an article that appeared shortly after the Moscow summit, Gorshkov drew an implicit comparison with the British acceptance of U.S. naval parity at the interwar naval conferences. Gorshkov claimed that: "Having agreed to 'parity' for the American Navy, England was no longer free to use 'diplomatic and propagandistic measures' to control the growth of American sea power."[66] If this quotation is applicable to the Incidents at Sea Agreement, it indicates that the Soviet navy may regard the agreement as a license for continued naval expansion or even eventual superiority. But the proper analogy may be between SALT and interwar naval arms control. The analogous U.S.–Soviet naval development would be mutual and equal limitations on naval forces. The agreement only accords parity to the Soviet Union in the sense that it subjects Soviet warships to the same rules as their U.S. counterparts. The Soviet response to current U.S. efforts to build a 600-ship navy will reveal far more of their attitude toward parity than the agreement does. Even if the Soviets believe the agreement entitles them to naval parity, it does not constrain U.S. efforts to maintain naval supremacy.

The most obvious reason for the success of the agreement is the mutual interest of the superpowers in avoiding dangerous incidents at sea. Neither country has any interest in accidentally triggering a naval conflict that could escalate or disrupt their political relations. Both navies want to prevent accidents that endanger ships and men. Although harassment at sea may sometimes serve a purpose, both sides have strong incentives to keep it limited and under control. But more than mutual interest is required for the successful negotiation and implementation of an agreement between the superpowers, as the history of arms control negotiations demonstrates. Inept diplomacy and U.S. domestic politics can frustrate even the most promising arms control initiatives.[67] The United States successfully overcame such obstacles in negotiating the Incidents at Sea Agreement.

---

65. Wolf, "Agreement at Sea," pp. 76–77.
66. Quoted in Abram N. Shulsky, "Gorshkov on Naval Arms Limitations: KTO KOGO?," in Murphy, ed., *Naval Power in Soviet Policy*, p. 250.
67. For a discussion of the U.S. domestic impediments to arms control, see Steven E. Miller, "Politics over Promise: Domestic Impediments to Arms Control," *International Security*, Vol. 8, No. 4 (Spring 1984), pp. 67–90.

The relative ease with which the text of the agreement was negotiated can be attributed to the careful and thorough U.S. preparations, the general political climate, which may have given each navy a stake in contributing to détente, the absence of any attempt to reduce force levels or to constrain deployments, and the asymmetries in naval capabilities and missions that enabled each side to derive different advantages from the agreement. In addition, the intensive involvement of the U.S. Navy in the preparations and actual negotiations may have helped by ensuring that the U.S. proposal and the memorandum of understanding were acceptable to the operational commanders of the service directly affected. The success of the agreement should not be traced to the simplicity of the issues. Although no thorny questions of verification were raised, the problems discussed were complex and often technical.

The nature of the 1972 agreement may also have reduced any domestic obstacles to its successful negotiation and implementation. As the agreement does not establish numerical limits on U.S. and Soviet weapons, it is less vulnerable to attacks from congressional and other critics who might allege that it codifies a U.S.–Soviet imbalance. It also does not require the U.S. Navy to forgo any ship-building programs. American domestic obstacles to arms control may not come into play when arms control is pursued through quiet efforts to limit potentially threatening activities instead of seeking formal treaties to control the level of armaments.[68] Although it might be argued that avoiding incidents at sea is inherently noncontroversial, it seems more plausible to suggest that the avoidance of quantitative limits reduces the potential for political controversy. Moreover, constraints on dangerous activities, like the Incidents at Sea Agreement, may offer greater potential for reducing the likelihood of war than small cuts in existing nuclear or conventional arsenals.

The actual implementation of the agreement may have been facilitated by the absence of publicity accorded it. The announcement of the agreement was overshadowed by the flurry of agreements that emerged from the 1972 Moscow summit, particularly SALT I. As the agreement is not a treaty, it was not subjected to public debate in the U.S. Senate. The Navy apparently believes that the lack of publicity has contributed to the success of the agreement, and has done little to call attention to it.

---

68. The SALT Standing Consultative Commission also appears to fit into this category. Ibid., p. 90.

The success of the agreement can also be attributed to the basic conceptual approach that underlay its negotiation and execution. The Incidents at Sea Agreement accepts the reality of U.S.–Soviet competition and competitive interaction. Unlike naval arms control measures that would impose geographic limitations on deployments, it implicitly assumes that U.S. and Soviet warships and aircraft will continue their rivalry at sea and engage in "gunboat diplomacy" to influence political outcomes in crises. Observance of the agreement makes U.S.–Soviet competition safer; it does not alter the basic terms of that competition.

In contrast to the U.S.–Soviet Basic Principles Agreement of 1972, the Incidents at Sea Agreement does not call for general political restraint by either superpower. Instead, it provides for specific measures to deal with a particular problem. It does not raise expectations as a more nebulous statement of principles might. The agreement serves to prevent crises that could lead to escalation as well as those with political ramifications, but it explicitly defines the behavior it attempts to prevent, rather than prohibiting undefined actions that one superpower perceives as leading to unilateral advantages for the other. Implementation of the agreement increases predictability in U.S.–Soviet relations, prevents possible crises that neither party intended, and controls the possibility of escalation in incidents.[69]

The agreement should be viewed as a confidence-building measure, an arrangement "designed to enhance such assurance of mind and belief in the trustworthiness of states and the facts they create."[70] No agreement can prevent the deliberate initiation of war, but agreements can reduce the possibility of unintentional conflict arising from mutual suspicion. The Incidents at Sea Agreement reduces the possibility of misinterpretation of potentially dangerous behavior at sea, thus increasing U.S. and Soviet confidence in the nonthreatening nature of each other's naval actions.

---

69. For an extended discussion of the Basic Principles Agreement of 1972 and the concepts of crisis prevention and escalation control, see Alexander L. George, ed., *Managing U.S.–Soviet Rivalry: Problems of Crisis Prevention* (Boulder, Colo.: Westview Press, 1983).
70. Johan Jørgen Holst, "Confidence-building Measures: A Conceptual Framework," *Survival*, Vol. 25, No. 1 (January/February 1983), p. 2. Numerous authors have offered various definitions of confidence-building measures. See Jonathan Alford, ed., *Confidence-Building Measures*, Adelphi Paper No 149 (London: International Institute for Strategic Studies, 1979); E.M. Chossudovsky, "Confidence Building and Confidence-Building Measures in East–West Interactions," *Coexistence*, Vol. 21, No. 1 (1984), pp. 23–36. Kevin N. Lewis and Mark A. Lorell, "Confidence-Building Measures and Crisis Resolution: Historical Perspectives," *Orbis*, Vol. 28, No. 2 (Summer 1984), pp. 281–306; and John Borawski, "Reorienting Arms Control: Confidence-Building Measures," *Arms Control Today*, March 1985, for differing perspectives on confidence-building measures.

Extensions and Applications of the Agreement

The Incidents at Sea Agreement could be amended to cover other naval activities, or it could serve as a model for similar agreements in other areas. The agreement itself could be improved in several ways. The provisions for notification of dangerous actions, for example, could be broadened to include mandatory notification of all naval exercises. U.S. interests in advance notification have increased as the Soviet navy has become more capable of actions at great distances from the Soviet Union. Advance notification would reduce U.S. or Soviet suspicions of any sudden, large-scale naval activities that appeared to be of a threatening nature.

The agreement might also be extended to include other countries. Collisions have occurred between Soviet warships and vessels of U.S. allies, including Great Britain. One collision even involved the British aircraft carrier *Ark Royal*.[71] Although a multilateral agreement would be more difficult to negotiate and apply, it would reduce the possibility of harassment by proxy and the dangers of "catalytic" incidents involving third parties. The potential for such incidents has existed in several superpower confrontations in the Middle East and the Indian Ocean, where Indian submarines of Soviet design operate.

Finally, the idea of a distance formula might be reconsidered. Although the United States rejected its inclusion in the 1971–1972 negotiations that led to the agreement, a distance formula appears to be consistent with U.S. arms control policies and might remove some ambiguities that could trigger incidents. The precise distance and the question of different distances for aircraft and surface ships could be the subject of negotiations.

The general approach of the Incidents at Sea Agreement could be applied to other, analogous areas of potential superpower conflict. A similar agreement might be negotiated to establish procedures for dealing with aerial incidents, including those involving civilian airliners. Earlier U.S.–Soviet communication and agreed-upon procedures might have prevented the tragic destruction of Korean Air Lines Flight 007.

Outer space is another area in which the principles of the Incidents at Sea Agreement seem particularly applicable. If, as seems likely, it is impossible

---

71. O'Connell, *The Influence of Law on Sea Power*, p. 178.

to negotiate the complete demilitarization of space, constraints on threatening activities could work to reduce the danger of unintended superpower conflicts above the atmosphere. Potentially threatening activities, such as the testing of antisatellite (ASAT) systems in conjunction with large-scale test launches of ballistic missiles, could be prohibited. This type of coordinated operational testing feeds U.S. fears that the Soviet Union is preparing for a first strike, even though most Soviet tests of this kind have not been very successful.

Application of the principles of the Incidents at Sea Agreement could also lead to prohibition of close high-speed passes or passes in geosynchronous orbit by satellites or spacecraft near the satellites of the other superpower. These actions could give rise to fears of imminent destruction of satellites by ASAT weapons as a prelude to a surprise attack. Close passes will obviously appear particularly threatening in the absence of an ASAT ban. Further applications of the agreement's principles may emerge as space technology continues to develop. If U.S.–Soviet military competition in space cannot be prevented entirely through arms control, it must at least be managed.[72]

More generally, additional confidence-building measures relevant to strategic forces could be sought. The agreement is, in many ways, analogous to the Hot Line Agreement and the 1971 agreement on the prevention of the accidental or unauthorized use of nuclear weapons. These measures could be expanded to require advance notification of all missile tests and limits on multiple missile launches within brief intervals.[73] Like simulated attacks at sea, the latter tends to increase fears of a surprise attack. The success of the agreement also suggests the utility of direct negotiations between the military services of the United States and the Soviet Union. Similar talks could improve mutual understanding and reduce suspicions in other areas.

There are, however, limits to the application of the principles contained in the Incidents at Sea Agreement. The agreement does nothing to reduce the instability created by forces that place a premium on striking first. Nor does it resolve any of the political differences that are the basic causes of U.S.–

---

72. See Daniel Deudney, "Unlocking Space," *Foreign Policy*, No. 53 (Winter 1983–84), pp. 91–113, for a discussion of emerging space weapons and issues. William Beecher, in "Soviets see talks as a way to avoid Space Arms Race," *The Boston Globe*, July 16, 1984, p. 7, notes that the U.S. position on space weapons may include the establishment of "ground rules" for space activities.

73. Holst, "Confidence-building Measures," p. 7.

Soviet hostility. Not all issues can be addressed without provoking domestic political controversy in the United States, even though confidence-building measures do not seem to be debated as heatedly as reductions in existing arsenals. Despite these limitations, the Incidents at Sea Agreement provides modest, yet encouraging evidence of the potential utility of confidence-building measures. Modesty in expectations may be a prerequisite for success in any U.S.–Soviet negotiations. The Incidents at Sea Agreement demonstrates that important results can emerge from modest expectations.